Advanced Wireless Communication Technologies

Advanced Wireless Communication Technologies: From Fundamentals to Future Applications explores the intricacies of modern and emerging wireless communication systems. It delves into fundamental theories, enabling technologies, network architectures, and practical applications to provide readers with a complete understanding of next-generation wireless communications.

The book addresses crucial aspects of advanced wireless communication systems, including RF circuits, network infrastructures, antenna systems, MIMO techniques, and beamforming methods. It also provides intelligent networking through the application of artificial intelligence and machine learning for performance optimization, resource management, and network automation. Furthermore, the book offers an in-depth exploration of cutting-edge advancements in wireless communication, encompassing technologies such as 5G and 6G, the Internet of Things (IoT), reconfigurable intelligent surfaces (RIS), drone and UAV communication, visible light communication, remote healthcare, and smart connectivity solutions, as well as energy-efficient and environmentally friendly green communication technologies.

Key Features:

- Thoroughly explores foundational principles and cutting-edge advancements in wireless communication technologies. It bridges the gap between theory and practice, making it a valuable resource for students, researchers, and professionals.
- Provides comprehensive coverage of 5G, 6G, IoT, and O-RAN, with the latest information on current technologies, and explores the application of AI and machine learning in wireless networks for intelligent communication.
- Presents in-depth discussions on antenna technologies, RIS, massive MIMO, and beamforming, enabling readers to understand the effective design and optimization of high-performance wireless systems.
- Presents real-world applications such as telemedicine, remote healthcare, smart connectivity, IoT ecosystems, and UAV and drone-based communication systems to demonstrate practical relevance.
- Covers green wireless communication, emerging technological trends, and potential breakthroughs, providing insights into sustainable next-generation wireless communication solutions and future research directions in the field.

Advanced Wireless Communication Technologies

From Fundamentals to Future Applications

Edited by
Sasmita Dash and Dilip Senapati

CRC Press
Taylor & Francis Group
Boca Raton London New York

CRC Press is an imprint of the
Taylor & Francis Group, an **informa** business

Designed cover image: Shutterstock AI Tool

First edition published 2027
by CRC Press
2385 NW Executive Center Drive, Suite 320, Boca Raton FL 33431

and by CRC Press
4 Park Square, Milton Park, Abingdon, Oxon, OX14 4RN

CRC Press is an imprint of Taylor & Francis Group, LLC

Library of Congress Cataloging-in-Publication Data
Names: Dash, Sasmita editor | Senapati, Dilip editor
Title: Advanced wireless communication technologies : from fundamentals to future applications / edited by Sasmita Dash and Dilip Senapati.
Description: First edition. | Boca Raton : CRC Press, 2027. | Includes bibliographical references and index.
Identifiers: LCCN 2026011901 (print) | LCCN 2026011902 (ebook) | ISBN 9781041067054 hbk | ISBN 9781041068976 pbk | ISBN 9781003637721 ebk
Subjects: LCSH: Wireless communication systems
Classification: LCC TK5103.2 .A384 2026 (print) | LCC TK5103.2 (ebook)
LC record available at https://lccn.loc.gov/2026011901
LC ebook record available at https://lccn.loc.gov/2026011902

ISBN: 978-1-041-06705-4 (hbk)
ISBN: 978-1-041-06897-6 (pbk)
ISBN: 978-1-003-63772-1 (ebk)

DOI: 10.1201/9781003637721

Typeset in Sabon
by KnowledgeWorks Global Ltd.

Contents

About the Editors

Dr. Sasmita Dash is working as Special Scientist at the IRIDA Research Centre for Communication Technologies, Department of Electrical and Computer Engineering, University of Cyprus, Cyprus. Since 2019, she has been associated with European Union FET-OPEN and European Research Council projects. She also holds the position of Associate Professor in the Department of Electronics and Communication Engineering at GIET University, Odisha, India. With over a decade of experience in teaching and research, Dr. Dash has expertise in the interdisciplinary domains, including electromagnetics, wireless communications, material science, and solid-state physics. Her research interests include graphene- and nanomaterial-based devices, terahertz and microwave technologies, antennas, absorbers, sensors, metamaterials and metasurfaces, and biomedical and wireless communication applications. Dr. Dash has authored more than 50 scientific publications in leading international journals and conferences, including publications in Frontiers, IEEE, IET, Elsevier, Wiley, Springer, and Nature Portfolio journals. She has also contributed to several edited books and holds patents in her field. She has presented her research internationally in USA, Japan, Germany, Finland, Greece, Cyprus, Latvia, and India. Dr. Dash received her PhD from the Indian Institute of Technology (IIT) Roorkee, India.

Dilip Senapati is currently working as Associate Professor in the Department of Computer Science, University of Delhi, India. He has more than 15 years of teaching and research experience. He has completed his PhD and MTech in Computer Science and Technology from Jawaharlal Nehru University, New Delhi, India. He also completed a postgraduate degree from Berhampur University, Odisha, India. He has received more than four research grants for conducting research-based projects as Principal Investigator (PI). He is the author of more than 40 scholarly research papers, which have been published in reputed journals from IEEE, Wiley, Elsevier, and Springer, as well as conferences and book chapters. He has been invited to give keynote lectures at international conferences and short-term courses. His research interests include wireless sensor networks, wireless communications, Internet of Things, entropy optimization, machine learning, and social networking.

Contributors

Supriya Aggarwal
Department of Information Technology
IIIT Bhopal
Bhopal, Madhya Pradesh, India

Ashni Chatterjee
Geetanjali Dental and Research Institute
Udaipur, Rajasthan, India

Satwik Chatterjee
Department of Forensic Science
Adamas University
Kolkata, West Bengal, India

Laxminarayan Dash
Department of Computer Science and Engineering
GITA Autonomous College
Bhubaneswar, Odisha, India

Sasmita Dash
Department of Electrical and Computer Engineering
University of Cyprus
Nicosia, Cyprus
and
Department of Electronics and Communication Engineering
GIET University
Gunupur, Odisha, India

Pimmy Gandotra
Department of Electronics Engineering
Jaypee Institute of Information Technology
Noida, Uttar Pradesh, India

Neetu Joshi
Department of Electronics Engineering
Jaypee Institute of Information Technology
Noida, Uttar Pradesh, India

Pritika Kujur
Department of Computer Science
Ravenshaw University
Cuttack, Odisha, India

Priyanka Kujur
Department of Computer Science
Ravenshaw University
Cuttack, Odisha, India

Vipin Kumar
Department of Information Technology
Babasaheb Bhimrao Ambedkar University
Lucknow, Uttar Pradesh, India

Gyana Ranjan Nanda
Department of Computer Science
Ravenshaw University
Cuttack, Odisha, India

Suvra Nayak
Department of Computer Science
University of Delhi
New Delhi, India

Kapila W. S. Palitharathna
Department of Electrical and Computer Engineering
University of Cyprus
Nicosia, Cyprus

Bijayalaxmi Panda
Department of Computer Science and Engineering
GITA Autonomous College
Bhubaneswar, Odisha, India

Manaswinee Madhumita Panda
Department of Computer Science and Engineering
GITA Autonomous College
Bhubaneswar, Odisha, India

Chhabi Rani Panigrahi
Department of Computer Science
Rama Devi Women's University
Bhubaneswar, Odisha, India

Bibudhendu Pati
Department of Computer Science
Rama Devi Women's University
Bhubaneswar, Odisha, India

Nikhil Kumar Rajput
Department of Computer Science, Ramanujan College
University of Delhi
Delhi, India

Rashmi R.
Clue4 Evidence Forensic Lab
Bangalore, Karnataka, India

Dilip Senapati
Department of Computer Science
University of Delhi
New Delhi, India

Amit Kumar Singh
Department of Information Technology
Babasaheb Bhimrao Ambedkar University
Lucknow, Uttar Pradesh, India

Baljeet Ram
Department of Information Technology
Babasaheb Bhimrao Ambedkar University
Lucknow, Uttar Pradesh, India

Neha Singh
Department of Electronics and Communication Engineering
IIT Bhopal
Bhopal, Madhya Pradesh, India

Jaya Yadav
Department of Information Technology
Babasaheb Bhimrao Ambedkar University
Lucknow, Uttar Pradesh, India

Yericharla Mary Asha Latha
Department of Electronics Engineering
IIT (ISM)
Dhanbad, Jharkhand, India

Preface

Wireless communication has become an essential part of modern society, shaping the way people communicate, access information, and interact with digital systems. From early mobile communication systems to today's intelligent and interconnected networks, the field has witnessed rapid advancements driven by innovations in hardware, software, and network architectures. The rapid evolution toward 5G, 6G, and intelligent networking systems in wireless communication technologies necessitates a thorough understanding of both the foundational concepts and the emerging technology.

This book, ***Advanced Wireless Communication Technologies: From Fundamentals to Future Applications***, presents the intricacies of modern and emerging wireless communication systems. It aims to provide readers with a thorough understanding of modern and next-generation wireless communication systems by combining theoretical foundations, enabling technologies, and real-world applications. The book covers a wide range of topics, including advanced antenna technologies, massive MIMO, beamforming, O-RAN, machine learning and AI, reconfigurable intelligent surfaces (RIS), IoT, UAV communication, visible light communication, telemedicine, and green communication technologies. The book also explores emerging trends, technological challenges, and research opportunities, encouraging readers to think critically about the future direction of wireless communication.

The book is intended for senior undergraduate and postgraduate students, researchers, academicians, and industry professionals in electronics, communication engineering, and related fields. This book will serve as a valuable resource for learning, teaching, and research, and will contribute to the advancement of knowledge in the field of wireless communication and inspire further innovation in this rapidly evolving field.

Abbreviations

1G	first generation
2D	two-dimensional
2G	second generation
3G	third generation
3GPP	third generation partnership project
4G	fourth generation
5G NR	5G new radio
5G	fifth generation
6G	sixth generation
7G	seventh generation
A2DP	Advanced Audio Distribution Profile
ADM	adaptive delta modulation
ADR	adaptive data rate
AE–ELM	autoencoder–extreme learning machine
AES	Advanced Encryption Standard
AGVs	autonomous guided vehicles
AI	artificial intelligence
ALMS	adaptive least mean square
AMC	adaptive modulation and coding
AoA	angle of arrival
AoD	angle of departure
AoI	age of information
AP	access point
APD	avalanche photodiode
API	application programming interface
APT	Asia-Pacific Telecommunity
AR	augmented reality
AR	autoregressive
ASK	amplitude shift keying
AUV	autonomous underwater vehicle
BCI	brain–computer interfaces
BD-RIS	beyond diagonal reconfigurable intelligent surface
BER	bit error rate
BLE	Bluetooth Low Energy
BPSK	binary phase shift keying
BS	base station
CAP	carrierless amplitude and phase modulation
CAPEX	capital expenditure

CCB/J	continuous class B/J
CCE	continuous class E
CCF	continuous class F
CCF^{-1}	continuous class F^{-1}
CCTV	closed-circuit television
CDMA	code-division multiple access
CE	continuous envelope
CGM	continuous glucose monitor
CMOS	complementary metal–oxide–semiconductor
CNN	convolutional neural network
CoAP	Constrained Application Protocol
cobots	collaborative robots
COPD	chronic obstructive pulmonary disease
COVID-19	coronavirus disease 2019
CR	cognitive radio
CRP	current source reference plane
CRS	cell-specific reference signals
CSI	channel state information
CSS	chirp spread spectrum
CU-CP	centralized unit-control plane
CU-UP	centralized unit-user plane
C-V2X	cellular vehicle-to-everything
D2D	device-to-device
DBPSK	differential binary phase shift keying
DC	direct current
DD	direct detection
DD	Doppler domain
DDoS	distributed denial of service
DDPM	denoising diffusion probabilistic models
DE	drain efficiency
DISHA	Digital Information Security in Healthcare Act
DMT	discrete multi-tone modulation
DNNs	deep neural networks
DPC	dirty paper coding
DRL	deep reinforcement learning
DRS	demodulation reference signals
DSRC	dedicated short-range communications
DSS	dynamic spectrum sharing
DSSS	direct sequence spread spectrum
EC	edge computing
ECCE	extended continuous class E
ECCF	extended continuous class F
ECCF^{-1}	extended continuous class F^{-1}
ECG	electrocardiogram
eDRX	extended discontinuous reception
EE	energy efficiency
EH	energy harvesting
EHR	electronic health record
EI	edge intelligence
EM	electromagnetic

eMBB	enhanced mobile broadband
eNB	evolved node base
ERCB/J	extended resistive continuous class B/J
EU	European Union
EV	electric vehicle
FAA	Federal Aviation Administration
FBMC-OQAM	Filter Bank Multicarrier with Offset Quadrature Amplitude Modulation
FD	frequency domain
FD	full-duplex
FDMA	frequency division multiple access
FHIR	Fast Healthcare Interoperability Resources
FM	frequency modulation
FoV	field-of-view
FPGAs	field-programmable gate arrays
GaN	gallium nitride
GDPR	General Data Protection Regulation
GFSK	Gaussian frequency shift keying
GHz	gigahertz
GIS	geographic information system
gNB	next generation node base
GNN	graph neural networks
GNSS	Global Navigation Satellite Systems
GPS	global positioning system
GSM	global system for mobile communications
HAPS	high-altitude platform stations
HCM	Hadamard coded modulation
HD	high definition
HF	high frequency
HFP	Hands-Free Profile
HID	human interface device
HIE	health information exchange
High-PHY	higher physical layer
HIPAA	Health Insurance Portability and Accountability Act
HL7	Health Level Seven (international standards organization)
HRM	hybrid reflection modulation
HSS	Home Subscriber Server
HVAC	heating, ventilation, and air conditioning
IBFD	in-band full-duplex
ICAO	International Civil Aviation Organization
ICD-10	International Classification of Diseases – 10th Revision
ICU	intensive care unit
IEEE	Institute of Electrical and Electronics Engineers
IIoT	industrial internet of things
IM	intensity modulation
IMSIP	Multimedia Subsystem
IoE	Internet of Everything
IoT	Internet of Things
IoV	internet of vehicles
IRS	intelligent reflecting surfaces
ISAC	integrated sensing and communications

ISM	industrial, scientific, and medical bands
IT	information technology
ITS	intelligent transport systems
ITU	International Telecommunication Union
JCAS	RIS-assisted joint communication and sensing
JSLS	*Journal of the Society of Laparoendoscopic Surgeons*
Kbps	kilobits per second
LAN	local area network
LC	liquid crystal
LD	laser diode
LED	light-emitting diode
LEO	low Earth orbit
LF	low frequency
LIDAR	light detection and ranging
Li-Fi	light fidelity
LOINC	Logical Observation Identifiers Names and Codes
LoRaWAN	low-power wide-area network
LoS	line-of-sight
LPD	low probability of detection
LPI	low probability of intercept
LPWAN	low-power wide-area network
LSTM	long short-term memory (neural network)
LTE	long-term evolution
LTE-M	long-term evolution for machines
MaaS	mobility-as-a-service
MAC	media access control
MAL	multi-agent learning
MBM	media-based modulation
Mbps	megabits per second
MEC	multi-access edge computing
MEO	medium earth orbit
MFA	multi-factor authentication
mHealth	mobile health
MHz	megahertz
MIMO	multiple-input multiple-output
ML	machine learning
mm	millimeter
MME	Mobility Management Entity
MMS	Multimedia Messaging Service
mMTC	massive machine-type communications
mm-wave	millimeter wave
MQTT	Message Queuing Telemetry Transport
MRC	maximum ratio combining
MRSS	multi-RAT spectrum sharing
MRT	maximum ratio transmission
MSF	metasurfaces
MTM	machine-to-machine communication
MU	multi-user
MU-MIMO	multi-user multiple-input multiple-output
NB-IoT	narrowband internet of things

Near RT RIC	near-real-time RIC
NFC	near-field communications
NFV	network function virtualization
NGC	next generation core
NLoS	non-line-of-sight
NOMA	non-orthogonal frequency-division multiplexing
Non-RT RIC	non-real-time RIC
NR	new radio
NTN	non-terrestrial networks
O-DU	open distributed unit
OFDM	orthogonal frequency-division multiplexing
OLED	organic light-emitting diode
OOK	on–off keying
OPAMP	operational amplifier
OPEX	operational expenditure
OQPSK	offset quadrature phase-shift keying
O-RAN	open radio access network
O-RU	open radio unit
OTA	over-the-air
PAM	pulse amplitude modulation
PAPR	peak-to-average power ratio
PD	photodiode
PDCP	packet data convergence protocol
PDF	probability density function
PETs	privacy-enhancing technologies
PGW	Painted Grey Ware
PHY	physical
PIN	personal identification number
PIN	positive-intrinsic-negative
PLS	physical layer security
PPM	pulse position modulation
PSM	power saving mode
PV	photovoltaic
PWM	pulse width modulation
QAM	quadrature amplitude modulation
QKD	quantum key distribution
QoS	quality of service
QPSK	quadrature phase shift keying
RADAR	radio detection and ranging
RAN	radio access network
RF	radio frequency
RFID	radio-frequency identification
RGB	red–green–blue
RIC	RAN Intelligent Controller
RIS	reconfigurable intelligent surface
RL	reinforcement learning
RLC	Radio Link Control
RNC	radio network controller
RNN	recurrent neural networks
RP	random process

RPM	remote patient monitoring
RRC	Radio Resource Control
RRM	radio resource management
RSA	Rivest–Shamir–Adleman
RWP	random waypoint
SAFER	Secure and Fast Encryption Routine
SC-FDMA	Single Carrier Frequency Division Multiple Access
SD	sphere decoding
SDAP	service data adaptation protocol
SDG	Sustainable Development Goal
SDMA	spatial division multiple access
SDN	software-defined network
SDR	software-defined radio
SE	spectral efficiency
SGW	Serving Gateway
SI	self-interference
SIC	successive interference cancellation
SIG	special interest group
SINR	signal-to-interference-plus-noise ratio
SiPM	silicon photomultiplier
SIW	substrate-integrated waveguide
SLIPT	simultaneous lightwave information and power transfer
SMO	service management and orchestration
SMS	short message service
SNOMED CT	Systematized Nomenclature of Medicine – Clinical Terms
SONs	self-optimizing networks
SRP	stochastic routing protocol
STAR	simultaneously transmit and receive
STAR-RIS	simultaneously transmitting and reflecting RIS
SUMO	Simulation of Urban MObility
SVR	support vector regression
SWIPT	simultaneous wireless information and power transfer
Tbps	terabit-per-second
TD	time domain
TDMA	time division multiple access
TDOA	radar and time difference of arrival
THP	Tomlinson–Harashima precoding
THz	terahertz
TIA	transimpedance amplifier
TLL	tunable liquid lens
TLS	Transport Layer Security
ToA	time of arrival
TV	television
TVWS	Television White Spaces
UAV	unmanned aerial vehicle
UE	user equipment
UHF	ultra high frequency
UMTS	Universal Mobile Telecommunications Systems
UNICEF	United Nations International Children's Emergency Fund
URLLC	ultra-reliable low-latency communication

USA	United States of America
UVLC	underwater visible light communication
UWB	ultra-wideband
UWC	underwater wireless communication
V2I	vehicle-to-infrastructure
V2N	vehicle-to-network
V2P	vehicle-to-pedestrian
V2V	vehicle-to-vehicle
V2X	vehicle-to-everything
VLC	visible light communication
VLP	visible light positioning
VoIP	Voice over Internet Protocol
VP	vector perturbation
VR	virtual reality
WAP	wireless access point
WCDMA	wideband code-division multiple access
WHO	World Health Organization
Wi-Fi	wireless fidelity
WLAN	wireless local area networks
WPT	wireless power transfer
WSN	wireless sensor networks
XR	extended reality
ZF	zero-forcing

Chapter 1

Advancement in wireless communication technology

Sasmita Dash and Neha Singh

1.1 INTRODUCTION

Wireless communication technologies have evolved from analog voice systems in the early 1980s to ultra-reliable low-latency fifth-generation (5G) and the sixth-generation (6G) networks, which increase data speed and network reliability and enable new paradigms such as the internet of things (IoT), machine-to-machine (MTM) communication, and ubiquitous mobile connectivity, driven by the exponential growth in mobile users and smart devices, and the increasing demand for real-time and data-intensive applications [1–3]. This is attributed to the increasing number of smart devices, the growing number of mobile users, and the rising demand for high-data-rate, real-time applications. Researchers have investigated higher frequency bands, such as the millimetre-wave (mm-wave) and terahertz (THz) bands, to support large bandwidths and high data rates [4–6]. Specifically, modern wireless communication is centred on smart cities, health, virtual worlds, and smart transportation [7].

The 5G and 6G networks are intended to create a smart and intelligent network architecture, with a focus on green and energy-efficient communication to minimise environmental impact while maintaining high performance [8]. Technologies such as THz communication, quantum encryption, and holographic beamforming will be utilised in 6G to support real-time applications, including brain–computer interfaces (BCI), extended reality (XR), and digital twins [9]. These modern techniques include software-defined radio (SDR), edge computing (EC) and network slicing, which improve the architecture of communication networks and promote adaptability, scalability, and service differentiation [10, 11]. Beamforming, reconfigurable intelligent surfaces (RISs), and massive multiple-input multiple-output (MIMO) systems also improve spectral efficiency and energy efficiency by exploring signal transmission and reception [12, 13]. In this chapter, the major technologies of the future, emerging trends, and applications in modern wireless communication are summarised, and the cellular generations from first generation (1G) to 6G are overviewed. The spectrum management strategies for next-generation wireless systems are also provided. The technological advancements, research challenges, and prospects of next-generation wireless communications are also discussed.

1.2 OVERVIEW OF CELLULAR GENERATIONS

The evolution of mobile communication, from basic voice calls to high-speed global connectivity, has enhanced human society. Mobile communication technology has evolved rapidly over the past few decades, and each generation, known as G, represents a major step in the evolution of wireless communication, defining advancements in capacity, data rate, latency, and service diversity. As shown in Figure 1.1 and summarised in Table 1.1, each generation of mobile technology has evolved with significant technological advancements.

DOI: 10.1201/9781003637721-1

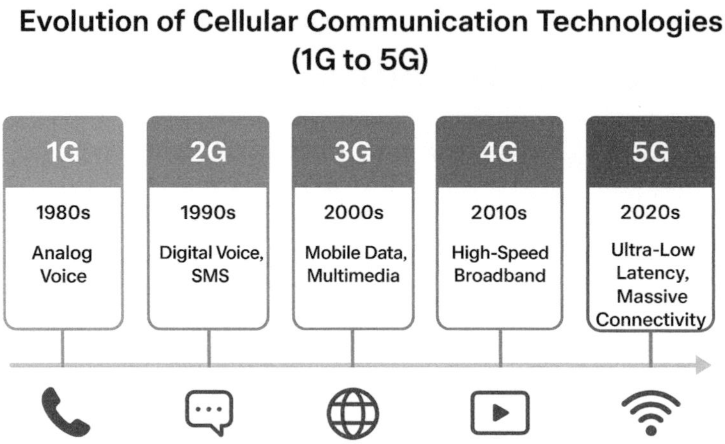

Figure 1.1 Performance evolution across cellular generations.

Table 1.1 The characteristics and technological advancements of each generation from 1G to 5G

Generation	*Time period*	*Core technology/ standard*	*Data rate*	*Key features*	*Typical applications*
1G	Early 1980s	Analog (FM), FDMA	Up to 2.4 kbps	Analog voice transmission, limited capacity, poor security	Basic voice telephony
2G	Early 1990s	GSM, CDMA, TDMA	Up to 64 kbps	Digital voice, SMS/MMS, improved security	Voice, text messaging
3G	Early 2000s	UMTS, WCDMA, HSPA	384 kbps – several Mbps	Mobile internet, multimedia, video calling	Internet access, video streaming
4G (LTE)	2010 onwards	LTE, OFDM, MIMO	100 Mbps – 1 Gbps	All-IP network, HD streaming, VoIP, low latency	Broadband internet, IoT, mobile apps
5G	2020 onwards	NR, mm-wave, massive MIMO	Up to 10 Gbps	Ultra–low latency, massive connectivity, high reliability	Smart cities, autonomous vehicles, remote surgery, AR/VR applications

Abbreviations: CDMA – code division multiple access; FDMA – frequency division multiple access; HSPA – high speed packet access; MIMO – multiple-input multiple-output; NR – new radio; OFDM – orthogonal frequency division multiplexing; TDMA – time division multiple access; UMTS – universal mobile telecommunications system; WCDMA – wideband code division multiple access.

1.2.1 First generation (1G): Analog voice communication

Mobile communication started with the 1G in the early 1980s, when making a call without wires seemed revolutionary. These were entirely analog systems that used frequency modulation (FM) techniques to carry communication [14]. The first networks, 1G, were designed exclusively for voice calls; there were no text messages, internet services, or multimedia capabilities. Different regions adopted their own standards: for example, AMPS was the main standard in the United States, TACS in the United Kingdom, and NMT in Northern Europe, which covered a wide geographical area. While 1G networks laid the foundation for mobile communication, several shortcomings existed: voice quality was often poor due to interference and noise, especially in crowded areas; the system lacked encryption; and security was a concern, as calls could be intercepted or cloned using simple radio equipment. The network provided extremely low data capacity, typically under 2.4 kbps, and no digital services were available. Nevertheless, 1G was a significant step towards a world where mobile communication was no longer a luxury but a necessity.

1.2.2 Second generation (2G): Digital voice and SMS services

The second generation (2G) mobile communication systems that began deployment in the early 1990s became fully digital for the first time and provided significantly enhanced performance over 1G systems with improved spectral efficiency, clearer voice quality, fewer dropped calls, lower background noise levels, and a more comfortable user experience; Short Message Service (SMS), originally developed as network notification messages but quickly adopted by users as personal messaging tools and text-messaging culture; Multimedia Messaging Service (MMS) for sharing pictures and simple multimedia files; data rates up to 64 kbps that provided basic internet access, email exchange, and very slow online services, which was the 1G of mobile data service beyond just voice calls; and higher level network security using encryption techniques during call connection or data transmission as compared with analog systems in 1G [15].

1.2.3 Third generation (3G): Mobile data and multimedia access

Mobile internet access has enabled basic communication devices to become intelligent, connected services through online services, available anywhere. By the turn of the century, third-generation (3G) mobile communications gave proper mobility for users and transformed simple phones into advanced, interconnected smart services; these features included higher data rates (384 kbps that later upgraded to several Mbps), faster browsing time and download speeds, video calling, social media in its early stages, mobile television (TV) service, and email use; 3G was technically innovative because it combined packet-switched data with traditional circuit-switched voice to allow simultaneous use [16].

1.2.4 Fourth generation (4G/LTE): High-speed broadband connectivity

Fourth generation (4G), commercialised from around 2010 onwards with technologies such as long-term evolution (LTE) and orthogonal frequency division multiplexing (OFDM), represented a major shift for mobile networks away from voice- and messaging-centric designs towards high-speed data access by typical users, allowing them to run demanding applications previously impractical or impossible on mobile. The migration towards an all-IP infrastructure provided more seamless support of the heavy load that these services entail and also enabled delivery of advanced multimedia functions such as HD video streaming, VoIP, mobile TV, online conferencing, and cloud-based apps, from basic connectivity to rich media consumption experiences with higher data rates, lower latency, greater spectral efficiency (e.g., via MIMO antennas), increased available bandwidth by carrier aggregation, and so on [17].

1.2.5 Fifth generation (5G): Ultra-low latency and massive connectivity

5G wireless communication is a next-generation mobile network technology being deployed globally which aims at delivering a vastly more diverse ecosystem of applications, devices, and industries by accommodating an extensive range of performance requirements for enhanced Mobile Broadband (eMBB), massive Machine-Type Communications (mMTC), and Ultra-Reliable Low-Latency Communication (URLLC) through the exceptional theoretical data rates exceeding 10 Gbps with latency less than 1 ms which enables near real-time responsiveness, due to key enabling technologies such as operating in mm-wave frequency bands, massive MIMO antenna systems, carrier aggregation, and advanced beamforming techniques that dramatically increase network throughput, spectral efficiency, and reliability, especially for dense urban and industrial environments [17]. Beyond raw speed and capacity, 5G will deliver billions of connected devices to accommodate large-scale IoT applications like smart city infrastructure, intelligent energy grids, environmental monitoring, and connected healthcare systems; massive device density so sensors, machines, vehicles, and automated systems can exchange data with minimal delay; as well as new and mission-critical uses that demand extreme reliability and low latency.

1.3 TECHNOLOGIES FOR ADVANCED WIRELESS COMMUNICATION

With the exponential growth in mobile traffic, connected devices, and data-intensive applications, wireless communication technologies have been compelled to evolve to meet the increasing demands for higher data rates, improved reliability, and lower latency. From advanced antenna architectures to intelligent network management, several enablers define the foundation of present- and next-generation wireless communication systems. These key enablers have played a major role in the rapid evolution of wireless communication from 1G to 5G and beyond, significantly enhancing data capacity, spectrum efficiency, energy utilisation, and network intelligence. The most important advanced technologies for wireless communication are described in the following subsections and illustrated in Figure 1.2.

1.3.1 Massive MIMO and beamforming

Massive MIMO is a technique that utilises an array of multiple base station antennas to serve multiple users simultaneously in the same time–frequency resources, thereby increasing spectral efficiency, energy efficiency, and link reliability by exploiting spatial multiplexing and diversity [18]. Massive MIMO is a closely related technique that enables directional transmission and reception of signals to intended users, rather than broadcasting energy in all directions. This reduces interference and maximises the Signal-to-Noise Ratio (SNR), especially for high-frequency communication, such as the millimetre-wave and THz bands.

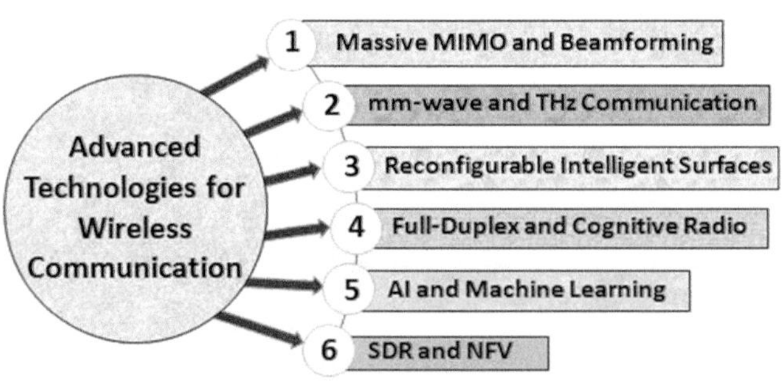

Figure 1.2 Advanced technologies for wireless communication.

Massive MIMO is the foundation of 5G and beyond networks, enabling high capacity and coverage with minimal energy waste. Hybrid analog–digital beamforming techniques are widely employed in 5G and beyond, striking a balance between performance and hardware complexity and facilitating high-capacity, low-latency communication for applications such as autonomous vehicles and smart cities.

Massive MIMO is a critical enabler of 5G and beyond networks. The technology enables step-change increases in network capacity, connection dependability, and spectrum efficiency [19]. The signals propagating through space are subject to degradation caused by multipath fading and interference. Security issues in massive MIMO and beamforming can result in erroneous channel estimation, which affects the confidentiality, availability, and integrity of wireless services [20]. Massive MIMO systems enable the massive collection of user locations and other personal information, which makes user privacy a critical issue beyond security. MIMO networks offer the potential for fine-grained user data, as they can serve multiple users simultaneously and may expose profiles and behavioural patterns if not properly encrypted. Stricter data privacy laws around the world have led to user privacy being disregarded and a loss of confidence. Massive MIMO systems are designed to protect privacy. In contrast to standard MIMO systems, which typically employ only a few antennas, massive MIMO systems utilise dozens or even hundreds of antennas to serve multiple customers simultaneously [21]. It controls interference, beam form, and spatial multiplex to increase data rates, improve signal quality, and increase network capacity.

1.3.2 mm-wave and THz communication

The conventional sub-6 GHz frequency spectrum is becoming increasingly congested due to the proliferation of wireless devices. To overcome spectrum scarcity, next-generation wireless systems are expanding into mm-wave (30–300 GHz) and THz (0.1–10 THz) frequency bands. Due to their ability to support exceptionally wide bandwidths and ultra-high data rates, high-frequency spectra in the mm-wave and THz frequency bands are key enablers of next-generation wireless communication systems that can provide multi-gigabit and even terabit-per-second (Tbps) transmission speeds for applications such as holographic communications, ultra-high-definition media delivery, immersive virtual and augmented reality (AR), tactile internet, and real-time digital twins, all of which are vital for future smart environments to connect autonomous vehicles, Industry 4.0 infrastructures, remote robotic systems, and intelligent healthcare platforms [22–24]. While mm-wave and THz signals offer numerous benefits, their extremely short wavelengths make them highly susceptible to propagation issues, including severe free-space path loss, atmospheric absorption (primarily from oxygen and water vapour), and scattering from environmental surfaces [25]. They have poor penetration through common barriers (e.g., walls, foliage, even the human body). Advanced techniques, such as adaptive beamforming, phased-array architectures, high-gain directional antennas, and RIS, are used to extend coverage, increase reliability, and maintain link robustness in dynamic and non-line-of-sight (NLOS) environments [25–30].

In addition to mm-wave, the THz spectrum (0.1–10 THz) is being considered a potential area of development for 6G communication, as THz wireless systems are expected to provide ultra-high data rates, sub-millisecond latency, high energy efficiency, and highly accurate localisation and sensing, which are key requirements for emerging applications such as space–terrestrial integrated networks, bio-inspired communication systems, and ultra-fast short-range wireless links for next-generation computing and semiconductor technologies. The recent advances in material science have also sped up the mm-wave and THz system development, such as graphene, metamaterials, and RISs that enable compact, tunable, and dynamically reconfigurable communication components (e.g., filters, modulators, transceivers, and antennas) with unique plasmonic properties (e.g., high carrier mobility and electrostatic tunability) that allow dynamic control of radiation patterns, operating frequencies, and beam direction, which will lead to programmable and adaptive wireless communication platforms that can meet the high-performance requirements for future 6G and beyond networks [26, 31–34].

1.3.3 Reconfigurable intelligent surfaces (RISs)

RISs are a promising technology for the future of wireless communication systems, enabling smart propagation environments in which electromagnetic waves can be dynamically manipulated to optimise communication performance [13, 32]. An RIS is a planar array of passive, semi-passive, or low-powered, tunable meta-atoms or unit cells that can be controlled by real-time signals to vary their phase, amplitude, polarisation, or scattering properties. RISs are usually designed with metamaterials, tunable Radio Frequency (RF) components, graphene, or CMOS (complementary metal-oxide-semiconductor)-integrated designs. RIS can reflect, refract, absorb, or redirect wireless signals to achieve desired communication outcomes, such as steering a communication beam towards a target user, suppressing interference, or improving spatial multiplexing [13, 35]. With the ability to make such configuration changes, RIS can create virtual Line-of-Sight (LoS) links in complex urban or indoor environments that would otherwise degrade signal quality, significantly enhancing network coverage, spectral efficiency, and communication reliability in NLOS environments.

A third advantage of RIS technology is that it requires very little power (only for configuration control, not for signal amplification) and has simple hardware compared to the active relays or conventional access points that need dedicated RF chains, continuous power supply, and complicated processing units, making RIS a scalable and sustainable solution for green communication infrastructures in alignment with energy-efficiency goals for future 6G and beyond systems. RIS can facilitate new communication paradigms such as cell-free massive MIMO, cooperative beamforming, and user-centric networking, where multiple surfaces coordinate to adjust wireless channels to meet user or service needs [26]. Due to the substantial propagation loss in the THz band, RIS is more attractive in this band and can be used to realise beam steering, focusing, and environmental reconfigurability via metasurface concepts such as holographic beamforming and programmable metastructures [13]. Furthermore, RIS-assisted localisation and sensing capabilities can enhance the accuracy of positioning systems, enabling applications in autonomous transportation, smart healthcare, and industrial automation. Besides the hardware innovations, implementing RIS in 6G networks also requires development in artificial intelligence (AI), optimisation algorithms, channel modelling, and control protocols.

Real-time adaptation to user mobility, interference patterns, and environmental dynamics will rely on machine learning (ML)-driven RIS control, which will also open new research directions such as hybrid active–passive RIS architectures, full-duplex (FD) RIS, RIS-aided MIMO, large-scale RIS deployment strategies, and joint communication-sensing systems. As communication networks move towards ubiquitous connectivity and intelligent environments, RIS is expected to be a key enabler by converting passive physical spaces into programmable and cooperative components of the communication system and will play an important role in cell-free architectures, URLLC, THz and mm-wave systems, holographic communications, and sustainable wireless network designs, which will lead to a paradigm shift in the thinking of the conception, deployment, and optimisation of wireless communication systems.

1.3.4 Full-duplex and cognitive radio systems

Traditional wireless communication systems are half-duplex systems, meaning that a device can transmit or receive but not both simultaneously, which may limit spectral efficiency and prevent simultaneous bidirectional communication. However, FD communication allows a wireless transceiver to transmit and receive signals simultaneously on the same time and frequency resource, which doubles the spectral efficiency and throughput, reduces latency, and is desirable for next-generation wireless systems, such as 6G and beyond [36]. Nevertheless, the main technical hurdle to practical FD wireless operation is self-interference (SI), which is the strong leakage of the transmitted signal into the receiver chain [37]. This can overwhelm the weak incoming signal, as the transmit signal power is much higher than the received signal power. FD systems are closer to practical implementation through multi-stage cancellation strategies, including antenna isolation, analog-domain cancellation, and digital signal processing-based interference suppression, as well as emerging techniques such as adaptive filtering, MIMO-based spatial cancellation, metamaterial antennas, and ML-assisted interference modelling.

In parallel with advancements in FD technology, the concept of cognitive radio (CR) technology has emerged as a promising technique for intelligent and flexible spectrum management [38–40]. The static spectrum allocation has become inadequate for the exponentially increasing number of wireless devices and services, resulting in underutilised licenced frequency bands and unlicenced bands becoming congested. CR introduces the concept of dynamic spectrum access, which allows secondary (unlicenced) users to opportunistically access the licenced spectrum that is idle or underused without causing harmful interference to primary users. CR systems can dynamically monitor and adapt to the radio environment through spectrum sensing, spectrum decision-making, learning-based optimisation, and adaptive transmission.

In addition to CR operation, ML and AI can further enhance CR operation by predicting spectrum allocation, avoiding interference, and making proactive decisions. This enables the transformation of traditional wireless systems into autonomous and intelligent spectrum frameworks that learn from usage patterns and environmental conditions. FD and CR technologies complement each other to shape the future of wireless communication systems, where FD operation enhances spectrum utilisation efficiency. CR ensures the available spectrum resources are identified, allocated, and utilised intelligently to support spectrum agility, enhanced network capacity, reduced latency, improved energy efficiency, and flexible resource management, which are all the characteristics of future dynamic communication environments, including 6G, smart cities, the IoT, autonomous vehicles, and mission-critical communication systems.

1.3.5 AI and ML in wireless networks

Traditional rule-based and static optimisation methods cannot meet the requirements of increasingly complex modern wireless communication networks, especially in cases where wireless systems scale to support billions of devices, diversified services, heterogeneous architectures, and dynamic user mobility. AI and ML have become fundamental to the automation, adaptability, and real-time decision-making throughout the communication stack. AI and ML are now deeply embedded in many layers of wireless systems, from the physical (PHY) layer to higher network and application layers [41–43]. While physical-layer tasks, such as modulation classification, channel estimation, and signal detection, have historically relied on mathematical approximations or pilot-based strategies, ML-based models are now being applied to these tasks. In addition, deep learning architectures like convolutional neural networks (CNNs), recurrent neural networks (RNNs), and graph neural networks (GNNs) are being used to learn channel characteristics, estimate fading conditions, enhance

MIMO detection, and enable intelligent beamforming and precoding strategies, especially for mm-wave and THz systems where the channel dynamics are very complex [44].

At the media access control (MAC) and network layers, AI enables adaptive resource management, mobility prediction, energy-efficient scheduling, and dynamic spectrum access, and reinforcement learning algorithms, such as deep Q-learning and multi-agent learning, facilitate autonomous and cooperative decision-making among distributed network nodes, which are particularly useful in highly dense environments such as smart cities, autonomous vehicular networks, and massive IoT ecosystems where resource contention and interference must be managed dynamically. Predictive network analytics using AI is also critical to identifying usage trends, traffic congestion, and system bottlenecks before they occur by learning from both real-time and historical data to predict demand surges, optimise bandwidth allocation, and balance load across edge, cloud, and fog computing infrastructures, and detect cyberattacks, system malfunctions, and abnormal traffic patterns [45].

AI-native networking will fundamentally alter the functioning of wireless communication systems, and AI will be an integral part of the network architecture, enabling self-configuring networks that adapt automatically to new devices and topology changes, self-optimising systems that improve performance through continuous learning, self-healing networks that can autonomously recover from faults or disruptions, and context-aware communication where networks adapt to user intent, environment, and mobility [46]. Furthermore, federated learning and distributed AI will be key to achieve distributed intelligence that maintains data privacy, especially in emerging applications such as healthcare, autonomous systems, and industrial automation. Therefore, the application of AI and ML turns wireless networks from passive and reactive systems into intelligent, scalable, and autonomous systems to manage massive connectivity, ultra–low latency communication, and energy-efficient operation, which are necessary to meet the various performance requirements of next-generation communication environments such as the metaverse, autonomous vehicles, tactile internet, holographic communications, and large-scale cyber-physical systems [47, 48].

1.3.6 Software-defined radio (SDR) and network function virtualisation (NFV)

SDR has evolved the hardware architecture of wireless communication systems from a fixed, hardware-dependent radio design to a fully programmable, software-driven radio system, which can be dynamically reconfigured and updated for frequency bands, communication protocols, bandwidths, and modulation formats, offering unprecedented flexibility and scalability, capable of multi-standard and multi-band operation for technologies such as LTE, Wi-Fi, Bluetooth, 5G new radio (5G NR), and emerging 6G waveforms on a single physical platform, while speeding up the research and development cycle through rapid prototyping, debugging, and field testing without needing to rebuild the hardware, reducing development cost and time. Another complementary approach to SDR is network function virtualization (NFV), which virtualises network services (such as firewalls, intrusion detection systems, load balancers, routers, and gateways) from proprietary hardware and runs them on commodity cloud servers for centralised management, elastic scaling, and resource optimisation. This dramatically reduces hardware dependency, operational complexity, and capital expenditure while increasing the agility and service innovation by allowing operators to deploy new services, update protocols, or scale capabilities using virtual machine or container orchestration instead of a physical upgrade or new appliance [49, 50].

Together, SDR and NFV are the fundamental technologies behind the softwarisation of the wireless networks, which allow 5G architectures to support cloud-native and virtualised cores, network slicing, distributed EC, real-time orchestration, and dynamic reconfiguration and programmability to satisfy different quality-of-service (QoS) requirements, latency constraints, and user mobility patterns for various use cases (e.g., smart manufacturing, autonomous vehicles, massive IoT, and immersive XR. SDR and NFV will also remain the foundational enablers for 6G networks, which will require even greater flexibility and intelligence to deliver ultra-high data rates, AI-driven automation, and ubiquitous sensing. Future SDR platforms will also be equipped with CR functionality to perform real-time spectrum sensing and adjust waveforms in a self-optimised manner. In contrast, NFV will be increasingly integrated with AI, edge cloud computing, and intent-based networking to create self-optimised, self-managed, and self-healing wireless infrastructures. Last but not least, the combination of SDR and NFV provides the technological foundation for highly agile, programmable, and sustainable communication ecosystems that can evolve from the current systems to the fully autonomous and connected paradigm envisioned for 6G smart networks [51].

1.4 EMERGING TRENDS AND APPLICATIONS

This transformation is fuelled by the convergence of new technologies and their increasing use in a variety of applications, and next-generation communication systems are specifically designed to provide seamless, intelligent, secure, and energy-efficient connectivity to support billions of devices, autonomous systems, and mission-critical services to fundamentally change the way networks are designed, optimised, and utilised to enable unprecedented levels of automation, mobility, and data exchange. These include the evolution to 6G and beyond, which will not only transmit data but also sense, localise, and provide environmental awareness. THz communication, holographic MIMO surfaces, and RIS are enabling ultra–high capacity wireless links and intelligent propagation control beyond traditional spectrum and coverage limitations. The use of AI-native architectures for complexity management is also increasingly at the forefront, encompassing intelligent resource allocation, predictive maintenance, and real-time autonomous network adaptation.

The other major trend is the scale of connectivity for the IoT. The performance profiles for IoT applications vary, with ultra–low latency, ultra-high reliability, and extremely low power requirements for industrial automation, smart agriculture, healthcare, and digital twins. These massive deployments must be supported efficiently with mMTC, narrowband IoT (NB-IoT), and EC technologies. At the same time, edge and cloud convergence is emerging as a critical technology to enable the distribution of computational resources and intelligence closer to end users and devices to minimise latency, maximise privacy, and provide better response times for time-sensitive applications such as remote surgery, autonomous vehicles, and immersive augmented reality (AR)/virtual reality (VR)/extended reality (XR) experiences. The convergence of digital twins, cyber-physical systems, and XR is expected to yield hyper-realistic simulations and interactive environments for training, design, and remote collaboration.

Moreover, the growing need for sustainable and energy-efficient communication has led to green networking, utilising renewable-powered base stations, energy-aware protocols, and intelligent sleep-mode mechanisms to minimise the environmental footprint of the exponentially growing data traffic, facilitated by technologies such as RIS, battery-free IoT, and low-power AI accelerators. These applications span smart cities and infrastructure automation, remote healthcare and telemedicine, autonomous transportation and smart mobility, precision agriculture and environmental monitoring, defence and space communication systems, Industry 4.0 and smart manufacturing, and immersive metaverse and multi-sensory communications. Wireless technology is evolving and will continue to do so, ushering in new efficiencies and applications in virtually every aspect of digital society. Figure 1.3 presents emerging applications of future wireless communication systems.

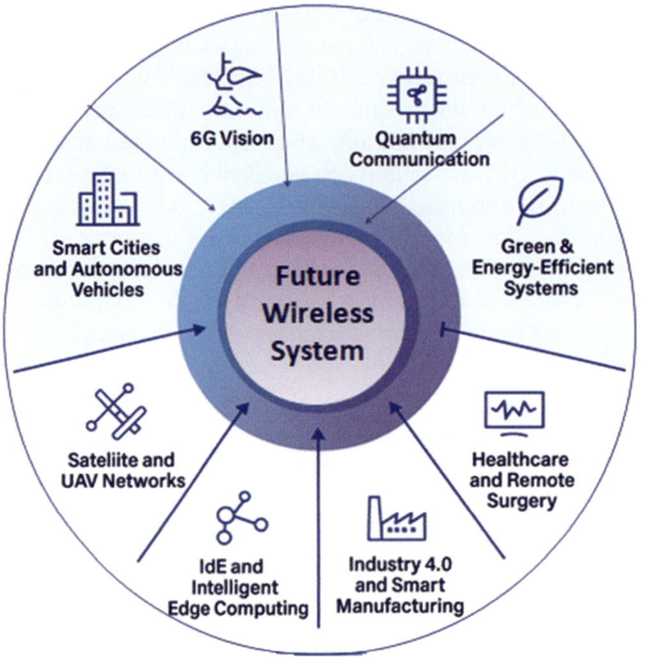

Figure 1.3 Emerging applications of future wireless communication systems.

1.4.1 6G communication

The transition to 6G wireless networks represents a paradigm shift from traditional communication technology to a cognitive, hyper-connected, and autonomous digital ecosystem that can deliver Tbps data rates, sub-millisecond latency, ultra-high reliability, and ubiquitous communication across terrestrial, aerial, and satellite platforms [52]. THz communication, with its vast unallocated spectral resources and ultra-high bandwidth, is expected to play a major role in 6G technology, supporting ultra-broadband applications with its high bandwidth [4, 25]. Holographic beamforming, IRS, and reconfigurable metasurfaces are incorporated into 6G to address the high propagation loss and atmospheric absorption and to realise the accurate focusing of energy and adaptive beam steering in dynamic environments [13, 26].

The other foundation of 6G is AI-native network design, in which AI is not just an enhancement layer. However, it is embedded into all aspects of the network, including resource allocation and routing, self-configuration and predictive maintenance, energy consumption, spectrum opportunities, traffic patterns, and real-time decision-making, which would allow for cognitive, context-aware communication to manage and maintain heterogeneous and massive device ecosystems and would enable new use cases such as full-scale XR (e.g., holographic communication, metaverse interaction, multi-sensory digital engagement), real-time digital twins (e.g., intelligent manufacturing, precision healthcare, smart energy grids, and disaster management), and high-fidelity telepresence and tactile internet (e.g., remote surgery, tele-robotics, autonomous transportation, and collaborative industrial automation with human-like precision) [53, 54].

Furthermore, 6G is predicted to play a major role in achieving sustainability goals through energy-aware protocols, green communication architectures, and energy-harvesting technologies, which will be required to have minimal environmental impact while enabling exponential growth in data traffic and device density [54].

1.4.2 Quantum communication

Quantum communication is a new technology for secure data transmission that leverages the laws of quantum mechanics to establish provable physical-layer security for future communication networks [55]. Quantum key distribution (QKD) is a key-enabling technology in this domain, as it enables two authenticated users to generate and exchange cryptographic keys that are protected against eavesdropping, because any attempt to eavesdrop on a quantum channel will introduce detectable disturbances due to the no-cloning theorem and the principle of measurement-induced collapse, which can be immediately detected, thus making quantum communication a cornerstone for secure transactions, national defence, confidential data transfer, and future autonomous systems [56, 57].

In addition to security, quantum entanglement-based communication enables ultra–low latency information exchange and may allow for instantaneous correlations over long distances without the use of traditional relay systems [55]. Quantum repeaters, quantum memories, and satellite-based quantum communication links are emerging to extend the operational range and develop a scalable quantum internet connecting distributed quantum processors and sensing platforms [67]. Quantum communication is anticipated to play a key role in the 6G and beyond ecosystem through secure interoperability, trusted frameworks, secure integration, and the application of quantum computing and sensing [58]. However, as research progresses towards practical deployment, current efforts are being made to address challenges, including photon loss in long-distance optical channels, environmental noise, and integration with existing network infrastructure. These challenges are being addressed with solutions, including entangled photon-based repeaters, hybrid optical-RF quantum links, and satellite-assisted communication frameworks. Hence, quantum communication is not just an add-on to traditional security protocols. However, a game-changer towards trusted, unbreakable, and future-proof communication systems is to be embedded in 6G networks, AI, and EC, thereby establishing a secure, reliable, and intelligent communication infrastructure for future digital societies [59].

1.4.3 Internet of everything and edge intelligence

The internet of everything (IoE) builds upon IoT by connecting not only devices but also people, processes, data, and intelligent devices in a deeply interconnected digital ecosystem where machines, humans, and digital services interact automatically, with context awareness, and with unified decision-making to optimise efficiency, safety, and user experience in real time [60]. As billions of devices are connected, edge intelligence (EI) is a necessary enabling technology to move computation and decision-making closer to the source of data, with localised intelligence, to minimise communication latency, accelerate response time, enhance privacy, and conserve energy, making it a key technology for mission-critical applications such as autonomous vehicles, industrial automation, remote surgery, public safety, and smart energy systems [61]. IoE with EI constitutes a distributed and self-organising communication architecture, in which devices sense and interpret the environment, edge nodes perform real-time processing, analytics, and control, and cloud platforms provide large-scale orchestration, long-term storage, and global optimisation.

It is also supported by new technologies like federated learning, neuromorphic computing, digital twins, and AI-enabled sensor fusion, which support collaborative intelligence without compromising privacy or security. As devices and systems learn, adapt, and optimise continuously, they do so without ever centralising data. The IoE and EI also provide the following advantages: Ultra–low latency: Real-time support for time-critical applications; Enhanced security and privacy: Data stays local, reducing exposure to cyber threats; Scalability: Enables large deployments of devices without overburdening the network infrastructure; Context-aware automation: Systems can adjust automatically to the behaviour of the users and the conditions of the environment [60, 61]. IoE and EI are also expected to deliver immersive, intelligent, and autonomous environments for 6G ecosystems, including smart factories, telemedicine, connected transportation, precision agriculture, and the metaverse infrastructure. AI-native architectures, ubiquitous connectivity, and distributed intelligence will characterise the future of resilient, sustainable, and human-centric communication systems.

1.4.4 Green communication

Green communication represents a significant paradigm shift in the design and implementation of modern wireless networks, driven by the need to reduce energy consumption, minimise electromagnetic pollution, and enhance overall environmental sustainability, particularly as the number of connected devices continues to grow [62]. The demand for data is growing exponentially, and to meet the requirements of future networks such as 6G, which aim to provide unprecedented performance while achieving low energy consumption, advanced technologies are required [63]. Green communication has implications at all layers of the communication stack, from hardware design to network architecture, resource allocation, and operational policies; at the infrastructure level, green communication incorporates energy-saving hardware such as low-power base stations, efficient power amplifiers, silicon photonics, and wide-bandgap semiconductor devices (e.g., GaN and SiC), and renewable energy sources such as solar, wind, and hybrid energy harvesting to make the network self-powered or energy-autonomous. On the network side, new technologies such as sleep-mode mechanisms, dynamic spectrum allocation, load balancing, and intelligent antenna beam management minimise unnecessary transmission power and increase energy efficiency when traffic is low. In contrast, advanced antenna technologies such as massive MIMO, RIS, and adaptive beamforming further increase energy efficiency by directing electromagnetic waves exactly where they are needed and minimising wastage and interference.

AI and ML techniques can be applied to green communication to enable predictive resource optimisation, context-aware scheduling, energy-aware routing, and proactive fault detection, which allow networks to autonomously adjust operational parameters according to real-time demand, user mobility, and system load, improving energy performance while maintaining QoS. Another enabler is EC, which minimises the need to transmit data by processing information closer to its source, lowering backhaul energy consumption and carbon emissions from centralised data centres. In addition to technological innovation, green communication also encompasses economic and social sustainability, such as the availability of affordable and accessible communication services in remote or underserved areas and the minimisation of the environmental impact of global digitalisation [62, 63]. Energy efficiency metrics, including energy per bit, carbon footprint per device, and network-level sustainability indices, will become important performance indicators alongside data rate, latency, and reliability as future mobile generations evolve.

1.4.5 Smart cities and infrastructure automation

Smart cities are one of the important application areas for new wireless communication technologies that use the interconnection of the infrastructure, the intelligent system, and data-driven services to increase the efficiency, sustainability, and quality of life in urban areas by leveraging advanced communication networks such as 5G, 6G, IoT, and EC to connect millions of devices (such as sensors, cameras, vehicles, wearable systems, and automated public services) to monitor, analyse, and make decisions in real time. Infrastructure automation is a key feature of smart cities, with essential public systems, such as transportation, power, water, and waste management, being digitised to optimise their operation with minimal human involvement. Intelligent transportation systems utilise vehicle-to-everything (V2X) communication to minimise traffic, accidents, and congestion, while also managing traffic flow and promoting autonomous mobility [64].

For example, smart traffic lights, dynamic tolling, and adaptive routing algorithms optimise transportation, minimise emissions, and enhance the commuter experience. Smart grids, such as those used in the energy and utilities sector, allow for the generation, distribution, and consumption of electricity to self-adjust load, detect failures, and reconfigure power distribution during outages, and smart water networks use real-time sensing and leak detection systems to reduce waste and improve resource allocation. Urban governance is also impacted by automation, with AI-driven smart surveillance and environmental monitoring systems capable of detecting anomalies such as decreases in air quality, fire risks, flooding, or security breaches, allowing authorities to respond quickly and enhance public safety. Digital twins of cities, which are a digital representation of physical urban infrastructure, provide real-time data for planning, resource optimisation, and scenario analysis to inform policymakers and administrators.

These applications require wireless communication because they involve low-latency data exchange, the connection of massive numbers of devices, secure communication frameworks, and distributed processing. Edge and cloud computing can be combined to process data at the edge more rapidly, reducing latency and enabling mission-critical applications to respond more quickly. Secondly, sustainability remains a key goal: AI-enabled analytics and low-power communication protocols can help conserve energy, enhance transportation, and manage services more efficiently, thereby conserving resources. Blockchain-enabled secure data sharing, autonomous drones for logistics, digital healthcare ecosystems, and intelligent buildings will further automate, personalise, and harden smart cities. In the end, the development of smart cities and the automation of infrastructure represents a leap towards living environments that are more efficient, safe, sustainable, and responsive to the needs of citizens, where future cities will transition from reactive systems to fully autonomous, adaptive, and context-aware ecosystems [65].

1.4.6 Remote healthcare and telemedicine

Advances in wireless communication, wearable devices, AI, and cloud computing have enabled remote healthcare and telemedicine, allowing patients to receive medical consultations, diagnosis, monitoring, and treatment without an in-person visit [66]. This is particularly beneficial for timely and accessible care in remote, rural, or underserved areas. These systems use next-generation communication technologies, such as 5G and emerging 6G networks, with ultra–low latency, high reliability, and massive connectivity, high-bandwidth communication, and, eventually, tactile internet and holographic communication to support high-resolution medical imaging, remote diagnostics, live patient data streaming, and immersive telepresence consultations, and, ultimately, remote surgeries and haptic feedback-assisted procedures whereby specialists can perform operations from anywhere in the world with precision and responsiveness [67].

Wearable and implantable medical devices take remote healthcare to the next level by continuously monitoring physiological data, including heart rate, body temperature, oxygen saturation, blood glucose levels, neural activity, and other biomarkers, which can be analysed with EI and AI-based analytics to generate automated alerts, detect anomalies, and make predictions for proactive healthcare that prevents hospitalisation and tailors treatment plans for better outcomes. Secure cloud infrastructures and blockchain-based medical data frameworks with privacy, integrity, and interoperability support telemedicine systems, including virtual healthcare assistants, language translation engines, and AI-based triage tools. These enhancements enhance accessibility, particularly for older patients and those who require ongoing health support [66–68].

The digital twins integrated with sensing networks in biomedicine together form an intelligent environment for medicine (smart hospital ecosystems) where physical and digital healthcare systems work hand-in-hand through real-time patient physiology simulations, remote rehabilitation via AR, and smart decision support based on medical AI. Beyond clinical efficiency improvements to telemedicine applications that reduce the cost of care delivery and decrease congestion in hospitals and transmission risks associated with infection spread, such as observed during the COVID-19 pandemic globally, we expect further advancements in wireless technology, AI, and biomedical engineering will allow remote healthcare to be an integral part of smart global medical systems designed for inclusive access and scalable implementation capable of democratising medical accessibility while enhancing the quality-of-life worldwide.

1.4.7 Autonomous transportation and smart mobility

In the era of smart cities, autonomous transportation and smart mobility are important elements of the future intelligent infrastructure, which is enabled by advances in high-speed wireless communication, AI, sensor fusion, and real-time data analytics. The transportation system is transitioning from manually driven vehicles to fully automated, intelligent, and connected mobility platforms as cities evolve into digitally connected ecosystems. The next-generation wireless networks, such as 5G, cellular V2X (C-V2X), and future 6G systems, will enable ultra-reliable and low-latency communication, allowing autonomous vehicles to operate safely and efficiently by seamlessly exchanging data among vehicles, road infrastructure, pedestrians, drones, and cloud networks. The end-to-end latency of next-generation wireless networks is expected to be potentially lower than 1 ms in 6G networks, which enables autonomous vehicles to process and respond to environmental changes almost instantaneously and thus perform collision avoidance, precise navigation, and traffic orchestration [69].

Autonomous transportation is based on an ecosystem of connected communication models: vehicle-to-vehicle (V2V) for cooperative movement, coordinated braking, lane switching, and platooning; vehicle-to-infrastructure (V2I) for communication with traffic signals, road sensors, and intelligent transport systems (ITS); vehicle-to-pedestrian (V2P) for safety enhancement by identifying vulnerable road users through wearables, smartphones, or embedded sensors; and vehicle-to-network (V2N) for cloud-based navigation, mobility analytics, and fleet management. Vehicles gather data from light detection and ranging (LIDAR), radio detection and ranging (radar), vision sensors, global positioning system (GPS), and communication networks and run AI-enhanced perception, localisation, and decision systems to analyse this data continuously. Using EC, they process critical safety decisions locally, while cloud platforms provide large-scale analytics, learning, and system-level optimisation. ML and digital twins can be used for predictive modelling of traffic flows, infrastructure planning, and preventive maintenance.

Beyond autonomous cars, smart mobility encompasses connected public transportation, delivery robots, autonomous drones, electric vehicle (EV) fleets, high-speed rail systems, and micro-mobility platforms, including smart bicycles and scooters [70]. Transportation is becoming an integrated mobility-as-a-service (MaaS) platform that offers real-time routing, energy management, dynamic pricing, and multimodal travel planning. Another primary motivator is sustainability. With EV charging networks powered by renewables, energy-aware routing, vehicle coordination, and smart traffic management, carbon emissions and congestion are lowered, while RIS, green wireless architectures, and optimised intelligent transport grids minimise the energy footprint of communication. The next generation of transportation and smart mobility will transform future cities into safer, less congested, more accessible, and environmentally sustainable mobility ecosystems over the coming decade, as advanced communication technologies combine with automation and intelligent infrastructure to deliver fully autonomous, seamlessly connected, and highly adaptive mobility solutions.

1.4.8 Precision agriculture and environmental monitoring

Modern advances in wireless communication, AI, and sensing are making it possible to manage natural resources with greater accuracy while feeding more people than ever before (precision farming) and protecting ecosystems on a grand scale against the expanding threat of hunger and rising climate-related environmental dangers from growing food demands. Precision agriculture utilises wireless sensor networks (WSN), IoT devices, unmanned aerial vehicles (UAVs), robotics, and satellite communication to capture data about the soil, crops, climate, and livestock, which can be paired with AI algorithms to inform decisions on irrigation scheduling, fertiliser distribution, pest management, and crop harvesting [71]. Fast communication networks, such as 5G and 6G, provide low-latency connectivity between field devices, drones, cloud analytics platforms, and automated farm machinery, enabling autonomous tractors, robot harvesters, and remote machinery control.

EC and federated learning technologies can process agricultural data locally, making them crucial in rural and remote farmland environments. Meanwhile, hyperspectral imaging from drones, combined with real-time wireless transmission, can be used for the early detection of crop diseases, soil nutrient deficiencies, and water stress. This transition from reactive to predictive and prescriptive farming leads to the most efficient use of inputs, reducing waste, lowering operational costs, and minimising environmental impact. Similar advances can be applied to environmental monitoring to monitor ecological variables such as air and water quality, biodiversity, forest health, aquatic ecosystems, and conditions of natural hazards continuously; low-power wide-area network (LPWAN), NB-IoT, satellite IoT, and future THz-based communication enable massive deployment of low-power sensors to observe climate patterns and environmental trends and thereby support early warning systems for wildfires, floods, droughts, and pollution events to protect ecosystems, agriculture, and human populations [72].

Digital twins, geographic information system (GIS)-based analytics, autonomous sensing drones, and blockchain-enabled traceability systems make agricultural value chains more transparent, compliant, and sustainable, while energy harvesting (such as solar-powered IoT devices, self-sustaining sensors, and low-power wireless architectures) supports long-term deployments with a low ecological footprint. The global sustainability goals of precision agriculture and environmental monitoring include enhanced food security, reduced greenhouse gas emissions, responsible land use, and natural resource conservation, which will be further facilitated by hyper-connected and intelligent ecosystems via fully autonomous farming systems and near-real-time environmental management at the planetary scale as 6G and beyond technologies mature.

1.4.9 Defence and space communication

As geopolitical complexities expand, space expansion increases, and information exchange becomes more critical to secure operations in these new areas of growth, modern communications technologies continue to evolve towards highly robust, flexible, adaptive, and intelligent architectures. Some of the most advanced and demanding communication environments are those required for defence and space applications, which must be highly reliable, with long-range coverage requirements, anti-jam capability, and robust cybersecurity protection. Next-generation wireless communication for defence applications enables real-time situational awareness, command and control systems, coordinated military operations, and unmanned platforms such as drones, autonomous ground vehicles, and naval assets and is also applicable for supporting high-speed, low-latency, and secure data transmission in defence applications, such as real-time situational awareness, command and control systems, coordinated military operations, and unmanned platforms such as drones, autonomous ground vehicles, and naval assets [73]. Technologies such as tactical 5G/6G, mm-wave), THz links, CR, and satellite-based communication networks are key enablers for high-speed, low-latency, and secure data transmission. In contrast, SDR and spectrum-agile communication systems provide the flexibility necessary to operate securely across multiple frequency bands and mitigate interference threats.

These advanced communication techniques include MIMO radar, low probability of intercept (LPI), low probability of detection (LPD), encrypted optical wireless communication, threat intelligence, and reinforcement learning, ensuring stealth, confidentiality, and operational resilience, real-time anomaly detection, adaptive beamforming, and dynamic spectrum reconfiguration in contested or hostile environments, and quantum communication for military-grade secure links through QKD for theoretically unbreakable encryption [74]. While traditional geostationary satellites have provided global broadband services, precision navigation, and earth observation, the future of space communication systems is a technological revolution that includes low earth orbit (LEO) and medium earth orbit (MEO) networks, as well as mega-constellations of satellites for global broadband services, coordinated communication links, inter-satellite laser communications, and AI network management to achieve global connectivity in complex orbits.

Ultra–long range links, advanced antenna arrays, radiation-hardened components, and extremely low-latency routing are required for emerging applications such as space-based IoT, lunar communication infrastructure, deep space exploration, and interplanetary internet. The combination of RIS, phased-array antennas, and graphene-based THz devices offers compact, adaptive, and high-bandwidth solutions that can withstand harsh aerospace conditions. Advancements in EC, federated AI, digital twins, and autonomous network management enable real-time decision-making with minimal ground support, which is also applicable to defence and space communication systems. This supports mission-critical services, including space traffic management, GPS augmentation, environmental surveillance, emergency response, and planetary science missions.

1.4.10 Industry 4.0 and smart manufacturing

Industry 4.0 describes the integration of digital technologies, such as intelligent automation and ML, combined with ubiquitous connectivity, to transform traditional production systems into smart factories or advanced manufacturing ecosystems [75]. In these ecosystems, connected next-generation communication systems coordinate between machines, sensors, control systems, and cloud-based analytics platforms. Smart manufacturing leverages 5G and next-generation 6G connectivity as well as Industrial IoT (IIoT), EC, WSN, and cyber-physical systems that allow real-time data exchange between production environments so machines can communicate with each other to diagnose performance issues and autonomously coordinate tasks for increased productivity, operational efficiency, and system reliability.

Predictive maintenance relies heavily on AI, ML, and data analytics to monitor critical assets for vibration levels, energy consumption, thermal profiles, and equipment stress, detecting anomalies and making informed decisions to prevent system failures before they occur, extend the life of critical assets, and minimise downtime. Robots, autonomous guided vehicles (AGVs), collaborative robots (cobots), AR, and digital twins improve automation, human–machine interaction, and production optimisation; digital twin models simulate factories, machines, or production processes in real-time, track performance, and optimise processes without physical interruptions; AR-based remote assistance and training enhance workforce productivity and minimise operational errors [75].

In addition to ultra–low latency and high-capacity data exchange, high-frequency communication bands, such as mm-wave and THz links, support deterministic and time-sensitive industrial applications, such as coordinated multi-robot systems and precision machining, via dedicated communication pathways with guaranteed QoS (network slicing). Cybersecurity is another crucial factor in Industry 4.0, as greater connectivity leaves industrial systems vulnerable to data breaches and cyberattacks. To protect network integrity and secure MTM communication, blockchain-based frameworks, encryption, secure authentication, and trusted execution environments are being deployed. In addition to achieving sustainability goals through energy efficiency, reduced waste generation (which is also important), supply chain traceability, and smart monitoring of consumption patterns about materials usage or carbon emissions in order to comply with green manufacturing practices as well as transitioning towards greener models of manufacture, Industry 4.0 facilitates a symbiotic relationship between humans and intelligent machines for the manufacturing industry through personalisation (as communication technologies evolve into AI-native ecosystems), resilience, and sustainable innovation.

1.5 CONCLUSION

The evolution of wireless communication technology from analog systems in 1G to the intelligent and adaptive networks of 5G demonstrates the progress of information transfer and sharing, with each generation bringing its own innovations: digital voice transmission, mobile internet, broadband connectivity, and ultra–low latency communication, which have combined to create the hyper-connected world of today. The recent developments of massive MIMO, beamforming, mm-wave and THz communication, RIS, FD radio, and AI-driven network optimisation have further enhanced the performance and efficiency of wireless networks, allowing higher data rates, improved reliability, and dynamic adaptability in various communication environments.

Future generations beyond 6G will fuse sensing, communications, and intelligence into one fabric with quantum communications, edge AI, holographic transmission, and energy-efficient green technologies embedded in future wireless systems to achieve Tbps data rates for applications such as XR digital twins, smart cities, and autonomous systems; it continues its evolution towards an intelligent, sustainable, immersive ecosystem. Those advances form the foundation of a world where communication networks are no longer connecting devices but become part of the human experience itself and a source of innovation.

REFERENCES

1. Jovović, I., Forenbacher, I. & Periša, M. (2015). Massive machine-type communications: An overview and perspectives towards 5G. In M. Mokrys & S. Badura (Eds.), *RCITD – Proceedings in Research Conference in Technical Disciplines 2015* (pp. 32–37). University of Žilina.
2. Al-Falahy, N. & Alani, O.Y.K. (2017). Technologies for 5G networks: Challenges and opportunities. *IT Professional*, 19(1), 12–20.
3. Ejaz, W., Anpalagan, A., Imran, M.A., Jo, M., Naeem, M., Qaisar, S.B. & Wang, W. (2016). Internet of Things (IoT) in 5G wireless communications. *IEEE Access*, 4, 10310–10314.

4. Akyildiz, I.F., Jornet, J.M. & Han, C. (2014). Terahertz band: Next frontier for wireless communications. *Physical Communication*, 12, 16–32.
5. Dash, S., Psomas, C., Patnaik, A. & Krikidis, I. (2022). An ultra-wideband orthogonal-beam directional graphene-based antenna for THz wireless systems. *Scientific Reports*, 12, 22145.
6. Dash, S., Psomas, C. & Krikidis, I. (2025). Graphene-based beam-reconfigurable liquid antenna for 5G mmWave wireless systems. *Frontiers in Communications and Networks*, 6, 1560311.
7. Gohar, A. & Nencioni, G. (2021). The role of 5G technologies in a smart city: The case for intelligent transportation system. *Sustainability*, 13(9), 5188.
8. Serrano, P., de la Oliva, A., Patras, P., Mancuso, V. & Banchs, A. (2012). Greening wireless communications: Status and future directions. *Computer Communications*, 35(14), 1651–1661.
9. Khan, L.U., Saad, W., Niyato, D.T., Han, Z. & Hong, C.S. (2021). Digital-twin-enabled 6G: Vision, architectural trends, and future directions. *IEEE Communications Magazine*, 60(1), 74–80.
10. Samdanis, K., Nait Abbou, A., Song, J. & Taleb, T. (2023). AI/ML service enablers and model maintenance for beyond 5G networks. *IEEE Network*, 37(4), 162–172.
11. Sanjalawe, Y.K., Fraihat, S., Al-E'mari, S.R., Abualhaj, M., Makhadmeh, S.N. & Alzubi, E. (2025). A review of 6G and AI convergence: Enhancing communication networks with artificial intelligence. *IEEE Open Journal of the Communications Society*, 6, 2308–2355.
12. Khaleel, A. & Başar, E. (2020). Reconfigurable intelligent surface-empowered MIMO systems. *IEEE Systems Journal*, 15(4), 4358–4366.
13. Dash, S., Psomas, C., Krikidis, I., Akyildiz, I.F. & Pitsillides, A. (2022). Active control of THz waves in wireless environments using graphene-based RIS. *IEEE Transactions on Antennas and Propagation*, 70(10), 8785–8797.
14. Jiang, W. & Han, B. (2024). Evolution to first-generation (1G) mobile cellular communications. In S. Sun & X. Zhang (Eds.), *Cellular Communication Networks and Standards* (pp. 7–21). Springer.
15. Garcia-Swartz, D.D. & Campbell-Kelly, M. (2022). New technologies, standards, customers, and markets in the world of 2G. In *Cellular: An Economic and Business History of the International Mobile-Phone Industry* (pp. 89–118). MIT Press.
16. Gomez, C., Catalan, M., Viamonte, D., Paradells, J. & Calveras, A. (2008). Web browsing optimization over 2.5G and 3G: End-to-end mechanisms vs. usage of performance enhancing proxies. *Wireless Communications and Mobile Computing*, 8(2), 213–230.
17. Rochman, M.I., Sathya, V., Fernandez, D., et al. (2023). A comprehensive analysis of the coverage and performance of 4G and 5G deployments. *Computer Networks*, 237, 110060.
18. Ma, T., Huang, J., Zu, J. & Li, W. (2025). Location-driven beamforming for massive multi-user MIMO systems. *Journal of Systems Engineering and Electronics*, 36(3), 609–622.
19. Pereira de Figueiredo, F.A. (2022). An overview of massive MIMO for 5G and 6G. *IEEE Latin America Transactions*, 20(6), 931–940.
20. Qi, Y. & Vaezi, M. (2020). Secure transmission in MIMO-NOMA networks. *IEEE Communications Letters*, 24(12), 2696–2700.
21. Yang, S. & Hanzo, L. (2015). Fifty years of MIMO detection: The road to large-scale MIMOs. *IEEE Communications Surveys & Tutorials*, 17(4), 1941–1988.
22. Vu, T.K., Bennis, M., Debbah, M., Latva-aho, M. & Hong, C.S. (2018). Ultra-reliable communication in 5G mmWave networks: A risk-sensitive approach. *IEEE Communications Letters*, 22(4), 708–711.
23. Chang, B., Tang, W., Yan, X., Tong, X. & Chen, Z. (2022). Integrated scheduling of sensing, communication, and control for mmWave/THz communications in cellular connected UAV networks. *IEEE Journal on Selected Areas in Communications*, 40(7), 2103–2113.
24. Saqib, N.U., Haroon, M.S., Lee, H.Y., Park, K., Song, H.-G. & Jeon, S.-W. (2023). THz communications: A key enabler for future cellular networks. *IEEE Access*, 11, 117474–117493.
25. Akyildiz, I. F., Kak, A. & Nie, S. (2020). 6G and beyond: The future of wireless communications systems. *IEEE Access*, 8, 133995–134030.
26. Othman, W.M., Ateya, A.A., Nasr, M.E., Muthanna, A., ElAffendi, M., Koucheryavy, A. & Hamdi, A.A. (2025). Key enabling technologies for 6G: The role of UAVs, terahertz communication, and intelligent reconfigurable surfaces in shaping the future of wireless networks. *Journal of Sensor and Actuator Networks*, 14(2), 30.
27. Dash, S., Soni, G., Patnaik, A., Liaskos, C., Akyildiz, I.F. & Pitsillides, A. (2021). Switched-beam graphene plasmonic nanoantenna in the terahertz wave region. *Plasmonics*, 16, 1855–1864.
28. Dash, S., Liaskos, C., Akyildiz, I.F. & Pitsillides, A. (2020). Graphene hypersurface for manipulation of THz waves. *Materials Science Forum*, 1009, 63–68.

29. Dash, S., Patnaik, A. & Kaushik, B.K. (2019). Performance enhancement of graphene plasmonic nanoantenna for THz communication. *IET Microwaves, Antennas & Propagation*, 13(1), 71–75.
30. Dash, S. & Patnaik, A. (2018). Material selection for THz antennas. *Microwave and Optical Technology Letters*, 60, 1183–1187.
31. Dash, S., Psomas, C. & Krikidis, I. (2023). Selection of metallic liquid in sub-6 GHz antenna design for 6G networks. *Scientific Reports*, 13, 20551.
32. Basar, E., Di Renzo, M., de Rosny, J., Debbah, M., Alouini, M.-S. & Zhang, R. (2019). Wireless communications through reconfigurable intelligent surfaces. *IEEE Access*, 7, 116753–116773.
33. Dash, S., Liaskos, C., Akyildiz, I.F. & Pitsillides, A. (2020). Nanoantennas design for THz communication: Material selection and performance enhancement. In *Proceedings of the 7th ACM International Conference on Nanoscale Computing and Communication* (NANOCOM 2020), College Park, MD, USA, September 23–25.
34. Dash, S. & Patnaik, A. (2018). Graphene plasmonic bowtie antenna for UWB THz application. In *Proceedings of the Twenty-Fourth National Conference on Communications (NCC 2018)*, IIT Hyderabad, February 25–28, pp. 1–4.
35. Dash, S., Liaskos, C., Akyildiz, I.F. & Pitsillides, A. (2019). Wideband perfect absorption polarization-insensitive reconfigurable graphene metasurface for THz wireless environment. In *2019 IEEE Microwave Theory and Techniques in Wireless Communications (MTTW)*, Riga, Latvia, October 1–2, pp. 93–96.
36. Chus, J. (2022). Exploring the past, present, and future of in-band full-duplex wireless systems. *IEEE Microwave Magazine*, 23(12), 16–19.
37. Liao, W.-S., Zhao, O., Li, K., Kawasaki, H. & Matsumura, T. (2023). Implementation of in-band full-duplex using software-defined radio with adaptive filter-based self-interference cancellation. *Future Internet*, 15(11), 360.
38. Xu, D., Yu, X., Sun, Y., Ng, D.W.K. & Schober, R. (2020). Resource allocation for IRS-assisted full-duplex cognitive radio systems. *IEEE Transactions on Communications*, 68(12), 7376–7394.
39. Towhidlou, V. & Shikh-Bahaei, M. (2018). Adaptive full-duplex communications in cognitive radio networks. *IEEE Transactions on Vehicular Technology*, 67(9), 8386–8395.
40. Xu, W., Tang, J., Wang, Z. & Guo, L. (2018). Capacity maximization in full-duplex cognitive radio systems with non-slotted primary user state change. *IEEE Communications Letters*, 22(9), 1890–1893.
41. Cao, G., Lu, Z., Wen, X., Lei, T., & Hu, Z. (2018). AIF: An artificial intelligence framework for smart wireless network management. *IEEE Communications Letters*, 22(2), 400–403.
42. Nawaz, S. J., Sharma, S. K., Wyne, S., Patwary, M. N., & Asaduzzaman, M. (2019). Quantum machine learning for 6G communication networks: State-of-the-art and vision for the future. *IEEE Access*, 7, 46317–46350.
43. Usman, Y., Oladipupo, H., During, A. D., Akl, R., & Chataut, R. (2025). AI, ML, and LLM integration in 5G/6G networks: A comprehensive survey of architectures, challenges, and future directions. *IEEE Access*, 13, 168914–168950.
44. Chen, A. C. H., Jia, W. K., Hwang, F. J., et al. (2022). Machine learning and deep learning methods for wireless network applications. *Journal of Wireless Communications and Networking*, 2022, Article 115.
45. Nguyen, D. C., et al. (2021). Enabling AI in future wireless networks: A data life cycle perspective. *IEEE Communications Surveys & Tutorials*, 23(1), 553–595.
46. Kaur, J., Khan, M. A., Iftikhar, M., Imran, M., & Emad Ul Haq, Q. (2021). Machine learning techniques for 5G and beyond. *IEEE Access*, 9, 23472–23488.
47. Salau, B. A., Rawal, A., & Rawat, D. B. (2022). Recent advances in artificial intelligence for wireless Internet of Things and cyber–physical systems: A comprehensive survey. *IEEE Internet of Things Journal*, 9(15), 12916–12930.
48. Ghadi, Y. Y., et al. (2024). Machine learning solutions for the security of wireless sensor networks: A review. *IEEE Access*, 12, 12699–12719.
49. Khater, A., Noohi, S. A., Hashemi, M. R., & Zali, Z. (2024). An NFV-based framework for autonomous deployment of new protocols in SDN networks. *IEEE Access*, 12, 148727–148740.
50. Erel-Özçevik, M., & Tekçe, F. (2023). Grant-free SCMA design using SDN/NFV with physical layer security. *IEEE Transactions on Network and Service Management*, 20(3), 2124–2135.
51. Ismail, T., & Mahmoud, H. H. M. (2020). Optimum functional splits for optimizing energy consumption in V-RAN. *IEEE Access*, 8, 194333–194341.
52. Wang, C.-X., et al. (2023). On the road to 6G: Visions, requirements, key technologies, and testbeds. *IEEE Communications Surveys & Tutorials*, 25(2), 905–974.

53. Kuruvatti, N. P., Habibi, M. A., Partani, S., Han, B., Fellan, A., & Schotten, H. D. (2022). Empowering 6G communication systems with digital twin technology: A comprehensive survey. *IEEE Access*, 10, 112158–112186.
54. Alwis, C. D., et al. (2021). Survey on 6G frontiers: Trends, applications, requirements, technologies and future research. *IEEE Open Journal of the Communications Society*, 2, 836–886.
55. Pan, D., et al. (2024). The evolution of quantum secure direct communication: On the road to the Qinternet. *IEEE Communications Surveys & Tutorials*, 26(3), 1898–1949.
56. Hasan, S. R., Chowdhury, M. Z., Sayem, M., & Jang, Y. M. (2023). Quantum communication systems: Vision, protocols, applications, and challenges. *IEEE Access*, 11, 15855–15877.
57. Yang, Z., Zolanvari, M., & Jain, R. (2023). A survey of important issues in quantum computing and communications. *IEEE Communications Surveys & Tutorials*, 25(2), 1059–1094.
58. Zeydan, E., et al. (2025). Quantum technologies for beyond 5G and 6G networks: Applications, opportunities, and challenges. *IEEE Open Journal of the Communications Society*, 6, 6383–6420.
59. Chehimi, M., Chaccour, C., Thomas, C. K., & Saad, W. (2024). Quantum semantic communications for resource-efficient quantum networking. *IEEE Communications Letters*, 28(4), 803–807.
60. Oliveira, F., Costa, D. G., Assis, F., & Silva, I. (2024). Internet of intelligent things: A convergence of embedded systems, edge computing and machine learning. *Internet of Things*, 26, 101153.
61. Cao, K., Liu, Y., Meng, G., & Sun, Q. (2020). An overview on edge computing research. *IEEE Access*, 8, 85714–85728.
62. Anpalagan, A., Ejaz, W., Sharma, S. K., Da Costa, D. B., Jo, M., & Kim, J. (2021). Editorial special issue on "Green Communication and Computing Technologies for 6G Networks." *IEEE Transactions on Green Communications and Networking*, 5(3), 1653–1656.
63. Kour, H., Jha, R. K., & Jain, S. (2024). Green and safe 6G wireless networks: A hybrid approach. *IEEE Transactions on Green Communications and Networking*, 8(4), 1729–1741.
64. Fortuna, C., Yetgin, H., & Mohorčič, M. (2023). Smart infrastructures: Artificial intelligence-enabled lifecycle automation. *IEEE Industrial Electronics Magazine*, 17(2), 37–47.
65. Dryjanski, M., Buczkowski, M., Ould-Cheikh-Mouhamedou, Y., & Kliks, A. (2020). Adoption of smart cities with a practical smart building implementation. *IEEE Internet of Things Magazine*, 3(1), 58–63.
66. Sri-Ganeshan, M., & Cameron, P. (2024). Remote monitoring in telehealth: Advancements, feasibility and implications. In *Biomedical Engineering*. IntechOpen.
67. Ahad, A., Ali, Z., Mateen, A., Tahir, M., Hannan, A., Garcia, N. M., & Pires, I. M. (2023). A comprehensive review on 5G-based smart healthcare network security: Taxonomy, issues, solutions and future research directions. *Array*, 18, 100290.
68. Gupta, R., Gamad, R. S., & Bansod, P. (2014). Telemedicine: A brief analysis. *Cogent Engineering*, 1(1), 966459.
69. Mahrez, Z., Sabir, E., Badidi, E., Saad, W., & Sadik, M. (2022). Smart urban mobility: When mobility systems meet smart data. *IEEE Transactions on Intelligent Transportation Systems*, 23(7), 6222–6239.
70. Prakash, A. (2021). Smart mobility solutions for a smart city. *IEEE Potentials*, 40(1), 24–29.
71. Ahmed, B., Shabbir, H., Naqvi, S. R., & Peng, L. (2024). Smart agriculture: Current state, opportunities, and challenges. *IEEE Access*, 12, 144456–144478.
72. Hashmi, A. U. H., et al. (2024). Effects of IoT communication protocols for precision agriculture in outdoor environments. *IEEE Access*, 12, 46410–46421.
73. Kaushal, H., & Kaddoum, G. (2017). Optical communication in space: Challenges and mitigation techniques. *IEEE Communications Surveys & Tutorials*, 19(1), 57–96.
74. Tedeschi, P., Sciancalepore, S., & Di Pietro, R. (2022). Satellite-based communications security: A survey of threats, solutions, and research challenges. *Computer Networks*, 216, 109246.
75. Abdullah, F. M., Saleh, M., Al-Ahmari, A. M., & Anwar, S. (2022). The impact of Industry 4.0 technologies on manufacturing strategies: Proposition of technology-integrated selection. *IEEE Access*, 10, 21574–21583.

Chapter 2

RF power amplifier circuits for wireless communications

Yericharla Mary Asha Latha

2.1 INTRODUCTION

2.1.1 RF power amplifiers for wireless communication

The exponential rise in mobile data traffic has necessitated the evolution of wireless communication standards for enhanced data rates. Modern 5G and beyond technologies use multi-band communication to provide wide-area coverage and support multiple devices. For example, the sub-6 GHz band of 5G NR uses nearly 60 bands covering from 410 to 7125 MHz. The key challenge while adapting these advanced technologies is to fully utilize the frequency spectrum, space, and the existing hardware. A transceiver is an essential component to transmit and receive the radio signal for wireless communication. Adapting a dedicated Radio Frequency (RF) transceiver for each of the communication bands increases system complexity and cost. Broadband RF transceivers are an effective solution to support various standards cost-effectively [1, 2]. They also have an additional capability even to meet the standards that come up in the future.

In an RF transceiver, the power amplifier (PA) plays a crucial role by increasing the output signal power required for effective long-distance communication. However, the PA is the most power-demanding block in an RF transceiver, significantly influencing the overall energy efficiency of the system. In a typical base station, a PA consumes 50–80% of the total power [3]. Therefore, there is a great need to reduce their power consumption by enhancing efficiency for a low operational cost of RF transceivers. Therefore, there is a great need for highly efficient PAs operating over a wide frequency range [4, 5]. This chapter discusses various design techniques such as continuous modes and extended continuous modes that can achieve high efficiency over an octave and a multi-octave bandwidth.

2.1.2 Classes of operation

The PAs can be classified based on their operation as biasing classes and waveform engineering modes. Traditionally, PAs are classified based on their current conduction angles, which are called biasing classes [6–12]. PAs are classified as Class A, AB, B, and C depending on the biasing level and the conduction angle of the transistor. In Class A operation, the amplifier is biased approximately halfway between the cutoff and saturation regions, resulting in a 360° conduction angle and a theoretical maximum efficiency of about 50%. In Class B mode, the transistor is biased at the cutoff point, leading to a conduction angle of 180°.This reduces power dissipation compared to Class A and therefore increases efficiency to 75%. In Class AB operation, the biasing point is positioned between those of Class A and B PAs – above cutoff and below the midpoint between cutoff and saturation. Consequently, both the conduction angle and efficiency fall between the two modes. The current conduction

DOI: 10.1201/9781003637721-2

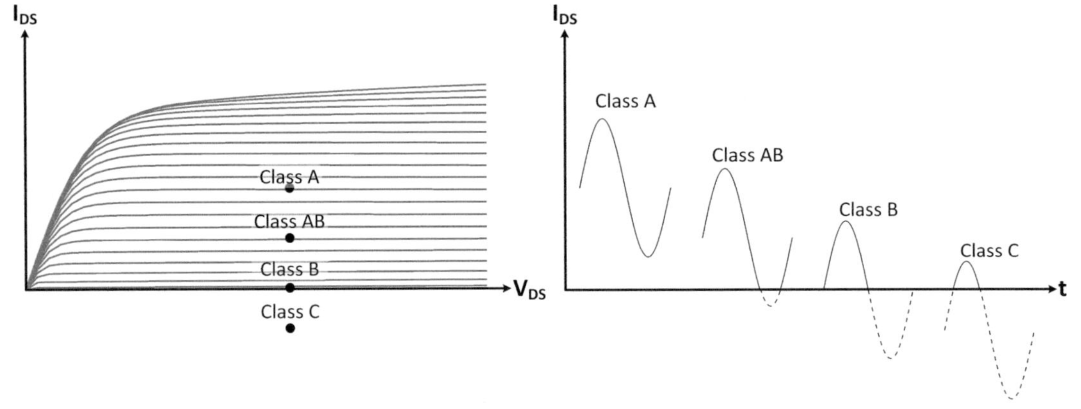

Figure 2.1 Biasing points and current conduction angles of the biasing class PAs.

angle typically ranges from 180° to 360°, while efficiency improves from about 50 to 75%. In Class C mode, the transistor is biased further below the cutoff region, resulting in a conduction angle of less than 180° and an efficiency exceeding 75%. Figure 2.1 depicts the bias points and conduction angles for each class. As the operation shifts from Class A to Class C, efficiency increases, but this comes at the expense of higher nonlinearity, which is generally undesirable in linear communication systems. Therefore, to achieve high efficiency, waveform engineering modes are used which will be discussed in detail in the rest of the chapter [13].

2.1.3 High-efficiency power amplifier modes

The efficiency of a PA is determined by the proportion of supplied DC power (P_{DC}) that is translated in the form of RF output power (P_{RF}). This relationship is expressed mathematically as follows [6]:

$$\text{Efficiency} = \frac{P_{RF}}{P_{DC}} = \frac{P_{RF}}{P_{diss} + P_{RF} + \sum_{n=2}^{\infty} P_{RF,n}} \tag{2.1}$$

In an ideal situation, all the supplied P_{DC} would be transformed into P_{RF}. However, practical losses due to transistor nonlinearity and parasitic effects reduce the overall efficiency. The total DC power delivered to the transistor is partly transformed into fundamental RF power, whereas the rest is either dissipated within the transistor (P_{diss}) or converted into harmonic output power ($P_{RF,n}$). The efficiency will be enhanced by reducing the power dissipation and harmonic power. The power dissipated across the transistor varies with the time domain voltage and current (V–I) waveforms. In essence, the power loss varies with the overlap between the time domain V–I waveforms and can be defined mathematically as follows [6]:

$$P_{diss} = \frac{1}{T} \int_{-T/2}^{T/2} v_{DS}(t).i_{DS}(t)\, dt \tag{2.2}$$

In general, a 180° phase shift exists between the time domain V–I waveforms across the transistor. In Class A operation, V–I waveforms are sinusoidal since the current conducts over the entire 360°, as illustrated in Figure 2.2. This full conduction causes a 50% overlap

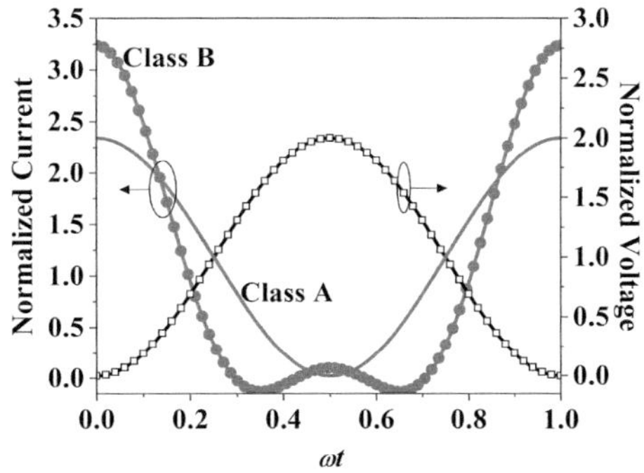

Figure 2.2 Voltage and current waveforms of Class A and Class B mode.

between the time domain V–I waveforms, resulting in equal proportions of power dissipation and useful output power, thereby limiting efficiency to around 50%. In contrast, for Class B mode, the transistor is "on" for only half the input cycle (180°), producing a half-sinusoidal current waveform, while the voltage waveform has sinusoidal characteristics as shown in Figure 2.2. The reduced overlap between the time domain V–I waveforms minimizes power loss, leading to an efficiency of approximately 78.5%. By further shaping and controlling the time domain V–I waveforms to minimize their overlap, efficiency can be improved beyond that achieved in Class B amplifiers.

As shown in (2.1), efficiency also depends upon the RF power at the harmonic frequencies. The PA is designed with the transistor which is a nonlinear component. Therefore, PA tends to amplify the power at harmonics as well. The mathematical expression of harmonic power is as follows [6]:

$$\sum_{n=2}^{\infty} P_{RF,n} = \frac{1}{2}\sum_{n=2}^{\infty} V_n I_n \cos\phi_n \tag{2.3}$$

The magnitude of the V–I at a given harmonic, and the phase angle between the time domain V–I waveforms, dictate the RF power contribution from that harmonic. The $P_{RF,n}$, i.e., (2.3) will be zero when $V_n = 0$, or $I_n = 0$, or $\cos\phi_n = \pm 90°$. Voltage V_n will be zero when there is a short circuit and current I_n will be zero when there is an open circuit. When the phase angle $\phi_n = \pm 90^{\circ}$, the value of $\cos(\phi_n)$ becomes zero, indicating a ±90° phase shift between the V–I components. In the case of an inductive load, current lags the voltage by 90°, while for a capacitive load, the current leads by 90°. On a Smith chart, inductive impedances are represented in the upper half, whereas capacitive impedances appear in the lower half. The outer boundary of the Smith chart corresponds to purely reactive terminations. Hence, by applying a short circuit ($V_n = 0$), an open circuit ($I_n = 0$), or a reactive load ($\cos(\phi_n) = 0$) at harmonic frequencies, the harmonic power contribution can be suppressed. In simpler terms, terminating harmonics with loads located along the periphery of the Smith chart effectively nullifies harmonic power.

Any load at the outer boundary of the Smith chart corresponds to total reflection. Therefore, by presenting a harmonic load surrounding the periphery of the Smith chart to the PA, the total harmonic power reflects toward the drain of the transistor. A sinusoidal waveform has one frequency component. When even harmonics are added to the fundamental component, it results in half-sinusoidal waveform and adding odd harmonics results in square waveform. Therefore, the reflected harmonic power can be added constructively to the fundamental component and engineer the shape of the time domain V–I waveforms to reduce their time overlap. The process of engineering the time domain V–I waveforms with the help of harmonic terminations to increase efficiency of PA is called the harmonic engineering [14–18]. Various harmonic engineering PA modes are described in the rest of this chapter.

2.2 WAVEFORM ENGINEERED MODES OF OPERATION

In waveform engineered modes, the harmonics are terminated with purely reactive loads. By properly adjusting the phase of harmonic components, they can constructively interact with the fundamental signal, allowing the time domain V–I waveforms to be shaped in such a way that their overlap is minimized. Using waveform engineering both P_{diss} and $P_{out,n}$ are made zero to enhance efficiency. Waveform-engineered PAs, often referred to as harmonic-tuned modes, include popular PA mode such as Class F, F^{-1}, J, and E [14–18]. Each of these modes is discussed in detail in this section. Class D and S also come under this category of waveform engineered PA modes; however, they are not commonly used [12]. The waveform engineered modes can be categorized as current modes and switched modes. Class F, F^{-1}, and J are example PAs operating in current mode, whereas Class D, E, and S are known as switching mode PAs.

2.2.1 Class F mode

In Class F, the time domain V–I waveforms are engineered to be non-overlapping in the time domain. This is accomplished by shaping the current as a half-sinusoid and the voltage as a square waveform [16]. The half-sinusoidal current results from biasing the transistor near pinch-off, similar to Class B operation. The square voltage characteristics are obtained with a short-circuit and an open circuit termination at even and odd harmonics, respectively, as depicted in Figure 2.3. Although achieving perfect harmonic terminations is impractical, terminating only three harmonics still produces an approximately square voltage characteristics, as shown in Figure 2.3. This configuration minimizes the voltage–current overlap and can theoretically yield an efficiency of about 90.7%.

The time domain V–I equations for Class F are as follows [12]:

$$v(\omega t) = V_{DD}\left(1 - \frac{2}{\sqrt{3}}\cos\omega t + \frac{1}{3\sqrt{3}}\cos 3\omega t\right) \tag{2.4}$$

$$i(\omega t) = I_m\left(\frac{1}{\pi} + \frac{1}{2}\cos\omega t + \frac{2}{3\pi}\cos 2\omega t - \frac{2}{15\pi}\cos 4\omega t + \frac{2}{35\pi}\cos 6\omega t + \dots\right) \tag{2.5}$$

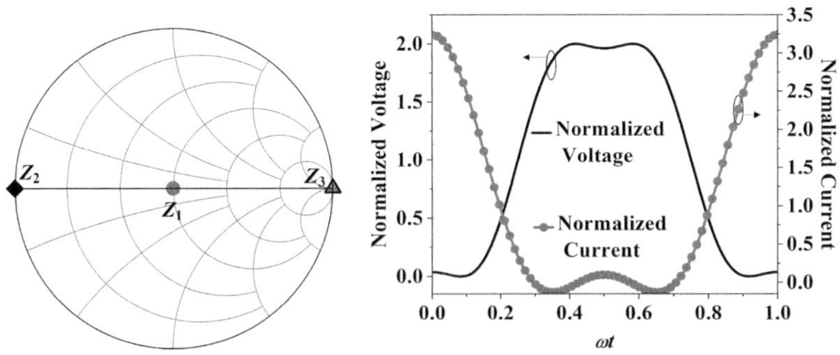

Figure 2.3 Load impedances and time domain V–I waveforms of Class F mode considering three harmonics.

The load terminations at fundamental and harmonics of Class F mode can be calculated from (2.4) and (2.5) as follows [12]:

$$Z_1 = \frac{2}{\sqrt{3}} R_{opt} \tag{2.6a}$$

$$Z_2 = 0 \tag{2.6b}$$

$$Z_3 = \infty \tag{2.6c}$$

$$R_{opt} = \frac{2V_{DD}}{I_m} \tag{2.7}$$

The term R_{opt} is the Class B load which can be calculated from the load line. The time domain V–I waveforms of Class F are shown in Figure 2.3 and its corresponding high efficiency can be achieved using any transistor by presenting the loads shown in (2.6) using an output matching network.

2.2.2 Class F^{-1} mode

The V–I characteristics are interchanged compared to Class F; therefore, it is named Class F^{-1} mode [17]. In this mode, the time domain V–I characteristics are half sinusoidal and square, respectively. These characteristics result in time domain V–I waveforms that do not overlap. As can be seen in Figure 2.4, the second harmonic load is open while the third harmonic load is at short in Class F^{-1} mode.

Ideally Class F^{-1} mode can achieve 100% efficiency like Class F. However, in practice when terminations up to the third harmonic are provided, it can achieve 82% efficiency. Class F mode can achieve 91.7% efficiency even with three harmonics since the Class B biasing makes it possible to present higher even harmonics to generate a half-sinusoidal characteristic of current. The Class F^{-1} mode produces a half-sinusoidal voltage characteristic, achieved by applying suitable load terminations. However, this approach naturally limits the termination of even harmonics in the circuit. Therefore, in Class F^{-1} nearly 10% less efficiency is achieved in comparison with Class F. However, the efficiency of 82% remains higher than that achieved in Class B operation. The time domain V–I waveforms when harmonics are terminated up to the third order are shown in Figure 2.4, which shows square and half-sinusoidal characteristics, respectively. It is observed that the overlap between time domain V–I waveforms remains minimal.

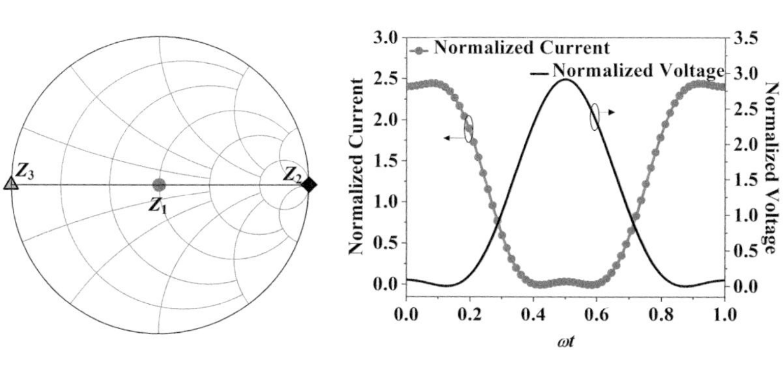

Figure 2.4 Load impedances and time domain V–I waveforms of Class F^{-1} mode considering three harmonics.

The time domain V–I in Class F^{-1} mode is given as follows [12]:

$$v(\omega t) = V_{DD}\left(1+\sqrt{2}\cos\omega t + 0.5\cos 2\omega t\right) \tag{2.8}$$

$$i(\omega t) = I_m\left(0.37 - 0.43\cos\omega t + 0.06\cos 3\omega t\right) \tag{2.9}$$

The load impedances can be determined using (2.8) and (2.9) for Class F^{-1} mode as follows [12]:

$$Z_1 = \frac{1}{0.43\sqrt{2}} R_{opt} \tag{2.10a}$$

$$Z_2 = \infty \tag{2.10b}$$

$$Z_3 = 0 \tag{2.10c}$$

The Class F^{-1} PA can be implemented with the impedances shown in (2.10) to achieve time domain V–I waveforms as can be seen in Figure 2.4. Minimizing the overlapping of time domain V–I waveforms allows the amplifier to reach Class F^{-1} efficiency.

2.2.3 Class J mode

Class J is an advanced PA configuration that evolved as an extension of the conventional Class B. Since the voltage characteristics of Class B mode are sinusoidal as depicted in Figure 2.2, the voltage doesn't have any harmonics. In other words, Class B has all the harmonic components short circuited, whereas the fundamental load R_{opt} behaves as shown in Figure 2.5. The efficiency of 78.5% like Class B may be realized by incorporating reactive terminations up to the second harmonic frequency. When fundamental components are made inductive, the capacitive second harmonic load needs to be provided, as shown in Figure 2.5. If only the fundamental is inductive, the phase of voltage waveform is shifted forward which increases the voltage–current overlap and reduces efficiency. Therefore, a capacitive second harmonic load, the voltage waveform becomes half sinusoidal, resulting in a flattened trailing edge and reduced overlap between time domain V–I waveforms, as can be seen in Figure 2.5. This operating mode, termed Class J, provides performance similar to Class B mode [7, 8].

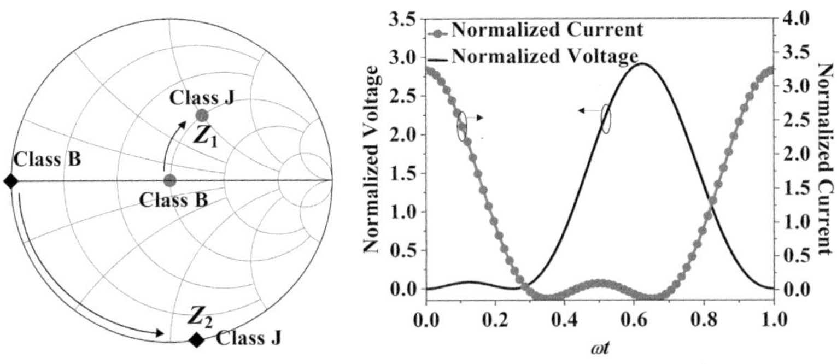

Figure 2.5 Load impedances and time domain V–I waveforms of Class J mode.

In Class J operation, the current characteristic is the same as that of Class B and F PAs, as per (2.5). The voltage waveform for this mode is determined by the relation presented below [12].

$$v(\omega t) = V_{DD}\left(1 - \cos\omega t - \sin\omega t + \frac{1}{2}\sin 2\omega t\right) \tag{2.11}$$

The load conditions for Class J operation are determined from (2.5) and (2.11) as follows [12]:

$$Z_1 = (1 + j)\ R_{opt} \tag{2.12a}$$

$$Z_2 = -j\frac{3\pi}{8}R_{opt} \tag{2.12b}$$

In Class J, both the time domain V–I waveforms have fundamental and second harmonic components. Therefore, both time domain V–I waveforms exhibit a half-sinusoidal, but with a slight phase shift, as can be seen in Figure 2.5. When Z_1 is capacitive and Z_2 is inductive, it is called Class J^{-1} where the voltage waveform moves toward the left. Class J is practically convenient to implement among the waveform-engineered modes, as it only requires terminations up to the second harmonic.

2.2.4 Class E mode

Class E is distinctive from other PAs where the time domain V–I waveforms emulate those of a switching controller; hence, it is classified under switched-mode PAs [18]. In this mode, the current waveform becomes zero when voltage exists across the transistor, and vice versa. This ensures non-overlapping time domain V–I waveforms, thereby reducing power loss. The Class E mode functions under two main principles: Zero Voltage Switching (ZVS) and Zero Voltage Derivative Switching (ZVDS). The ZVS condition guarantees that the voltage drops to zero just before the current begins conduction, while ZVDS maintains zero drain-to-source voltage during the transition from the off to on state. These behaviors are realized through the charge and discharge mechanism of the drain shunt capacitance. Therefore, with suitable fundamental and capacitive harmonic terminations as illustrated in Figure 2.6, ZVS and ZVDS conditions can be met. This results in non-overlapping time domain V–I waveforms as can be seen in Figure 2.6 and can achieve theoretical efficiency of 100%. In other words, the transistor remains "on" for 50% duty cycle and "off" for the rest of the duty cycle.

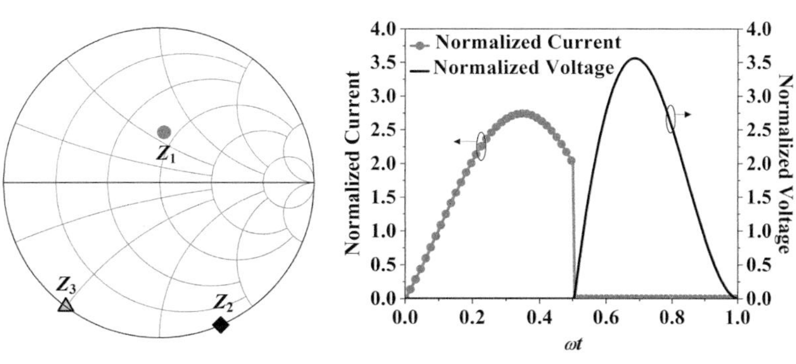

Figure 2.6 Load impedances and time domain V–I waveforms of Class E mode.

The expression of time domain voltage for "off" and current for "on" conditions in Class E mode are as follows [12, 18]:

$$v(\omega t) = \pi V_{DD}\left(\omega t - \frac{3\pi}{2} - \frac{\pi}{2}\cos\omega t - \sin\omega t\right) \tag{2.13}$$

$$i(\omega t) = I_{max}\left(1 + \frac{\pi}{2}\sin\omega t - \cos\omega t\right) \tag{2.14}$$

The load conditions for a Class E PA are defined by the following equations:

$$Z_1 = (R + j\omega L)\left\|\frac{1}{j\omega C}\right. \tag{2.15a}$$

$$Z_n = -j\frac{5.45R}{n} \tag{2.15b}$$

$$R = \frac{0.5768V_{DD}^2}{P_{OUT}} \tag{2.16}$$

$$L = \frac{1.1525R}{\omega} \tag{2.17}$$

$$C = \frac{0.1836}{\omega R} \tag{2.18}$$

Class E can achieve 100% efficiency theoretically. But, in practice, a transistor has finite "on" resistance which cannot achieve ideal ZVS and ZVDS conditions. The drain-to-source capacitance (C_{ds}) inherently present in the transistor provides a capacitive termination of higher-order harmonics, improving waveform shaping and efficiency.

2.3 CONTINUOUS MODES OF OPERATION

The waveform-engineered modes presented earlier are derived at the current source reference plane (CRP). In practical implementation, operation of a transistor deviates from ideal current source behavior due to inherent device and package parasitic elements, as shown in the equivalent circuit of a packaged GaN HEMT in Figure 2.7. The equivalent model of GaN HEMT shown in Figure 2.7 consists of reactive elements which are frequency dependent components. In a particular waveform engineered mode, required fundamental and harmonic loads need to be presented at CRP to achieve high efficiency.

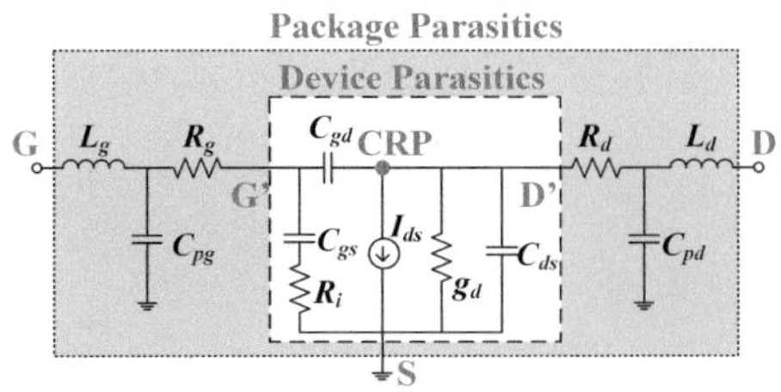

Figure 2.7 Model of a packaged GaN HEMT.

When the effect of parasitic elements such as g_d, C_{ds}, R_d, C_{pd}, and L_d are de-embedded, appropriate loads that need to be presented at the package plane can be obtained. Although CRP loads are frequency independent, the termination at the package plane is frequency dependent. Usually, passive matching network presents foster loads, i.e., clockwise rotating loads with frequency on the Smith chart. Sometimes these de-embedded loads at the package plane results in active impedance loads which is not attainable using a passive matching network [19–28]. Therefore, it is quite challenging to design broadband PA using conventional waveform engineering modes. Continuous modes of operation are an extension of conventional waveform engineering modes, which have the wideband capability [19–28]. This section provides an overview of different continuous modes.

2.3.1 Continuous Class B/J mode

As discussed in Section 2.2.3, Class J/J^{-1} modes can achieve efficiency of 78.5% like Class B by introducing a reactive element into the loads up to second harmonic frequency. Therefore, as presented in [19–22], a continuum of impedance space exists from Class J to B and to J^{-1}, as can be seen in Figure 2.8. Therefore, this operation mode is termed Continuous Class B/J (CCB/J) mode. Each mode exhibits distinct voltage waveforms, as illustrated in Figure 2.8, but interestingly can give similar output power and efficiency. The current characteristics in CCB/J mode is half sinusoidal similar to that of Class B, as per (2.5). The voltage waveform is as per the following expressions [19, 20].

$$v(\omega t) = V_{DD}\left(1-\cos\omega t\right)\left(1-\alpha\sin\omega t\right),\ -1\le\alpha\le1 \tag{2.19}$$

The design parameter α controls the phase angle between the time domain V–I. The time domain V–I waveforms of Class B can be obtained when $\alpha=0$, whereas $\alpha=1$ and -1 are associated with Class J and J^{-1}, respectively. The load terminations of CCB/J mode can be derived from (2.5) and (2.19) as shown below [19, 20].

$$Z_1 = \left(1+j\alpha\right)R_{opt} \tag{2.20a}$$

$$Z_2 = -j\alpha\frac{3\pi}{8}R_{opt} \tag{2.20b}$$

By changing the design parameter α from -1 to 1, a continuum of loads as shown in Figure 2.8 will be obtained. Each of these loads can be mapped to different frequencies to

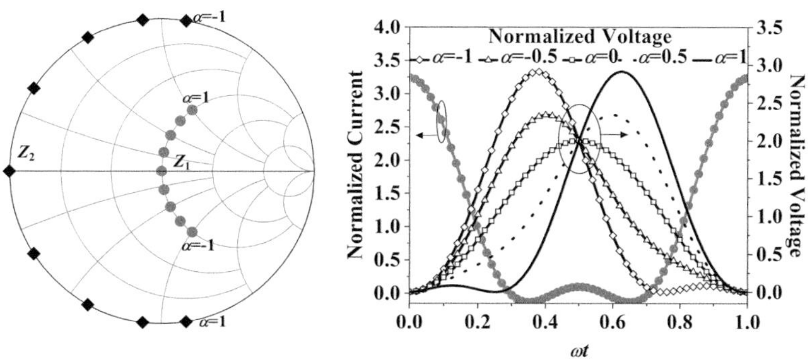

Figure 2.8 Load impedances and time domain V–I waveforms of CCB/J mode.

design a broadband matching network. Since Class J serves as the foundation for continuous modes, it is often referred to as their parent class.

2.3.2 Continuous Class F mode

Like CCB/J, the Class F continuum can also be realized by introducing a design parameter δ with respect to the voltage waveform in Class F operation as shown below [23–26].

$$v(\omega t) = V_{DD}\left(1+\sqrt{2}\cos\omega t+\frac{1}{2}\cos 2\omega t\right)(1-\delta\sin\omega t) \tag{2.21}$$

The design parameter δ, which ranges from –1 to 1, controls the phase difference between the time domain V–I waveforms. A unique voltage waveform can be obtained for each value of δ and can achieve 90.7% efficiency like Class F mode. The load terminations of Continuous Class F (CCF) mode determined from (2.21) and (2.5) are as shown below [23].

$$Z_1 = \left(\frac{2}{\sqrt{3}}+j\delta\right)R_{opt} \tag{2.22a}$$

$$Z_2 = -j\delta\frac{\pi}{2}R_{opt} \tag{2.22b}$$

$$Z_3 = \infty \tag{2.22c}$$

The impedances and the time domain V–I characteristics of the CCF mode are as shown in Figure 2.9. When $\delta = 0$, the operation corresponds to Class F mode. The loads shown in Figure 2.9 need to be presented at the CRP to achieve high efficiency. Therefore, the effect of parasitic elements needs to be de-embedded from CRP loads to obtain packaged loads. Once finding the packaged loads corresponding to δ, each of these loads can be considered at different frequencies and design a broadband PA.

The broadband PA designed with CCF mode was presented in [26]. This PA is designed with a packaged 10 W GaN HEMT and operates over a broadband from 1.6 to 2.3 GHz. A fractional bandwidth of 35.9% is achieved by the PA. This PA achieved DE of 72–78% and output power varying from 41.6 to 42.7 dBm over a broadband of 1.6–2.3 GHz in measurements. This hardware prototype exhibits the capability of CCF mode to achieve high efficiency over broadband [26].

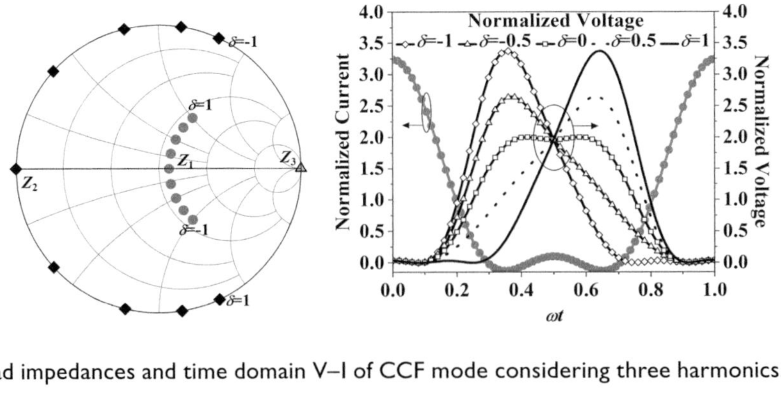

Figure 2.9 Load impedances and time domain V–I of CCF mode considering three harmonics.

2.3.3 Continuous Class F^{-1} mode

The concept of the Continuous Class F^{-1} (CCF^{-1}) mode was presented in [28–31], which enables the configuration of a broadband PA with the Class F^{-1} efficiency. The voltage characteristics of CCF^{-1} mode remain half sinusoidal, like the Class F^{-1} mode. The current equation for the CCF^{-1} mode are as follows [28]:

$$i(\omega t) = I_m(0.37 - 0.43\cos\omega t + 0.06\cos 3\omega t)(1 - \xi\sin\omega t) \tag{2.23}$$

In this expression, the design parameter ξ varies from –1 to 1, which determines current waveform. The performance is the same as Class F^{-1} mode for any value of the design parameter ξ. The load terminations of the CCF^{-1} mode are determined from (2.8) and (2.23) as follows [28]:

$$Z_1 = \frac{\sqrt{2}R_{opt}}{(0.43 + j\xi 0.37)} \tag{2.24a}$$

$$Z_2 = -\frac{R_{opt}}{0.49\xi} \tag{2.24b}$$

$$Z_3 = 0 \tag{2.24c}$$

The load impedances and the time domain V–I waveforms of CCF^{-1} for various values of ξ can be seen in Figure 2.10, respectively. Impedances corresponding to different ξ can be mapped to different frequencies to realize a highly efficient broadband PA. When $\xi = 0$, the load impedances and time domain V–I waveforms corresponding to Class F^{-1} mode are obtained.

2.3.4 Continuous Class E mode

The Continuous Class E (CCE) mode extends Class E for the design of broadband PA. The time domain V–I equations of CCE mode are as follows [32, 33]:

$$v_C(\omega t) = \frac{V_{DD}\tan\phi_1}{\pi}\begin{bmatrix}(4\pi\cot\phi_1 - 8)\sin^2\omega t - 2\pi\sec\phi_1\sin(\omega t + \phi_1) \\ +(\pi - 2\tan\phi_1)\sin 2\omega t - 6\pi\tan\phi_1 + 4\omega t\tan\phi_1\end{bmatrix} \tag{2.25}$$

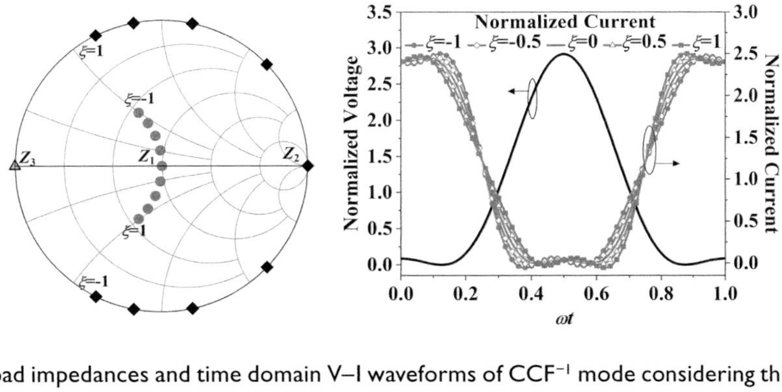

Figure 2.10 Load impedances and time domain V–I waveforms of CCF^{-1} mode considering three harmonics.

$$i_S(\omega t)=\frac{4V_{DD}}{\pi^2 R}\begin{bmatrix}\cos\phi_1\left(\pi\cos\left(2\omega t-\tan^{-1}(2\cot\phi_1)\right)\right)\times\sqrt{1+4\cot^{-1}\phi_1}\\ -4\sin 2\omega t\cos\phi_1-\pi\cos(\omega t+\phi_1)+4\sin\phi_1\sin^2\omega t\end{bmatrix}\sin\phi_1 \quad (2.26)$$

where

$$R=\frac{8V_{DD}^2\sin^2\phi_1}{\pi^2 P_{out}} \quad (2.27)$$

The load impedances of CCE are as follows [32, 33]:

$$Z_1=\frac{(1+jx_1)\pi R}{\pi+j2(1+jx_1)\cos^2\phi_1} \quad (2.28a)$$

$$Z_2=\frac{jx_2\pi R}{\pi-4x_2\cos^2\phi_1} \quad (2.28b)$$

$$Z_n=\frac{-j\pi R}{2n\cos^2\phi_1} \quad (2.28c)$$

where

$$x_1=\frac{16\pi\cos^2\phi_1\cot\phi_1+2\pi\sin 2\phi_1+3\pi^2-32}{12\pi\cos^2\phi_1} \quad (2.29a)$$

$$x_2=\frac{\pi\sec^2\phi_1}{24(\pi-2\tan\phi_1)}\begin{bmatrix}4\sin\left(2\tan^{-1}(2\cot\phi_1)\right)+3\pi\\ +2\left(\cos\left(2\tan^{-1}(2\cot\phi_1)\right)-2\right)\tan\phi_1\end{bmatrix} \quad (2.29b)$$

Provided that V_{DD} and P_{out} are known, the impedances and voltage and current equations of CCE can be calculated from (2.25) to (2.29) for a particular ϕ_1. In CCE mode, ϕ_1 is the phase angle parameter in the fundamental current component and serves as a design variable. The valid range of ϕ_1 is from 43.3° to 78°, which yields the performance comparable to that of Class E. The load impedances and time domain V–I waveforms of the CCE are as can be seen in Figure 2.11, respectively. Distinct time domain V–I waveforms are obtained for different values of ϕ_1 as can be observed from Figure 2.11. When ϕ_1 = 57.5°, time domain

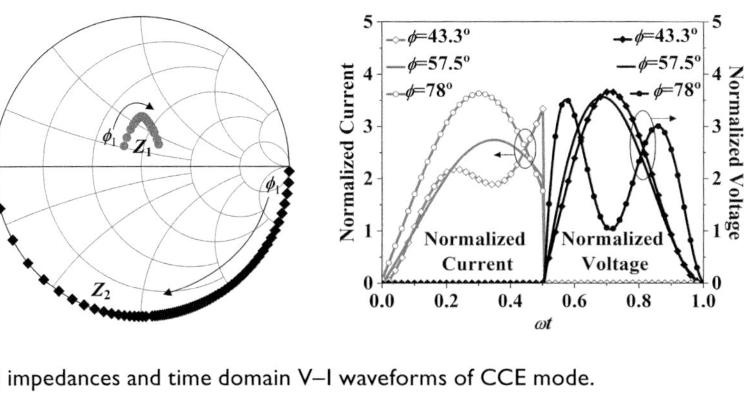

Figure 2.11 Load impedances and time domain V–I waveforms of CCE mode.

V–I characteristics of a conventional Class E amplifier are illustrated in Figure 2.11. When $\phi_1 = 43.3°$, Z_2 behaves as an open circuit similar to that in Class F^{-1} operation; therefore, it is called Class E/F_2 mode. One can see from Figure 2.11 that the current waveform exhibits square-like characteristics similar to those of Class F^{-1}, when $\phi_1 = 43.3°$. The Z_2 behaves as a short circuit when = 78°, like in Class F mode. Therefore, this mode is known as Class EF_2 mode, and the voltage waveform exhibits square-like characteristics as can be seen in Figure 2.11.

The broadband PA, designed with CCE theory and employing the 15 W packaged GaN HEMT CGH27015F, was presented in [33]. This PA is designed to operate over a broadband of 1.5–2.9 GHz, which has a fractional bandwidth of 63.6%. The PA has achieved DE of 60.2–76.5%, PAE of 56.5–73.8%, and the output power varies from 39.4 to 41.9 dBm.

2.4 EXTENDED CONTINUOUS MODES OF OPERATION

The continuous classes discussed in Section 2.3 are suitable to achieve high efficiency over wider bandwidth. However, their bandwidth is constrained to one octave because of the requirement of purely reactive harmonic loads to achieve high efficiency. For example, let us consider a multi-octave bandwidth from 1 to 4 GHz, where the first octave is from 1 to 2 GHz and the second octave is from 2 to 4 GHz. To design a multi-octave PA with traditional continuous modes, the fundamental load needs to be either purely resistive or complex to achieve the required output power. However, to boost efficiency, the second harmonic frequency from 2 to 8 GHz range should be terminated with a purely reactive impedance. There is a spectral overlap of fundamental and second harmonic frequencies. Moreover, the second octave's fundamental band and the first octave's second harmonic band both occupy the 2–4 GHz range. If conventional continuous modes are used, the fundamental load requires a purely resistive or complex impedance while the second harmonic demands for a purely reactive load from 2 to 4 GHz. However, it is not possible to achieve two different loads at the same frequencies. Therefore, the conventional continuous modes fail for multi-octave bandwidths. The extended continuous modes solve this problem [34–48]. This section provides a detailed explanation of the extended continuous modes derived from various waveform engineering techniques.

2.4.1 Extended Resistive Continuous Class B/J mode

The Extended Resistive Continuous Class B/J (ERCB/J) was introduced with a complex second harmonic load to achieve an overlap between the loads at the fundamental and second harmonic frequencies. In order to achieve resistive second harmonic load, design parameters β and γ are introduced in the voltage waveform as follows [40]:

$$v(\omega t) = V_{DD}(1-\cos\omega t)(\gamma-\alpha\sin\omega t)(1+\beta\cos\omega t) \tag{2.30}$$

Like Class B/J mode, the drain current exhibits a half-sinusoidal characteristic. The load terminations for the ERCB/J mode are calculated using Equations (2.5) and (2.30) as given below.

$$Z_1 = R_{opt}\left(\gamma(1-\beta)+j\alpha(1-0.25\beta)\right) \tag{2.31a}$$

$$Z_2 = \frac{3\pi R_{opt}}{8}\left(\gamma\beta - j\alpha(1-\beta)\right) \tag{2.31b}$$

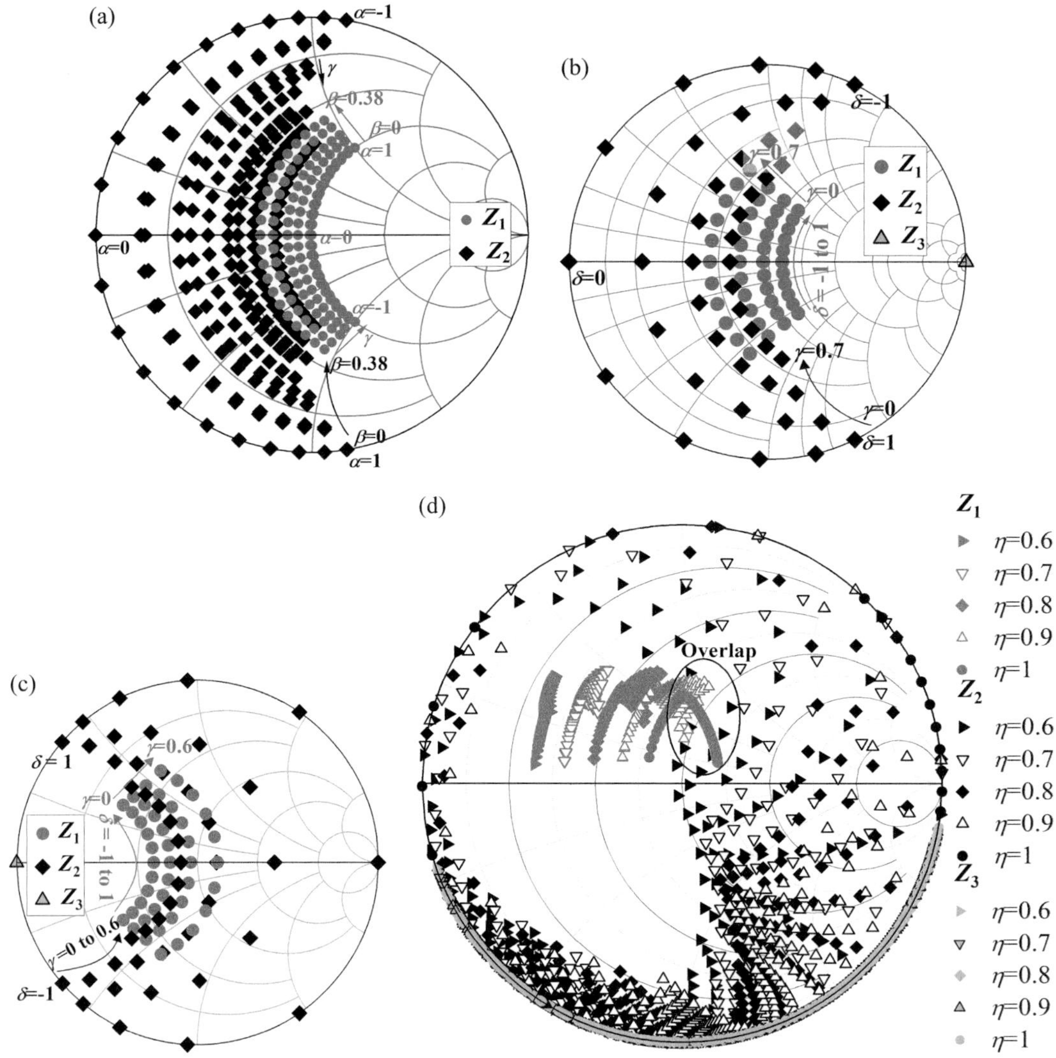

Figure 2.12 Load impedances of (a) ERCB/J mode, (b) ECCF mode, (c) $ECCF^{-1}$ mode, and (d) ECCE.

The load impedances for the ERCB/J mode are illustrated in Figure 2.12(a). The design parameters α varies from –1 to 1, β from 0 to 0.38, and γ from 1 to $1/(1-\beta)$. When $\beta=0$ and $\gamma=1$, the load impedances corresponding to CCB/J mode are obtained. One can see from Figure 2.12(a) that the loads at the fundamental and second harmonic frequencies overlap. These overlap loads can be mapped to the frequencies in higher octaves to achieve multi-octave bandwidth. The ERCB/J PA can achieve efficiency from 60 to 78.5%.

The PA presented in [40] employs a commercial 10 W GaN HEMT CGH40010F and is designed to operate using ERCB/J mode. This PA operates from 0.5 to 3.25 GHz in a multi-octave bandwidth, with a very high fractional bandwidth of 146.7%, covering nearly three octaves. The DE of 60–70.1% and the output of 39.1–41.75 dBm are achieved in the measurements for the ERCB/J PA. The measured PAE varies from 52.6 to 68.7%. The ERCB/J mode enabled the design of a highly efficient multi-octave PA [40].

2.4.2 Extended Continuous Class F mode

An extended version of the CCF, appropriate for multi-octave bandwidth, is achieved by incorporating a complex second harmonic impedance [35]. The voltage equation of the Extended Continuous Class F (ECCF), incorporating additional design parameter γ, is given below.

$$v(\omega t) = V_{DD}\left(1+\sqrt{2}\cos\omega t+\frac{1}{2}\cos 2\omega t\right)(1-\delta\sin\omega t)(1+\gamma\cos\omega t) \tag{2.32}$$

The design parameter δ varies from –1 to 1 and γ varies from 0 to 0.7. The load terminations of ECCF mode derived from (2.32) and (2.5) are as follows [35]:

$$Z_1 = \left(\left(\frac{2}{\sqrt{3}}-\gamma\right)+j\delta\left(1-\frac{7}{12\sqrt{3}}\right)\right)R_{opt} \tag{2.33a}$$

$$Z_2 = \frac{3\pi R_{opt}}{8}\left(\frac{5}{3\sqrt{3}}\gamma - j\delta\left(\frac{7}{3\sqrt{3}}-\gamma\right)\right) \tag{2.33b}$$

$$Z_3 = \infty \tag{2.33c}$$

The load impedances of ECCF mode are as shown in Figure 2.12(b) where an overlap is achieved between the load terminations at fundamental and harmonic frequencies. When $\gamma = 0$, CCF waveforms and impedances can be obtained where the efficiency of 90.7% can be achieved. As γ increases from zero, resistance of second harmonic impedance increases, which increases the harmonic power and reduces the efficiency from 90.7%. However, the resistive term is crucial to obtain interaction between fundamental and second harmonic load impedance in multi-octave operation. When γ is decreased below 0.7, efficiency drops from 60%. Therefore, γ is varied from 0 to 0.7 to achieve efficiency from 60 to 90.7%.

2.4.3 Extended Continuous Class F^{-1} mode

The extension of ECCF^{-1}, i.e., Extended Continuous Class F^{-1} (ECCF^{-1}) mode, was introduced to enable the design of multi-octave PAs [36, 37]. Similar to other extended continuous modes, ECCF^{-1} also introduces complex second harmonic load, resulting in an overlap between the load terminations at fundamental and harmonic frequencies. The overlapped loads can be mapped to frequencies in higher octaves to design highly efficient PA operating over multi-octave bandwidth. Like Class F^{-1} mode, the voltage exhibits a half-sinusoidal characteristic as per (2.8). The time domain current expression in ECCF^{-1} mode is as follows [37]:

$$i(\omega t) = I_m\left(1-1.16\cos\omega t+0.16\cos 3\omega t\right)\left(1-\gamma\sin\omega t\right)\left(1+\delta\cos\omega t\right) \tag{2.34}$$

The design parameter γ varies from –1 to 1, while δ varies from 0 to 0.6 to maintain DE from 60 to 82%. The load terminations for the ECCF^{-1} mode are derived using equations (2.8) and (2.34) as shown below [37].

$$Z_1 = \frac{R_{opt}}{\sqrt{2}\left((1.16-\delta)-j\gamma\left(0.33\delta-1\right)\right)} \tag{2.35a}$$

$$Z_2 = \frac{R_{opt}}{2\left(\delta-j\gamma\left(1.32-\delta\right)\right)} \tag{2.35b}$$

$$Z_3 = 0 \tag{2.35c}$$

The load impedances of the ECCF-1 mode for various values of γ and δ are shown in Figure 2.12(c). One can see that an overlap is achieved between load terminations at fundamental and harmonic frequencies which can be utilized to design multi-octave PAs.

2.4.4 Extended Continuous Class E mode power amplifiers

The Extended Continuum of Class E (ECCE) mode was introduced in [48]. The voltage and current equations below were solved by considering a complex second harmonic load [48].

$$v_C(\omega t) = V_{DD}\frac{\left\{-4\pi\begin{bmatrix} I_{01}I_{02}(\pi-\omega t)+I_{02}\sin\phi_1-0.5I_{01}\sin\phi_2 \\ +I_{02}\sin(\omega t+\phi_1)+0.5I_{01}\sin(2\omega t+\phi_2)\end{bmatrix}\right\}}{(4I_{02}\cos\phi_1+\pi I_{01}\sin\phi_2)} \tag{2.36}$$

$$i_S(\omega t) = I_m\left\{1-\left[\frac{(\cos(\omega t+\phi_1))}{I_{01}}\right]-\left[\frac{(\cos(2\omega t+\phi_2))}{I_{02}}\right]\right\} \tag{2.37}$$

Here

$$I_{01} = \frac{(2\sin\phi_1)}{\pi} \tag{2.38a}$$

$$I_{02} = \frac{(2\sin\phi_1\cos\phi_2)}{(2\sin\phi_1-\pi\cos\phi_1)} \tag{2.38b}$$

$$I_{21} = \frac{(2\sin\phi_1-\pi\cos\phi_1)}{(\pi\cos\phi_2)} \tag{2.38c}$$

The load impedances of ECCE, with a complex second harmonic, can be expressed as follows:

$$Z_n = Z_{ni}\|X_{Cn},\ n>0 \tag{2.39}$$

$$Z_{1i} = \frac{1}{12\pi\omega C}\begin{bmatrix}\begin{bmatrix}12I_{01}(\pi\sin\phi_1+2\cos\phi_1)-24\sin^2\phi_1 \\ +8I_{21}(\cos\phi_1\cos\phi_2+2\sin\phi_1\sin\phi_2)\end{bmatrix} \\ +j\begin{bmatrix}6\pi+12I_{01}(\pi\cos\phi_1-2\sin\phi_1) \\ +8I_{21}(2\cos\phi_1\sin\phi_2-\sin\phi_1\cos\phi_2)\end{bmatrix}\end{bmatrix} \tag{2.40a}$$

$$Z_{2i} = \frac{1}{12\pi\omega CI_{21}}\begin{bmatrix}-4\,(2\cos\phi_1\cos\phi_2+\sin\phi_1\sin\phi_2) \\ +j\,(3\pi I_{21}+4\,(2\cos\phi_1\sin\phi_2-\sin\phi_1\cos\phi_2))\end{bmatrix} \tag{2.40b}$$

$$Z_{ni} = \infty,\ n>2 \tag{2.40c}$$

$$X_{Cn} = \frac{1}{jn\omega C} = \frac{2\pi I_{02}R}{jn\eta I_{01}[4I_{02}\cos\phi_1+\pi I_{01}\sin\phi_2]} \tag{2.41}$$

where

$$R = \frac{\left[8\left(\eta\left(V_{DD} - V_k\right)\sin\phi_1\right)^2\right]}{\left(\pi^2 P_{out}\right)} \tag{2.42}$$

For a particular transistor, V_{DD} and P_{out} are usually known, therefore, the impedances and voltage and current equations of ECCE can be calculated from (2.36) to (2.42) for particularly η, ϕ_1, and ϕ_2. The valid range of ϕ_1 are from 39° to 79°, whereas the range of ϕ_2 conditionally depends upon ϕ_1 as per the following relation.

$$\phi_2 \in [\phi_a, \phi_b],\ 39° \leq \phi_1 \leq 57.5° \tag{2.43a}$$

$$\phi_2 \in [\phi_b, \phi_a],\ 57.5° < \phi_1 \leq 79° \tag{2.43b}$$

where $\phi_a = -\tan^{-1}\left(2\cot\phi_1\right)$ and $\phi_b = -\tan^{-1}\left(\Delta_2/\Delta_1\right)$
$\Delta_1 = 64\ \sin^3\phi_1 + 8\pi\sin 2\phi_1\left[\pi\cos\phi_1 - 4\ \sin\phi_1\right]$
$\Delta_2 = 8\pi^2\cos^3\phi_1 + 4\sin 2\phi_1\left[16\sin\phi_1 - 10\pi\cos\phi_1\right]$

The load impedances for η varying from 0.6 to 1, and ϕ_1 varying from 39° to 79°, are shown in Figure. 2.12(d). An overlap is achieved between the loads up to second harmonic, as can be seen in Figure 2.12(d). In ERCB/J and ECCF modes, Z_{3i} is a short circuit, while in ECCF^{-1} mode it is an open circuit. While targeting a multi-octave bandwidth, the range of third harmonic load would be large. Designing a matching network that provides a single load, i.e., short/open circuit, at the third harmonic across a multi-octave band is challenging. However, in ECCE mode third harmonic load is a range of impedances at the periphery of the Smith chart as can be seen in Figure 2.12(d). Therefore, there is an additional freedom with ECCE mode in terms of third harmonic load termination while designing matching network for PA.

A highly efficient multi-octave PA presented in [48] was designed with commercial 10 W GaN HEMT CGH40010F using ECCE mode. The PA is designed from 0.45 to 2.9 GHz covering multi-octave bandwidth, resulting in a high fractional bandwidth of 146.27%. The measured DE was from 60 to 72.9% and the output ranged from 39.6 to 41.7 dBm are achieved in the measurements for the fabricated ECCE PA. The measured PAE varies from 52 to 67%. The ECCE mode enabled the design of a highly efficient multi-octave PA.

2.5 CONCLUSION

This chapter has summarized advanced RF PA design techniques aimed at enhancing efficiency and bandwidth for modern wireless communication systems. Emphasis was placed on waveform engineering approaches, including high-efficiency modes such as Class B/J, E, F, and F^{-1}, which offer efficiencies between 75 and 91%. The concept of continuum modes was explored for achieving octave bandwidth, while extended continuum modes demonstrated the potential for multi-octave operation. Through discussion of these modes and their practical implementations, this chapter concludes that waveform engineering plays a vital role in enabling wideband, energy-efficient PAs suitable for next-generation wireless technologies.

REFERENCES

1. Raab, F.H., Asbeck, P., Cripps, S., Kenington, P.B., Popovic, Z.B., Pothecary, N., Sevic, J.F. and Sokal, N.O., 2002. Power amplifiers and transmitters for RF and microwave. *IEEE Transactions on Microwave Theory and Techniques*, **50**(3), pp.814–826.
2. Raab, F.H., Asbeck, P., Cripps, S., Kenington, P.B., Pothecary, N., Sevic, J.F. and Sokal, N.O., 2004. RF and microwave power amplifier and transmitter technologies – part 5. *High Frequency Electronics*, **3**, pp.46–54, January.
3. Fang, C., Yu, F.R., Huang, T., Liu, J. and Liu, Y., 2015. A survey of green information-centric networking: research issues and challenges. *IEEE Communications Surveys & Tutorials*, **17**(3), pp.1455–1472, February.
4. Wright, P., Lees, J., Tasker, P.J. and Benedikt, J., 2008. GaN power transistors in narrow and wide bandwidth applications. *IET Seminar on Wideband Receivers and Components*, pp.1–1, May.
5. Musser, M., Walcher, H., Maier, T., Quay, R., Damman, M., Mikulla, M. and Ambacher, O., 2010. GaN power FETs for next generation mobile communication systems. *European Microwave Integrated Conference (EuMIC)*, pp.9–12, September.
6. Colantonio, P., Giannini, F. and Limiti, E., *High efficiency RF and microwave solid state power amplifier*. Wiley Series in Microwave and Optical Engineering, United Kingdom: John Wiley & Sons, 2009.
7. Cripps, S.C., 2002. *Advanced techniques in RF power amplifier design*. 1st ed. Norwood, MA: Artech House.
8. Cripps, S.C., 2006. *RF power amplifiers for wireless communications*. 2nd ed. Norwood, MA: Artech House Publishers.
9. Pozar, D.M., 2000. *Microwave and RF design of wireless systems*. 1st ed. New York, NY: John Wiley & Sons.
10. Grebennikov, A., Sokal, N. and Franco, M., 2011. *Switchmode RF power amplifiers*. New York, NY: Newnes.
11. Gonzalez, G., 1996. *Microwave transistor amplifiers: analysis and design*. Englewood Cliffs, NJ: Prentice-Hall.
12. Rawat, K., Roblin, P. and Koul, S.K., 2020. *Bandwidth and efficiency enhancement in radio frequency power amplifiers for wireless transmitters*. Switzerland: Springer Nature.
13. Tasker, P.J., 2009. Practical waveform engineering. *IEEE Microwave Magazine*, **10**(7), pp.65–76.
14. Al Tanany, A., Bruner, D., Sayed, A. and Boeck, G., 2010. Highly efficient harmonically tuned broadband GaN power amplifier. *European Microwave Integrated Circuits Conference (EuMIC)*, pp.5–8, September.
15. Grebennikov, A., Sokal, N.O. and Franco, M.J., 2012. *Switchmode RF and microwave power amplifiers*. London, UK: Academic Press.
16. Raab, F.H., 1997. Class-F power amplifiers with maximally flat waveforms. *IEEE Transactions on Microwave Theory and Techniques*, **45**(11), pp.2007–2012.
17. Wei, C.J., DiCarlo, P., Tkachenko, Y.A., McMorrow, R. and Bartle, D., 2000. Analysis and experimental waveform study on inverse class-F mode of microwave power FETs. *IEEE MTT-S International Microwave Symposium Digest*, Boston, MA, pp.525–528, June.
18. Raab, F.H., 1977. Idealized operation of the class E tuned power amplifier. *IEEE Transactions on Circuits and Systems*, **24**(12), pp.725–735.
19. Wright, P., Lees, J., Benedikt, J., Tasker, P.J. and Cripps, S.C., 2009. A methodology for realizing high efficiency Class-J in a linear and broadband PA. *IEEE Transactions on Microwave Theory and Techniques*, **57**(12), pp.3196–3204.
20. Saxena, S., Rawat, K. and Roblin, P., 2017. Continuous Class-B/J power amplifier using a nonlinear embedding technique. *IEEE Transactions on Circuits and Systems II: Express Briefs*, **64**(7), pp.837–841.
21. Asha Latha, Y.M. and Rawat, K., 2019. Methodology to realize wideband high efficiency power amplifier using active harmonic injection. *IEEE MTT-S International Microwave and RF Conference (IMARC)*, Mumbai, India, pp.1–4, December.
22. Asha Latha, Y.M., Rawat, K., Helaoui, M. and Ghannouchi, F.M., 2019. Broadband continuous mode power amplifier with on-board harmonic injection. *IET Microwaves, Antennas & Propagation*, **13**(9), pp.1402–1407.
23. Carrubba, V. *et al.*, 2011. On the extension of the continuous Class-F mode power amplifier. *IEEE Transactions on Microwave Theory and Techniques*, **59**(5), pp.1294–1303.

24. Carrubba, V., Lees, J., Benedikt, J., Tasker, P.J. and Cripps, S.C., 2011. A novel highly efficient broadband continuous Class-F RFPA delivering 74% average efficiency for an octave bandwidth. *IEEE MTT-S International Microwave Symposium Digest*, pp.1–4, June.
25. Benedikt, J., Tasker, P.J. and Cripps, S.C., 2012. Continuous Class-F3 power amplifier mode varying simultaneously first 3 harmonic impedances. *MTT-S Microwave Symposium Digest*, pp.1–3, June.
26. Asha Latha, Y.M., Nunes, L.C., Barradas, F.M. and Pedro, J.C., 2023. The role of nonlinear Cout in continuous Class-F PAs. *IEEE MTT-S International Microwave Symposium Digest*, pp.579–582, June.
27. Asha Latha, Y.M., Nunes, L.C., Barradas, F.M. and Pedro, J.C., 2024. Theory and design of Doherty power amplifier with active harmonic injection. *IEEE Access*, **12**, pp.41631–41642.
28. Carrubba, V. *et al.*, 2010. Exploring the design space for broadband PAs using the novel continuous inverse Class-F mode. *41st European Microwave Conference*, Manchester, pp.333–336, October.
29. Asha Latha, Y.M., Shariatifar, M. and Rawat, K., 2022. Performance estimation of continuous Class-F–1 PA with non-ideal harmonic terminations. *IEEE MAPCON*, pp.1–4.
30. Asha Latha, Y.M. and Rawat, K., 2023. Continuous Class-F^{-1} Ku-band GaN MMIC power amplifier with an effect of nonlinear output capacitance. *IEEE Transactions on Circuits and Systems II: Express Briefs*, **70**(10), pp.3887–3891.
31. Asha Latha, Y.M. and Rawat, K., 2023. Methodology to design a CCF–1 Ku-band GaN MMIC PA with nonlinear output capacitance. *IEEE Microwaves, Antennas, and Propagation Conference (MAPCON)*, pp.1–5.
32. Ozen, M., Jos, R. and Fager, C., 2012. Continuous Class-E power amplifier modes. *IEEE Transactions on Circuits and Systems II: Express Briefs*, **59**(11), pp.731–735.
33. Asha Latha, Y.M., Rawat, K. and Roblin, P., 2019. Nonlinear embedding model-based continuous Class E/F power amplifier. *IEEE Microwave and Wireless Components Letters*, **29**(11), pp.714–717.
34. Shi, W., He, S., Li, Q., Qi, T. and Liu, Q., 2016. Design of broadband power amplifiers based on resistive-reactive series of continuous modes. *IEEE Microwave and Wireless Components Letters*, **26**(7), pp.519–521.
35. Zheng, S.Y., Liu, Z.W., Zhang, X.Y., Zhou, X.Y. and Chan, W.S., 2018. Design of ultrawideband high-efficiency extended continuous Class-F power amplifier. *IEEE Transactions on Industrial Electronics*, **65**(6), pp.4661–4669.
36. Carrubba, V., Akmal, M., Quay, R., Lees, J., Benedikt, J., Cripps, S.C. and Tasker, P.J., 2012. The continuous inverse Class-F mode with resistive second harmonic impedance. *IEEE Transactions on Microwave Theory and Techniques*, **60**(6), pp.1928–1936.
37. Tang, Q., Li, Y. and Li, W., 2017. Over second octave power amplifier design based on resistive–resistive series of continuous Class-F/F^{-1} modes. *IEEE Microwave and Wireless Components Letters*, **27**(5), pp.494–496.
38. Friesicke, C., Quay, R. and Jacob, A.F., 2015. The resistive–reactive Class-J power amplifier mode. *IEEE Microwave and Wireless Components Letters*, **25**(10), pp.666–668.
39. Li, Q., He, S., Shi, W., Dai, Z. and Qi, T., 2016. Extend the Class-B to Class-J continuum mode by adding arbitrary harmonic voltage elements. *IEEE Microwave and Wireless Components Letters*, **26**(7), pp.522–524.
40. Asha Latha, Y.M. and Rawat, K., 2022. Design of ultra-wideband power amplifier based on extended resistive continuous Class B/J mode. *IEEE Transactions on Circuits and Systems II: Express Briefs*, **69**(2), pp.419–423.
41. Asha Latha, Y.M. and Rawat, K., 2021. Methodology to design multi-octave power amplifier using extended Class B to Class J continuous mode. *IEEE MTT-S International Microwave and RF Conference (IMARC)*, Kanpur, India, pp.1–4.
42. Pang, J., He, S., Dai, Z., Huang, C., Peng, J. and You, F., 2016. Design of continuous-mode GaN power amplifier with compact fundamental impedance solutions on package plane. *IET Microwaves, Antennas & Propagation*, **10**(10), pp.1056–1064.
43. Ma, C., Liu, Y., Pan, W. and Tang, Y., 2015. 0.4–3.0 GHz highly efficient harmonic-tuned power amplifier. *Electronics Letters*, **51**(23), pp.1911–1913.
44. Huang, C., He, S., Shi, W. and Song, B., 2018. Design of broadband high-efficiency power amplifiers based on hybrid continuous modes with phase shift parameter. *IEEE Microwave and Wireless Components Letters*, **28**(2), pp.159–161.

45. Zhang, Z. and Cheng, Z., 2019. A multi-octave power amplifier based on mixed continuous modes. *IEEE Access*, **7**, pp.178201–178208.
46. Chen, X., Chen, W., Feng, Z. and Ghannouchi, F.M., 2009. A 10–3100 MHz nested-mode highly efficient power amplifier for multi-octave applications. *IEEE MTT-S International Microwave Symposium (IMS)*, Boston, MA, pp.702–705.
47. Huang, H., Zhang, B., Yu, C., Gao, J., Wu, Y. and Liu, Y., 2017. Design of multioctave bandwidth power amplifier based on resistive second-harmonic impedance continuous Class-F. *IEEE Microwave and Wireless Components Letters*, **27**(9), pp.830–832.
48. Asha Latha, Y.M. and Rawat, K., 2022. Extending the design space of Class-E mode to design a multi-octave power amplifier. *IEEE Transactions on Circuits and Systems II: Express Briefs*, **69**(12), pp.4829–4833.

Chapter 3

O-RAN architecture and interfaces

Jaya Yadav, Baljeet Ram, Vipin Kumar, and Amit Kumar Singh

3.1 INTRODUCTION

Wireless communication technologies have significantly revolutionized our daily lives in a manner never seen before. Users are no longer confined to connecting and interacting with information from homes, schools, to offices everywhere worldwide. This transformation was accomplished through the growth of cutting-edge services and applications since the beginning of the twenty-first century, transforming billions of customers globally. The Radio Access Network (RAN) has facilitated numerous changes over time and has sped up the information age revolution, therefore becoming mobile telecommunications systems' most important element. The notable alteration in RAN architecture was the use of packet-switched data to increase 3G data traffic throughput. Another evolution resulted when the Long-Term Evolution (LTE) RAN was implemented [1]. Because 5G and beyond 5G are gaining popularity, there is an increasing demand and need for dynamic, adaptable, and intelligent RANs to foresee and satisfactorily fulfill the user demands [2]. As the move toward 5G and B5G raises telecom providers' revenues from businesses other than traditional cell operations, it also presents difficulties for large, traditional enterprises. This shift entails operational and cultural transformations that are far more profound than the surface technological changes. Generally, network infrastructure was expected to be operational for 10–15 years with multiyear return on investment. Moreover, RAN components were interchangeable every decade, and mobile operators scheduled system upgrades in a three-to-four year timeframe [3]. Each of these components was designed intricately and specifically for Network Equipment Manufacturers (NEM). These components were very tightly coupled to specially designed high-performance hardware. Each new generation of wireless technology requires very extensive and expensive redesigns of hardware, because of inflexibilities in the basic structural hardware. Components of most RANs are also based on proprietary vendor-specific protocols for the same reason that a few NEMs in the RAN market competed for larger market shares [4].

The RANs architecture has evolved from static, vendor-dependent, and inflexible systems to intelligent, open, and more flexible solutions. This evolution can be seen in three main phases: traditional RAN, first-level disaggregation, and Open RAN (O-RAN), as shown in Figure 3.1. Traditional RAN architecture based on a distributed RAN (D-RAN) model, in this Remote Radio Head (RRH) and Baseband Unit (BBU) were closely coupled and integrated in a proprietary manner, restricting operators into single-vendor silos [5]. The lack of open interfaces in monolithic system designs forced traditional telecom providers to cope with high CAPEX, high OPEX, and time-to-market delays. Even the services offered in D-RAN architecture do not support dynamic User Equipment (UE) allocation among cell sites and follow a closed proprietary system.

DOI: 10.1201/9781003637721-3

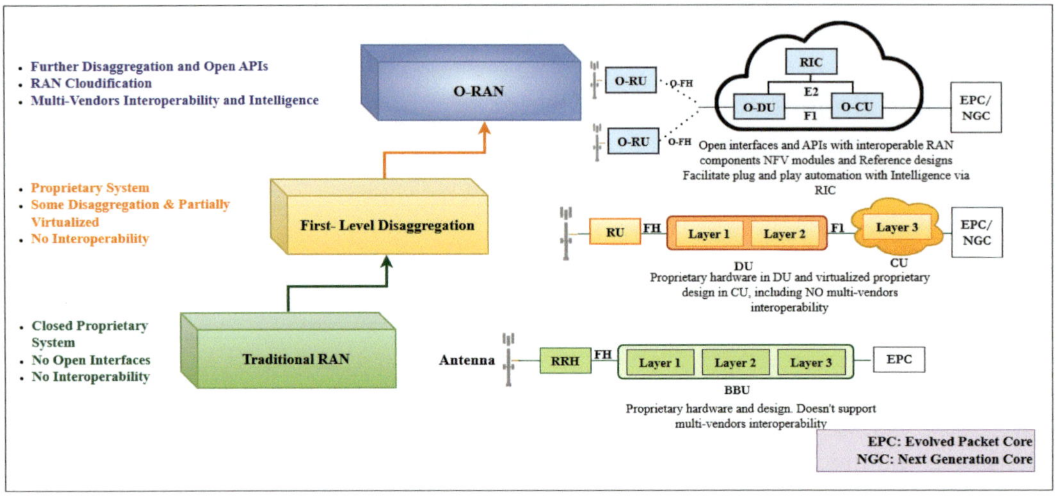

Figure 3.1 Evolution of O-RAN.

At the advent of the first-level disaggregation, the RU, Distributed Unit (DU), and Central Unit (CU) were split. In this split, Layer 3 processing was migrated to a virtualized CU, while Layers 1 and 2 processing were assigned to the DU. Although there was intent to design this in a disaggregated manner, there were many problems in achieving openness and excessive dependence on proprietary inputs. Such problems led telecom operators to rethink their network design and consider migrating to an O-RAN as an alternative. The intent was to eliminate lock-in by NEMs, increase competition in the market, and inspire new developments. RANs have undergone a remarkable transformation—one that embraces the cloud, automation, and AI/ML (Machine Learning), while also facilitating new, standardized, and open interfaces to create a more open and adaptable architecture.

The O-RAN was established to promote this initiative and implement the O-RAN vision focused on integrating open interfaces in a disaggregated network architecture. O-RAN was created when two separate entities—the C-RAN Alliance and the X-RAN Forum—united together. The C-RAN Alliance was mainly a consortium of China Mobile vendors, while the X-RAN Forum comprised US, European, Japanese, and South Korean merchant operators. The initial five founding members of the O-RAN concept include AT&T, NTT Docomo, China Mobile, Telekom, and Orange, which are illustrated in Figure 3.2 [7]. Since its inception, the O-RAN network has grown into a global network of diverse and varied O-RAN ecosystem constituents, including mobile network operators, distributors, academia, and research. This new approach is also improved and enhanced by combining the RAN Intelligent Controller (RIC) component, which has both near-real-time and non-real-time layers. This lets AI/ML algorithms automate and optimize things in new ways. O-RAN is a major driver which is prominent for the future of wireless communications as it makes network utilization better and provides multi-vendor options.

As the evolution of 6G wireless, which seeks to expand on the foundation built by 5G, is another noteworthy advancement in wireless communication. This expansion increased speed, reduced latency, and provided the ability to run a broad range of novel applications. Whereas 6G will exploit all the strengths of 5G so far, it dreams of a future when the wireless network carries not only transistors and screens but full-blown technologies such as smart cities, innovative healthcare solutions, autonomous vehicles, and many more other services. Throughout this chapter, the primary focus is on O-RAN architecture and its

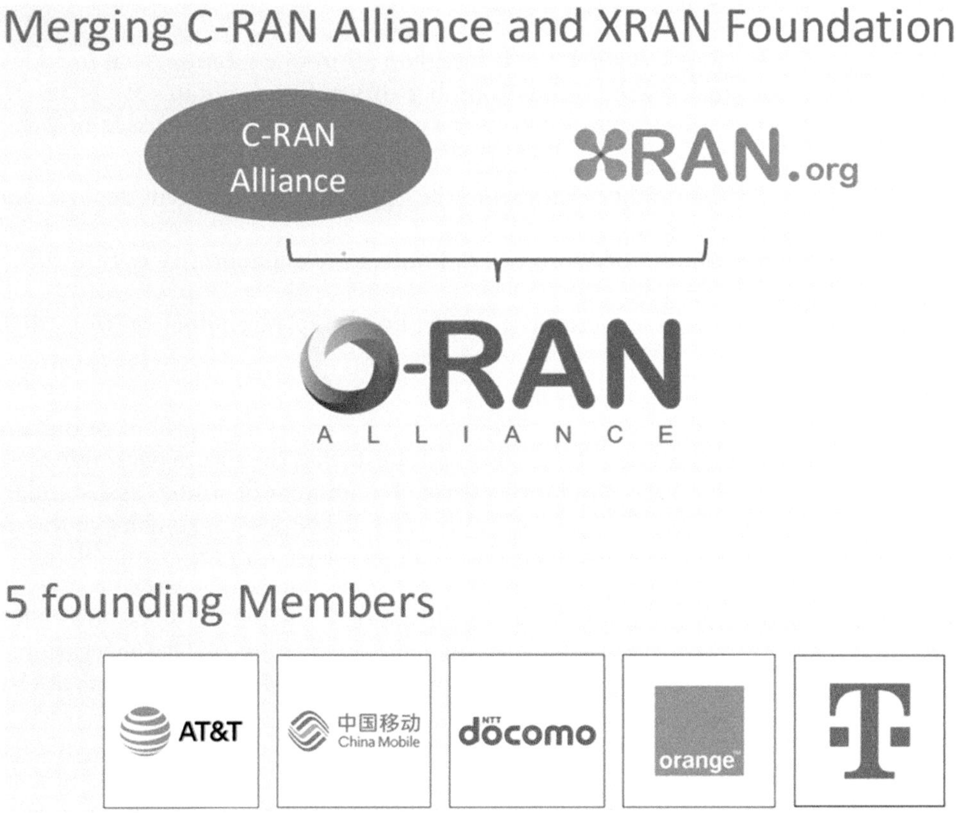

Figure 3.2 Founding operators of O-RAN [6].

interfaces. Additionally, we have covered various use cases and future possibilities. As a result, transitioning to O-RAN not just overcomes the existing architecture limitations but also makes network administration more erudite and profitable for next-generation wireless systems, restores cost, shortens time to market, aids innovation, and addresses other significant issues.

3.2 O-RAN ARCHITECTURE

O-RAN Alliance enhances the existing approach, by introducing more openness and intelligence than their predecessors. It improved network functionality and used real-time analytics based on back-end models of artificial intelligence and embedded ML systems. O-RAN proposed reference designs add more hyper-virtualized network elements and will have disaggregated components that interconnect through open and standardized interfaces. These designs mainly depend on open-source and white-box network technologies for both hardware and software components [8, 9]. Additionally, O-RAN architecture also supports open multi-vendor interoperability, allowing for the programmatic optimization of components, including the use of a data-driven closed control loop. For these reasons, it is important for all researchers and practitioners in the telecom industry to understand O-RAN, along with its architecture, interfaces, and workflows. This chapter aims to offer valuable insights into the comprehensive architecture of O-RAN.

3.2.1 Vision of O-RAN architecture

The existing monolithic RAN architecture is based on all-in-one solutions that include every cellular protocol stack tier. As the limited number of vendors facilitates RAN services, it causes several issues due to reliance on black-box systems which are discussed below:

i. Limited RAN configurability that cannot be optimized for different deployment scenarios and traffic patterns;
ii. Poor coordination among network nodes hinders the collaborative optimization and control of RAN components; and
iii. Vendor lock-in prevents operators from deploying RAN equipment from different vendors.

These challenges show that real-time adaptation is necessary for optimal resource management and spectrum usage. O-RAN has been established as the future paradigm for RAN systems in order to get over these limitations through a number of studies and standardization efforts. O-RAN implementations seek to modify RAN designs to achieve the above-mentioned goals. Extensive research spanning multiple years on programmable and open networks is the cornerstone of the O-RAN concept. These principles have been critical to the software-defined networking (SDN) transformation in wired networks over the past 15 years and now reshape to build their way into wireless networks. Disaggregation, open interface, RIC and closed-loop control, and virtualization are the four key pillars of O-RAN that will govern the initiatives of O-RAN [4, 10].

Disaggregation: For New Radio (NR) Next-Generation Node Bases (i.e., gNBs), 3GPP introduced a functional disaggregation model. This approach revolutionizes traditional base station architecture by splitting it into different functional parts: CU, DU, and Radio Unit (RU) are the three components of the gNB (known as O-CU, O-DU, and O-RU as per O-RAN specifications). The CU comprises two logical components for the control plane (CP) and user plane (UP), known as CU-CP and CU-UP. This logical segregation enables the deployment of unique functionalities on different hardware platforms and network locations. For instance, multi-vendor interoperability, where operators can mix and match elements from multi-vendors.

As earlier discussed, to accomplish this, at the network edge, white-box servers can be used to virtualize CUs and DUs, with hardware acceleration for some physical layer functions [11]. Conversely, RUs are commonly designed with radio frequency (RF) antennas and are mounted close to application-specific integrated circuits (ASICs) and field programmable gate arrays (FPGAs). However, the 3GPP has offered several distinct RU/DU split options, depicted in Figure 3.4, provided by the O-RAN consortium which are discussed later in functional components.

Open Interface: Without openness, incorporating cloud-scale economics and service agility into the RAN will be challenging. For smaller vendors and operators, implementing open interfaces is vital to offering differentiated services and customizing the network to meet their needs. Leveraging these open interfaces will enable the use of multiple vendors, fostering greater flexibility and competition regarding the procurement of network elements while ensuring the interoperability of the O-RAN architecture. The O-RAN acknowledges the importance of open-source communities and supports their missions and objectives. There is a wide array of ongoing initiatives; numerous architectural elements of O-RAN will be made available as open-source, encompassing components such as the virtualization platform, protocol stack, PHY layer processing, and RIC. The O-RAN open-source software framework is designed to provide a reference blueprint for the future generation of Radio Resource Management (RRM) with embedded intelligence to enhance the capabilities of

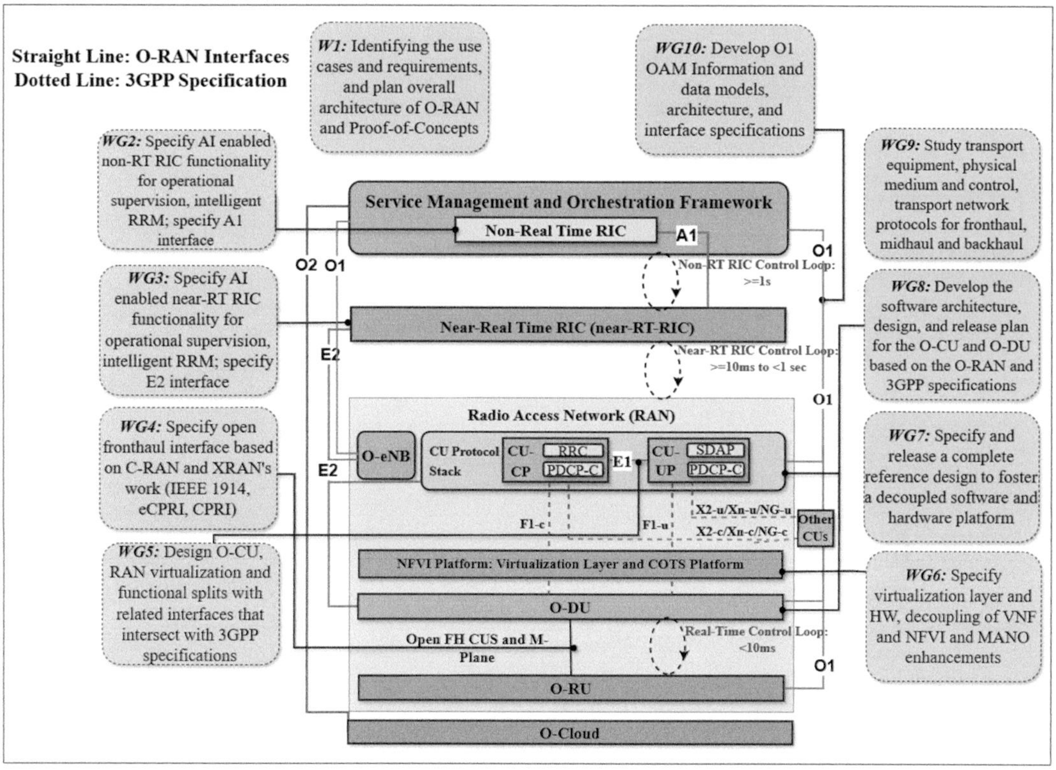

Figure 3.3 O-RAN architecture and specific workgroups [23].

the RIC [10]. A comprehensive set of essential interfaces connects various decoupled RAN components via 3GPP-defined interfaces (E1, F1, X2, Xn, and NG), forming the basis of the O-RAN reference design. However, the O-RU and O-DU components of O-RAN interconnected with Open Fronthaul (FH) (OFH) interface are depicted in Figure 3.3. In contrast, the O-RAN interfaces assist in overcoming the conventional RAN black box approach by providing the RICs with access to data analytics and telemetry and enabling various automation and control activities such as RAN real-time control, virtualization, and deployment optimization. The interfaces defined by O-RAN are the E2 interfaces, which link the near-RT RIC to the RAN-enabled near-real-time control loops. These control loops stream the telemetry reports from the RAN nodes and enable the near-RT RIC to take control decisions. Moreover, A1 interface facilitates communication between near-RT RICs and non-RT RICs by enabling communication in non-RT and allowing for policy deployment and updation in policy definition through intelligent models. To monitor and orchestrate network functions, the non-RT RIC also terminates the O1 interface, which links to all other RAN components. Additionally, the O2 interface—discussed in Section 3.3—connects the non-RT RIC and SMO to the O-Cloud.

RIC and Closed-Loop Control: The networks become increasingly complex, with the rapid evolution of 5G technology, along with increased network densification and more sophisticated applications. There is no way to manage this complexity using traditional, manual network deployment, optimization, and operation processes. To automate these operational network processes and reduce OPEX, networks must adopt new learning-based technologies that enable them to function autonomously [8]. The O-RAN partnership aims to integrate intelligence into each component of the RAN architecture by employing

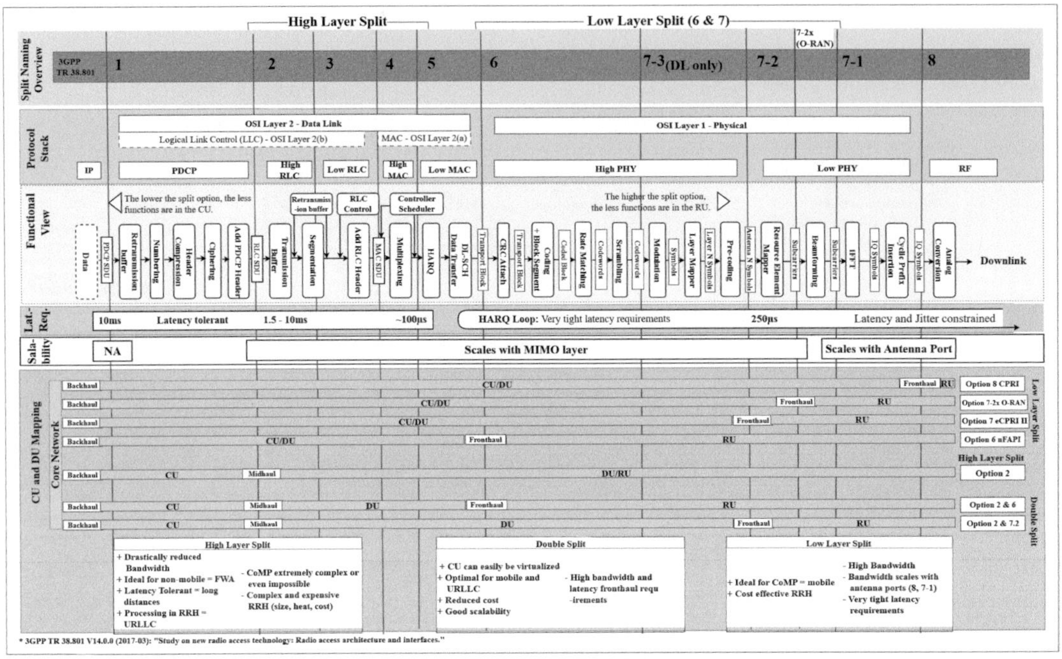

Figure 3.4 O-RAN Split based on 3GPP TR 38.801.

advanced AI/ML techniques. When applied at each component and network level, this embedded intelligence enhances overall network efficiency and allows for dynamic allocation of local radio resources. Converging AI-optimized closed-loop automation with open interfaces can usher in a new era of network operations. This principle enhances the SDN concept within the existing architecture by splitting the CU into two components: CP and UP, as further discussed, developed by 3GPP through the E1 interface. It also embedded the intelligence component known as the RIC via the A1 and E2 interfaces. The O-RAN specifications are defined for two types of RIC: a near-RT RIC that drives control loops with RAN nodes on a time scale of 10 ms to 1 sec, and a non-RT RIC that integrates with the network orchestrator, i.e., SMO, and operates on a time scale of more than 1 s which will be covered in Section 3.2.2.2. However, the real-time extension in O-RAN architecture is considered for future work that operates below 10 ms for RAN node-level, such as RRM as well as less than 1 ms for UE management and optimization. Moreover, scheduling, beam management, and feedback-less parameter detection (such as interference recognition, modulation, and coding scheme (MCS)) are typical instances of real-time control in physical layer [8, 12].

RAN Virtualization: Cloudification is another principle that advances virtualization platforms by optimizing power consumption and accommodating various splits, such as a low-layer split within the PHY and a high-layer split between Radio Link Control (RLC) and Packet Data Convergence Protocol (PDCP). This supports the development of significant solutions, including programmable hardware and software accelerators, real-time processing, and lightweight virtualization technologies. O-Cloud is a hybrid cloud computing platform that may be used to install all of the elements of the O-RAN architecture. The O-Cloud is a collection of virtualization infrastructure and computing resources aggregated in one or more physical data centers. In addition to specialized virtualization for O-RAN, this platform integrates physical nodes, software components (such as operating systems and virtual machine hypervisors), and management and orchestration features [13, 14]. This principle facilitates isolation between hardware and software components; standardized hardware capabilities can be shared by several tenants, and RAN functionality can be automatically deployed and instantiated. In addition, virtualization of the RAN components and computing elements is anticipated to culminate in power consumption optimization and savings. Power consumption is restricted to the necessary network functions since virtualization enables the simple and dynamic scaling up or down of the computing resources needed to fulfill user requirements. The RAN's virtualization and the closed-loop control features discussed above allow for more sophisticated and dynamic sleep cycles for the base stations and RF components, which typically account for most cellular networks' power consumption [15, 16].

3.2.2 Functional and intelligent components of O-RAN architecture

The functional components of the O-RAN architecture include RAN, the Operations Support System (OSS), and open interfaces. The 5G and beyond 5G RAN architecture specified by 3GPP is compatible with the O-RAN architectural design. Within this architecture, the various functions of the RAN protocol stack are defined, including Radio Resource Control (RRC), Service Data Adaptation Protocol (SDAP), PDCP, RLC, Medium Access Control (MAC), and the Physical (PHY) layer. The O-RU, O-DU, and O-CU provide these functionalities. Additionally, to manage the RAN protocol stack in an optimized manner, intelligence is incorporated at the top of the RAN protocol stack via the RIC. This architecture embraces the concept of openness, allowing different functions to be integrated more flexibly. The various functions of the O-RAN architecture, including the associated workgroups with specific O-RAN specifications based on 3GPP for LTE and NR RANs, are depicted in Figure 3.3 [17, 18].

3.2.2.1 Functional components

The specifications of O-RAN build upon the 3GPP LTE and NR RANs, enhancing the 3GPP NR 7.2 split for base stations. This component is crucial for a distributed and virtualized RAN architecture. The 7.2× split that has been specified balances the RU's simplicity with the high data rates and low latency need on the RU–DU interface. As the RU in split 7.2× carries out Fast Fourier Transform (FFT) and cyclic prefix (CP) addition/removal operations, it is cheap and straightforward to implement. Afterward, the DU guides the remaining physical layer functions as well as those of the MAC and RLC layers [19, 20]. These three components' operations are typically closely synchronized as the MAC layer decides scheduling and multiplexes MAC SDUs (Service Data Units) from one or different logical channels using data stored at the RLC layer to create Transport Blocks (TBs) for the physical layer. Eventually, the CU-CP and CU-UP components implement the higher layers of the 3GPP stack, including the PDCP layer, which handles encryption, header compression, and retransmission (if needed) for the air interface while the RRC layer, which controls the connection setup and management, mobility, and security configuration [21], and the SDAP layer, which controls the quality of service (QoS) of the traffic flows (also known as bearers) [22]. This paragraph shows multiple split options defined by 3GPP TR 38.801 with multiple objectives, including low latency, bandwidth efficiency, and deployment flexibility.

3.2.2.1.1 O-RU: Radio unit

The RU plays a critical function within the O-RAN architecture, managing RF operations, digital beamforming, the digital front end (DFE), and components of the PHY layer. This functional component is positioned at the edge and links with the O-DU via the OFH interface, enacting split 7.2×, as a crucial functional separation in O-RAN. O-RU primarily handles the actual transmission and receiving analog radio signals, whereby BBU signals are transformed into radio frequencies that are broadcast over the air and vice versa. This logical node hosts PHY processing tasks such as lower PHY functions, such as digital beamforming and FFT/Inverse FFT (iFFT) while upper PHY and higher functions, such as channel coding and modulation, remain within the O-DU which is interconnected with the FH interface. Although considerations of size (compact and lightweight miniature), weight (used lightweight materials like aluminum alloys), and power consumption are vital in O-RU design, 5G RU designs are intended to be inherently intelligent.

In summary, split 7.2× delineates PHY processing responsibilities as follows:

- Lower PHY roles are adding CP, FFT/iFFT, digital beamforming resides in the O-RU.
- Upper PHY and higher roles are responsible for channel coding, modulation, scrambling, rate matching, and coding block segmentation which are managed by the O-DU.

3.2.2.1.2 O-DU: Distributed unit

The O-RAN DU is responsible for managing mid- and lower-layer radio functions, including real-time processing and RRM. It communicates with O-CU to coordinate higher-layer protocols and operates alongside O-RU to control the physical layer. O-DU is responsible for efficiently rerouting traffic between O-CU and O-RU using low-latency functions and high speed, including channel scheduling and HARQ (Hybrid Automatic Repeat Request) techniques. It operates various protocol layers, including the following:

- RLC: This layer segments, reassembles, and corrects errors, involving elements from the PHY layer.
- MAC: This layer schedules tasks, performs HARQ, and allocates resources. The MAC is situated near the RU. The CU regulates the operations of this logical node, which encompasses a subset of the eNB and gNB functions depending on the chosen functional split.
- High-Physical Layer: This layer performs modulation, coding, layer mapping, and accomplishing physical transmission procedures.

The O-DU is linked to the O-CU through the F1 interface and processes UP and CP data. While O-CU is responsible for centralized decision-making and mobility management, DU mainly performs scheduling and low-layer processing near the radio edge to achieve low-latency response.

3.2.2.1.3 O-CU: Central unit

The O-RAN CU is in charge of controlling the upper-layer protocols that are part of the O-RAN architecture. This central component is in charge of RRC and PDCP management. Here, the gNB is composed of the CU and the DU, which are connected by means of Fs-U (UP) interface and Fs-C (CP). A single CU can support several gNBs, which are just connected to multiple DUs. Regarding midhaul interference availability and network topology, a 5G network may apply flexible policy in the distribution of protocol stacks between CU and DU due to this architecture split. Such a "split architecture" results in a functional separation of O-CU and O-DU, allowing for programmable allocation of protocol stacks. It communicates with SMO for orchestration and fetches optimal configurations and policies from RIC. The CU and DU exchange processed packets through standardized interfaces that are transmitted to the rest of the network nodes.

In addition, the CU is a logical node that contains many gNB operations, such as mobility control, RAN sharing (MORAN), traffic steering, location services, session management, and user data transfer. This makes sure that users stay connected and that data transfers go smoothly. The midhaul interface lets the CU control the work of several DUs at once. The O-CU can be broken down into different logical parts.

The O-CU is also classified into a number of logical segments:

- CU-CP: It is responsible for signaling-related functions like RRC, mobility management, and connection control.
- CU-UP (UP): Controls user data transport and PDCP functionality and maintains QoS per traffic type with the help of SDAP.

This control and user separation enables more flexibility and scalability thereby supporting multiple network slices and services.

3.2.2.2 Intelligence Components

In the O-RAN architecture, another core component is the "Intelligent" component, which is responsible for managing and orchestrating all automation, administration, and monitoring processes for RAN components in both near-RT and non-RT environments.

Non-RT RIC: The non-RT RIC is located at the SMO which provides a suite of interfaces (elaborated later in the next section) that facilitate communication across network components, as well as enable data collection capabilities to sustain AI/ML-driven network monitoring and control. Within the O-RAN, non-RT RIC is a key entity, and it works for a duration longer than 1 s. It also enables third-party applications (rApps) to operate within SMO. These applications offer value-added functionalities, including policy guidance, enrichment information, data analytics, and configuration management to support and improve RAN-optimized operations. The R1 termination, part of the non-RT RIC, provides rApps access to the internal message infrastructure and inter-rApp communication (as well as to A1/O1/O2 interfaces' AI/ML capabilities or data management and exposure services). While rApps intend to exert control at a more abstract layer that impacts a broader range of users and network nodes, they can also provide the same control functionalities as xApps, including scheduling control, traffic steering, and mobility (handover (HO)) management, albeit over longer control windows. Network slicing, performance diagnostics, RAN sharing, frequency and interference management, and end-to-end Service Level Agreement (SLA) assurance are examples of rApps applicable to non-RT RIC control [24]. The non-RT RIC facilitates two advanced management and orchestration services discussed below that support a versatile architecture, allowing real-time modification of network components' behavior and functionality to align with operators' intent and objectives [25]:

- Intent-Based Network Management: This feature allows operators to articulate their intentions through a human–machine interface using a high-level language. These intentions are expected to align with the procedures employed by existing SMOs and be formatted as YAML or XML configuration files. The non-RT RIC then interprets these intents and identifies which policies need to be enforced, and what the associated rApps and xApps must be installed and executed to comply with them.

- Intelligent Orchestration: Managing network is critical due to the multiple rApps and xApps networks have that run across various RICs and network locations. To resolve such complexity, we need technologies, enabling the efficient management and coordination of these applications. The primary responsibility of the non-RT RIC is to orchestrate network intelligence to ensure that the chosen rApps and xApps (i) are aligned with operator intents and specifications, (ii) are instantiated at the accurate RIC location to assure control over the specific RAN elements, (iii) provide relevant data, and (iv) prevent conflicts when several apps concurrently manage the same parameters and/or functionalities. For instance, the service provider has set multiple network slices and intends to manage radio resources and optimize scheduling policies for selected set base stations, considering traffic density and geographical locations in real time for each slice. The non-RT RIC can automatically decide which xApp running at a distinct near-RT RIC can meet the stringent timing requirement for scheduling decisions. However, non-RT RIC selects only those xApps that may make optimal scheduling decisions, which will then send them to the near-RT RICs that are in charge of the base stations set up in the region of interest and can provide data to feed to the xApps [4].

Near-RT RIC: The near-RT RIC is positioned at the network edge and operates a control loop with a periodicity ranging from 10 ms to 1 s. This intelligent component serves as the CU for RAN control and optimization, leveraging the capabilities of E2 interfaces. The near-RT RIC communicates with RAN O-DUs, O-CUs, and legacy O-RAN-compliant eNodeBs. As the near-RT RIC is typically integrated with numerous RAN nodes, its closed control loop can significantly impact the QoS experienced by hundreds or even thousands of user equipment (UEs). The near-RT RIC is composed of multiple applications, called xApps, which contain custom logic with all services to run them. An xApp is basically a microservice in the near-RT RIC with a specific role in provisioning radio resources using standard service models and interfaces. It is primarily developed as a plug-and-play module that provides custom functionalities to create RAN statistics and control. This includes obtaining information from the RAN, like slice, user, or cell-level Key Performance Indicators (KPIs), and, these information are considered for computing and transmitting control actions. The near-RT RIC extends the functionality of xApps in several ways: (i) it contains a database with RAN information, such as lists of connected RAN nodes or users, that acts as shared data among xApps; (ii) it offers a messaging infrastructure that facilitates communication between the platform's various components, enabling xApps to subscribe to RAN elements; (iii) it supports terminations for open interfaces and APIs; and (iv) it includes conflict resolution mechanisms allowing multiple xApps to collaboratively manage the same RAN feature [4].

The near-RT RIC platform consists of three interface terminations (E2, O1, and A1), the necessary xApps, and essential components for efficient operation and management. O-RAN specifications do not stipulate how to implement such components, but they explicitly define the functionalities and prerequisites for the ready and available RIC components. It's based on the assumption that all implementations which are compliant with the standard will deliver a consistent range of services. Moreover, the O-RAN provides a complete reference model of a fully functional near-RT RIC that satisfies their designed requirement as well. This model is available to developers through the O-RAN Software Community (OSC) for creating O-RAN solutions. The near-RT RIC in the OSC deployed with a microservice set of software within the Kubernetes framework laid down a good foundation from an architectural perspective. Key microservice includes dynamic resource allocation, QoS and QoE

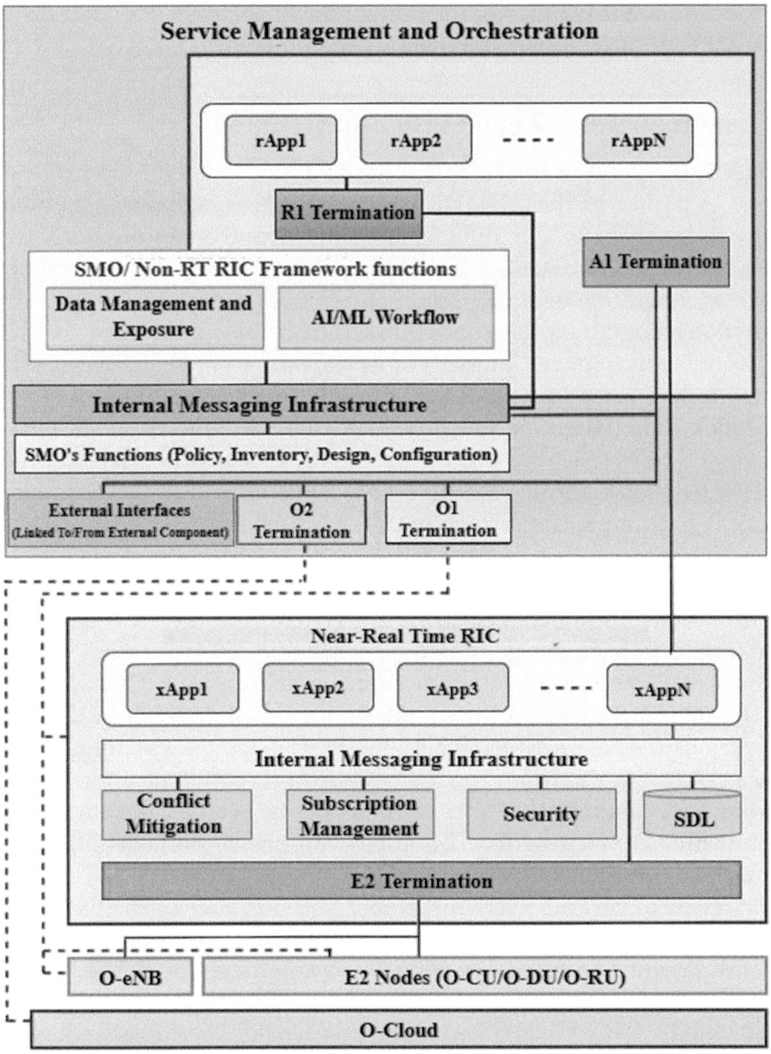

Figure 3.5 Non-RT RIC and Near-RT RIC components.

management, RAN slicing, traffic steering, and congestion control. The non-RT RIC and near-RT RIC components are presented in Figure 3.5.

3.3 INTERFACES OF O-RAN ARCHITECTURE

The O-RAN architecture introduces a new set of standardized interfaces to make all wireless networks more flexible, transparent, and intelligent. These include interfaces like E2, O1, A1, F1, and more, which are the backbone to promoting a modular structure, real-time management, and practical cooperation among the decentralized components of the RAN. Services like telemetry reporting from the RAN are made possible by each interface with a clearly defined and integrated procedure. In the following paragraphs, we elaborate on these interfaces and examine the processes, including logical abstractions, that define these interfaces like how they are functions to the overall system of the O-RAN ecosystem.

3.3.1 E2 interface: Near-Real-Time and RAN control

The E2 interface is the point of interface that act as a RAN closed loop between the near-RT RIC and E2 nodes named as the CUs, DUs, and O-RAN compliant LTE eNBs [26]. The E2 interface allows the near-RT RIC to dynamically control and configure RAN functions in real time, including RRM, interference control, radio parameter control, and response to network conditions such as congestion or traffic bursts. Additionally, this interface makes it possible to gather metrics which contained KPIs for UE and cell from the E2 node and send them back to near-RT RIC either at frequent intervals or in response to pre-established trigger events. In case there is any failure between near-RT RIC and E2 interface, it does not interrupt the procedure of E2 nodes; it shall function independently. To collect metrics data and controlling the action, data may be collected from one or more cells, slices, QoS classes, or individual UEs. This ensures that user information remains secure while upholding a consistent and standardized user identity across the entire system. In O-RAN architecture, the E2 interface is logically categorized into two protocols: E2AP (Application Protocol) and E2SM (Service Model).

E2AP: This protocol is a simple procedural CP protocol deployed at the application layer that controls communication between the near-RT RIC and the E2. Perhaps this E2AP can also embed multiple E2SMs which deal with a specific set of services such as controlling RAN parameters, KPM monitor, cell configuration and control (CCC), or reporting RAN metrics [27]. The SCTP protocol is the foundation for the E2 interface [28], where every E2 node unveils various RAN functions, including its services and capabilities.

For instance, DUs from various vendors might have different control knobs based on the functions and parameters that can be adjusted, as well as how well they can gather and report performance information. E2 nodes can utilize publish-subscribe mechanisms to disseminate their information, while the xApps on the near-RT RIC can employ the E2 interface to subscribe to various RAN functions. The existence of the E2 interface facilitates the deployment of xApps within the near-RT RIC, which are modular applications at the application level designed to carry out distinct functions about RAN control. The xApps use the E2 interface to enforce control mechanisms on the RAN nodes, hence increasing the efficiency and responsiveness of the network. As a result, the capabilities of each node can be easily distinguished, and the xApps' interactions with the RAN may be specified.

In addition, the E2AP is also responsible for near-RT RIC service updates (i.e., exchanging the list of RAN functions that the E2 node supports) and interface management (i.e., setup, reset, and fault reporting for the E2 interface itself) at the lowest level. E2AP manages several procedures with RIC such as RIC control, RIC query, RIC service load/update status (REPORT, INSERT, CONTROL, POLICY, QUERY) to implement an E2SM [28].

E2SM: E2SM is used to determine the content structure of E2 interface messages. These models specify how data, events, and control actions are exchanged in a standardized way between the near-RT RIC and E2 nodes. According to O-RAN WG3 version 07.00, multiple service models have been standardized, namely, E2SM-NI (network interface), E2SM-KPM (key performance metric), E2SM-RC (radio control), E2SM-CCC, and E2SM-LLC (Lower Layer Control) [29].

i. *E2SM-NI:* This interface specifies how a RAN function forwards the contents of both outgoing and incoming network interface messages with network interface services. The received and transmitted interface messages across the E2 interface to the near-RT RIC domain observe and influence the CP and UP traffic behavior. Moreover,

E2SM-NI facilitates the implementation of AI- or policy-based decisions that could alter the general behavior of the network, including traffic rerouting, HO management, or QoS adaptation. E2SM-NI enables the RIC to dynamically optimize RAN operations, hence supporting the objectives of automation, agility, and service-specific adaptation in a 5G network that complies with O-RAN.

ii. *E2SM-KPM:* The E2SM KPM v1 reports the periodic KPM from the RAN such as O-DU populates cell-related performance IEs (Information Elements) and O-CU-CP reports for cell/UE-associated performance IEs, including bearer-related performance IEs, are exposed by O-CU-UP. These KPMs are requested by xAPP in near-RT RIC, where the relevant report is triggered or periodically generated based on subscription message. KPM v2 is an exposure of available measurements like cell-specific and user-specific performance containers from E2 nodes via the RAN function definitions and maintains periodic report of measurements subscribed from near-RT RIC.

iii. *E2SM-RC:* The E2SM RC domains were initially unveiled by the O-RAN Alliance in July 2021, utilizing E2 interfaces to carry out diverse control functionalities. E2SM: The E2SM specifications' initial edition covers a wide range of control domains, such as radio bearer control, RAC, radio resource allocation, and carrier aggregation control. Moreover, E2SM RC provides functionalities for UE information reporting and identification. The standardized parameters established by 3GPP pertain to the control actions and policies outlined by E2SM [27].

iv. *E2SM-CCC:* E2SM-CCC focuses on exposing configuration and control interfaces at both the node level (e.g., base stations) and cell level (e.g., individual sectors or carriers). This model provides an interface for xApps to get the current RAN configuration parameters and apply changes dynamically. For instance, this includes data related to frequency allocation, bandwidth, transmission power, HO thresholds, and scheduling policies. E2SM-CCC enables both monitoring (observability) and control (actuation), rendering itself useful for intelligent RAN adaptation. One common application of it is load balancing. In case one cell encounters traffic congestion, the RIC can control power or hand in users to neighboring cells via E2SM-CCC for resource utilization and QoE enhancement. Energy saving is another application, where the RIC may switch off some underused cells in low-traffic hours. Standardizing access and modification of configuration parameters, E2SM-CCC directly supports the integration of AI/ML-based RIC applications that may analyze the real-time data and act according to need without use-case-specific vendor APIs.

v. *E2SM-LLC:* E2SM-LLC allows access to and control of the L1 and L2 of the RAN protocol stack. First layer (L1)—Physical Layer: At the physical layer, signal modulation, antenna handling, and power control are involved. Layer 2 (L2) consists of MAC, RLC, and PDCP. The RIC is, with E2SM-LLC, able to collect fine-grained runtime metrics such as SINR (Signal to Interference plus Noise Ratio), HARQ processes, and buffer occupancy—and also take actions like adjusting MCS, allocating PRBs (Physical Resource Blocks), or switching retransmission strategies. This model is important for real-time RAN optimization such as URLLC (Ultra-Reliable Low Latency Communications). For example, E2SM-LLC could allow the RIC to respond at deteriorating radio conditions by dropping L1 parameters for continued service quality. Also, L2 control can assist in resource prioritization in sudden traffic peaks to meet QoS demands.

3.3.2 O1 interface: Operations, administration, and management

The O1 interface is the underlying interface to interoperate between the SMO platform and RAN functional units (CU, DU, and RIC) [30]. It is primarily used to support the FCAPS functions, namely, Fault Management that supports providers in detecting, isolating, and preventing network alarms; Configuration Management to identify if a network is configured correctly as well as make changes in a controlled manner; Accounting Management to monitor the consumption of resources within networks like how resources are consumed; Performance Management that ensures networks meet performance standards required; Security Management to safeguard the network against unauthorized access. Further, the O1 interface manages vital capabilities such as software upgrades, log retrieval, and file handling to keep the network running smoothly. These services provided by the SMO platform ensure that it has an efficient interface to verify the operating state and performance of network elements, update new parameters, and process any failure in a timely manner, assuring operation stability of the network. In addition, maintenance and operation protocols already in use are included and integrated for such an O1 interface. It manages the whole lifecycle of O-RAN components through Management services (MnS), which spans from startup and configuration to high availability and heartbeat services.

The interface ensures optimal performance by monitoring KPI reports and facilitating software and file management. In addition, the O1 interface accommodates multiple MnS services:

Provisioning MnS: This MnS enables the SMO to start, configure, and stop configuration from managed nodes to the SMO and push configuration back to managed nodes. O1 interface uses this to monitor the configuration of SMO and RAN components via REST/HTTPS APIs and NETCONF [29].

Fault Supervision MnS: Here, SMO receives error and event report via MnS by following 3GPP-defined fault events [31, 32]. The RAN nodes can use REST APIs to report faults (using standardized JSON payloads).

Heartbeat MnS: SMO can manage both virtual and physical network functions (PNFs) and provision heartbeats on the devices under its control via the Heartbeat MnS. Heartbeat messages track the availability and state of nodes and services.

Performance Assurance MnS: This MnS facilitates data collection for AI/ML by enabling real-time performance data streaming or batch reporting through file transfer to the SMO. The SMO has the right to decide which KPIs—also known as counters—will be reported and how often. This decision is guided by KPI reporting requirements and formats outlined by either the VNF Event Stream (VES) project or the 3GPP [32, 33]. In addition, the performance metrics may be specific to the vendor, standardized by various O-RAN Alliance working groups, or aligned with 3GPP documentation.

Software and File Management: To transfer files the SFTP protocol is used, which enables SMO to download files and monitor trace-based events via the Trace MnS.

The O1 interface is well known for its standard management, wherein standardized components from several vendors are operated in accordance with a standard coherency. It also has automated RAN control for increasing its overall efficiency and performance. The main emphasis of the O1 interface on standardization provides the interface with strong confidence in terms of interoperability [34].

3.3.3 O2 interface

O2 interface in the O-RAN architecture is an enabler for the communication between the SMO platform and the O-Cloud infrastructure. It is essentially enabling the effective management of cloud resources and the lifecycle of network functions in an elastic, multi-vendor environment. The O2 interface has two major service components: Infrastructure

Management Services (IMS) and Deployment Management Services (DMS) [35]. IMS oversees both physical and virtualized infrastructure (e.g., compute, storage, and network resources) and performs tasks such as fault detection, performance monitoring, and configuration management. In contrast, DMS focuses on the deployment, scale, and termination of O-RAN network functions to enable dynamic orchestration based on real-time network needs. Leveraging RESTful APIs for the O2 interface provides standardization and simplicity to integrate between SMO and cloud infrastructure for automation, interoperability, and scalability. Consequently, the O2 interface is critical for scalable and transparent 5G RAN deployments [36].

3.3.4 A1 interface: Policy guidance and intelligence sharing

The A1 interface is critical to facilitating the coupling of the non-RT RIC, a fundamental functionality of SMO with the near-RT RIC. It acts as a communication channel through which policies, enrichment data, and ML models are exchanged. This exchange is enabled by standard mechanisms that allow communicating high-level intent and policies using a specific syntax via a specific language, primarily the JSON schema. Through policy recommendation, the A1 interface enables the near-RT RIC to take proactive decisions that are consistent with the operator's specific intentions (i.e., improving user experience and/ or reducing energy consumption). The A1AP (A1 Application Protocol) is a fundamental part of the A1 interface over 3GPP and fuses REST APIs with HTTP to move JSON objects [37, 38]. RICs leverage HTTP client and server capabilities to control services, data exchange, and notifications. The A1 Policy Management from the non-RT RIC is based on how the near-RT RIC works to meet the high-level goals of RAN. This can be performed by using QoS or KPI metrics for all users or just certain groups of users (like a slice). The O1 interface's reporting features and A1 feedback keep track of it. Policies are defined by the non-RT RIC, after which they are subsequently put into action over A1 [25]. The near-RT RIC is ultimately responsible for utilizing their APIs to create, update, retrieve, and delete these policies. The near-RT RIC is also tasked with managing and ensuring that the policies' lifecycle persists sufficiently. Additionally, the A1 enrichment information service seeks to empower the RAN with the capability of boosting its performance. However, it does this by providing access to data and insights that the RAN can't access, such as capacity projections, data from other sources, or joint analytics. Consequently, the transfer of enrichment data to the near-RT RIC over an A1 interface provides it with useful insights, like traffic predictions or user behavior patterns, which it can use to make intelligent decisions.

3.3.5 F1 interface: Communication between CU and DU

It is a standardized communication interface between the DU and the CU of the split RAN architecture. The F1 has two sub-interfaces: F1-C, which supports communication of the CP, and F1-U for UP data transfer [39]. To put it in perspective, the F1-C interface handles signaling procedures such as bearer setup, mobility management, and configuration updates. These signaling procedures are user-oriented. The F1-U interface, on the other hand, is for forwarding actual user data packets from the network to the final users. Moreover, F1-C signaling procedures are considered to be the most stringent traffic due to their stringent delay constraints. But then, the F1 has unique attributes that allow it to serve several roles. For instance, network operators can position CU and DU elements in real time with no degradation of performance or interoperability. This has an exceptional advantage where network operators can have DUs and CUs from different manufacturers; hence, this leaves a window for optimally tuning the entire RAN for the best results. In addition, the F1 interface is a critical linkage for smooth cooperation between different network elements and contributes to delivering service in a seamless manner for end users. In summary, the F1 interface maximizes network performance, supports smooth connectivity, and allows flexible deployment of network elements in a split RAN setup. The segmentation of the F1 interface is split into two sub-interfaces: F1-C and F1-U in which CP signaling and UP traffic can be clearly separated to facilitate processing of network traffic.

3.3.6 Open Fronthaul

According to O-RAN specifications, the O-RAN FH interface connects the DU to one or more RUs in the same gNB [40, 41]. The O-RAN FH interface lets the RU and the DU share physical layer tasks, which lets the DU control what happens on the RU. The O-RAN has chosen a specific configuration (split 7.2×) from a number of several options proposed by the 3GPP [42] for the optimized way to split the physical layer. In split 7.2×, the low-PHY in RU does OFDM phase frequency-to-time conversion through iFFT and CP in the downlink, while for the uplink it removes CP and FFT. Also, more competent RUs (i.e., category B RUs) can do precoding. This feature is available at the DU for RUs that are not as capable (category A). To enhance the DU's capabilities, "category A" RUs must be able to handle low PHY processing for at least eight data streams. Moreover, the DU's physical layer (high PHY) does layer mapping, resource element mapping, modulation, and scrambling. This interface is designed in such a manner to meet URLLC requirements, including modulation compression techniques to reduce jitter and load in the network. The OFH protocol encompasses four unique functionalities, also called planes:

CP (C-Plane): Transmits beamforming and scheduling commands from DU to RU.
UP (U-Plane): The U-Plane is in charge of transmitting I/Q samples for data transfer.
Synchronization Plane (S-Plane): The S-Plane allows the RU and DU to be synchronized in terms of time, frequency, and phase.
Management Plane (M-Plane): The M-Plane is in charge of fault monitoring, software updates, and RU configuration.

3.3.7 Other interfaces: N2/N3 and R1

N2 and N3 interfaces are essential parts of the 5G system architecture as they are the main interface points between the gNB—whether it is composed of both CU and DU functions or not—and the 5G core network. N2, a CP communication management facility, is in charge of processing activities such as session management and signaling to the AMF. Simultaneously, N3 deals with the UP and ensures the smooth flow of data between the UPF and the gNB. Others include an R1 interface that is crucial in promoting communication between RAN applications and either SMO or non-RT RIC. These rApps are assigned tasks, including long-term network optimization, policy-making, and service reliability. The R1 interface is the bridge that facilitates registration, service discovery, and communication between these applications and the orchestration elements, ultimately enabling a highly flexible and scalable network intelligence layer [42].

3.4 O-RAN USE CASES

O-RAN is an innovative idea that transforms the conventional mobile network structure by bringing flexibility, disaggregation, and intelligence. It enables the deployment of various

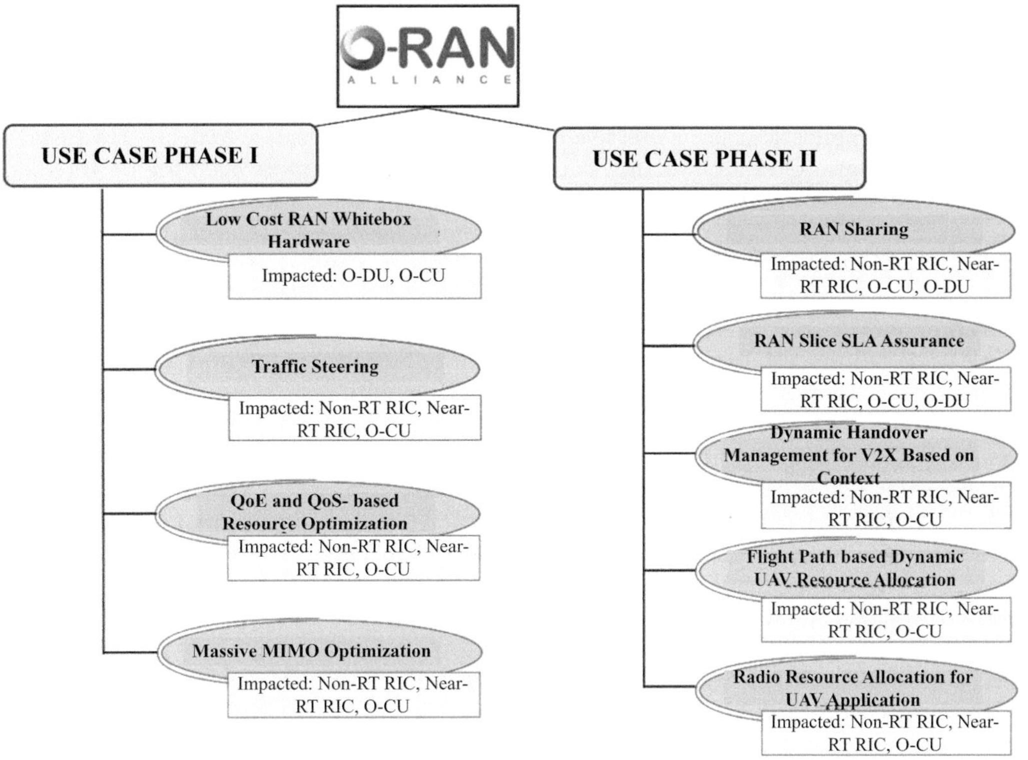

Figure 3.6 O-RAN Alliance use case phases.

advanced applications across industries and telecom sectors. Use cases are mainly based on needs or the enablement of new technologies. The O-RAN Alliance defined a set of use cases split into two phases: Use Case Phase 1 and Use Case Phase 2 [43, 44], as shown in Figure 3.6 and discussed in the following paragraph.

Low Cost RAN White-Box Hardware: The existing monolithic RAN architecture is rigid and has high OPEX/CAPEX. To overcome this limitation, a white-box hardware concept is presented. The main aim is to lower the cost of 5G deployment and achieve high performance, spectral, and energy efficiency. In addition, releasing the reference designs encourages more small and medium-sized businesses to enter the telecom sector, fosters innovation and industrial success, and cuts R&D expenses and challenges. The primary focus of this design is on O-DU, O-RU, and FH gateway (if needed), such as for DL and UL baseband processing, converting the signal to an RF signal, or a power supply for O-RU.

Traffic Steering: Traffic steering is a prominent network solution that optimizes traffic according to predetermined objectives. The complexity of mobile networks has increased along with the network capacity and traffic. It can be challenging to figure out how to direct traffic to boost network performance and user experience efficiently. Conventional network traffic control optimization involves a lot of manual intervention, which is passive and insufficient, and the feedback response is slow after processing. Furthermore, existing cellular networks' RRM features are cell-centric rather than UE-centric. There are variations in the radio environment even within various parts of a cell, including the coverage of neighboring cells, signal intensity, interference status, etc. The O-RAN architecture facilitates dynamic traffic steering and load balancing by deploying xApps and rApps that assess live network data via RICs. These innovative applications enhance user satisfaction by directing traffic to less-utilized cells and effectively managing network load. Utilizing AI technology, the xApps can forecast cell load levels and modify traffic distribution as needed to ensure seamless network performance.

QoE and QoS-Based Resource Optimization: Cloud VR and other extremely demanding 5G native applications use a massive bandwidth and are sensitive to delay. However, the current semi-static QoS architecture could not effectively support such highly interactive and traffic-intensive applications. To overcome this limitation, proactive closed-loop control is required in real-time and vertical applications for QoE-based network optimization and prediction. Multi-dimensional data can be obtained and analyzed using ML methods to facilitate QoE prediction and traffic pattern recognition, and ultimately inform closed-loop QoS enforcement choices. Whereas near-RT RIC will conduct model inference in real time, non-RT RIC allows ML models to be trained offline. To impose QoS for QoE optimization, the interface A1 enforces the policy to near-RT RIC and interface E2 assists the feedback to AI models, and RAN control. Moreover, the O1 interface accumulates information for non-RT RIC training.

In the same manner to achieve QoS, the non-RT RIC can employ the QoS-based resource optimization policy. The analytics functions in non-RT RIC determine that it is preferable to prioritize a specific user or group using the same service in a given scenario after closely examining the congestion situation. When this occurs, the near-RT RIC can outline the service details to the non-RT RIC model via the A1 interface where policies are upgraded or deleted based on the congestion state. Later, the upgraded policies are enforced via near-RT RIC on RAN components via the E2 interface to redistribute RAN resources among customers.

Massive MIMO Optimization: The objective of this use case is to enhance spectral efficiency and alleviate network congestion by directing a high-gain beam antenna on one or more subscribers. Filtering out interference from neighboring subscribers results in improved spectral efficiency and reduced network congestion. This key feature boosts network capacity and is ideal for URLLC applications such as the following: autonomous driving, cloud VR, and industrial automation. MIMO optimization intent is to proactively enhance user-centric QoE or cell-centric network QoS in a multi-cell with multiple transmission/reception points. Non-RTRIC and near-RT RIC continuously monitor real-time network conditions such as traffic, coverage, and interference, which help in intelligent decision-making. At the same time, ML used such data to train the model to proactively take decisions and provide this enriched information to near-RT RIC for real-time dynamic and adaptive adjustment. Toward this direction, the O1, A1, and E2 interface control of the process at architectural components. The O1, A1, and E2 framework governs the seamless coordination between its architectural elements.

RAN Slice SLA Assurance: Network slicing is a prominent feature supporting the customization of virtual networks to meet the service or customer demand. According to 5G and 6G standards, the UE services should be fulfilled with low latency, where these KPIs—number of users, data throughput, service availability, traffic capacity, latency, and reliability—are monitored. The mobile operator and the business customer have an SLA that specifies these capabilities and ensures that it is set up for specified usage, with guaranteed high-quality bandwidth and low latency for self-driving cars, remote surgery, and other mission-critical applications [45]. In addition to preventing possible SLA violations, the O-RAN architecture supports dynamic RAN slice SLA assurance using AI/ML in RIC and open interfaces. Operators thus can ensure maximizing efficiency while serving different slice-specific requirements.

Dynamic HO Management for V2X Based on Context: V2X is vehicle-to-vehicle, vehicle-to-infrastructure, or whatever other combination of those two is possible. It enables fast-paced environments to boost opportunities around road safety, reduce emissions, and tailor travel time through real-time traffic orchestration. Traffic synchronization and user preferences adaptive application (considering the current status of road conditions, traffic, driving intent, etc.). However, V2X connectivity also brought a problem with it and so when a vehicle is moving at high speed and through multiple networks, the HO anomalies are disclosed as follows: Short Stay HO (periodic switching between base stations); Ping-Pong HO (fast oscillation between two cells); Remote Cell HO (connection to a far suboptimal cell). These can degrade V2X performance, connectivity, and robustness.

The O-RAN architecture handles such circumstances efficiently with RIC components and leveraged AI/ML for real-time monitoring and advanced network configurations to enhance the reliability and efficiency of the V2X ecosystem.

Flight Path-Based Dynamic UAV Resource Allocation: UAVs have been used extensively in civil applications such as pollution monitoring, border/forest surveillance, base station inspection, and agricultural plant protection. The domains in which they are used are also growing quickly. Usually, traditional UAVs rely on point-to-point communication links with a ground station, but at high altitudes or remote locations signals are weak and disrupted. As a result, there could be a disconnect issue that causes a long-standing UAV to land or return. O-RAN architecture allows for the acquisition of multi-dimensional data, and by integrating UAV flight data with O-RAN's AI-driven resource allocation, drones may maintain longer connections, minimize interference, and dynamically adjust to real-world scenarios. For example, using the UE's measurement report and SMO, non-RT RIC can obtain the metrics linked to aerial vehicles from the network. This includes information such as the aircraft's flight path, meteorological data, and space load information sourced from applications like UTM (Unmanned Traffic Management). These are the metrics used in the development and training of AI/ML models that will be finally embedded in the RAN architecture. AI/ML models developed by non-RT RIC will operate effectively with the assistance of the near-RT RIC component. As such, near-RT RIC can allocate radio resources to UAVs whenever necessary, considering factors such as flight path data, radio channel conditions, and other pertinent application information.

UAV Radio Resource Allocation: The O-RAN architecture is ideal for managing unmanned aerial vehicles when overseeing wireless resources. The Rotor UAV operates at variable altitudes and speeds, outfitted with various devices like cameras and sensors. The terminal uses a 5.8 GHz frequency to control the UAV remotely, and some instances involve monitoring Rotor UAVs for pollution gas measuring, along border/forest surveillance, high voltage/base station inspections, terrain mapping, and high-definition live broadcasting. They also integrate with anti-UAV defense system to check for unauthorized Rotor UAVs, thus maintaining the safety of each restricted area's low-altitude airspace.

The UAV requires very minimal data exchange between terminal and network manipulation. Still, there is a high demand for uploading HD video with large bandwidth requirements. The UAV Control Vehicle utilizes non-RT RIC, near-RT RIC, O-CU, O-DU, and application servers. The non-RT RIC provides policy guidance and analytics, whereas the near-RT RIC module manages radio resources on the O-CU and O-DU sides to dynamically handle UAV services.

5G and O-RAN applications introduction in wireless communication technology has brought substantial changes. The most critical factor is enabling vendor-independent flexibility and intelligent RAN deployment. Network slicing, traffic steering, massive MIMO optimization, and service-level QoS assurance are just a few of the numerous use cases as earlier discussed. O-RAN's cloud-native features, modular architecture, and AI integration through RIC platforms have made it possible to accomplish these use cases. However, since user demands and technological advances are shifting quickly, especially by 2030, next-generation services will need even more power than current O-RAN and 5G can offer.

With themes of intense connectedness, intelligent automation, and highly immersive communication, the 6G vision began to take shape in this respect. Organizations like the International Telecommunication Union's-Telecommunication Standardization Sector (ITU-T) and the Next-G Alliance are leading international forums to find and prioritize use cases that will influence the design of 6G systems in the future. The ITU-T has identified seven essential use cases for 6G networks [46]: Holographic Type Communications (HTC: to make ultra-realistic communication that goes beyond traditional video calls), Tactile Internet for Remote Operations (TIRO: to facilitate real-time controlling for remote users by focusing on low latency and high reliability), Intelligent Operation Network (ION: AI-enabled automation to handle dynamic allocation and resource management), Network and Computing Convergence (NCC: to integrate cloud and networking services together for real-time processing), Digital Twin (DT: to fabricate virtual replicas of physical systems majorly used in smart cities, industry, and many more areas), Space-Terrestrial Integrated Network (STIN: to facilitate global connectivity via the amalgamation of satellite and terrestrial networks), Industrial IoT with Cloudification (IIoT: to enable smart factories through industrial automation by exploiting cloud services). In addition, the Next-G Alliance classifies future 6G applications into four more general categories: Network-Enabled Robotic and Autonomous Systems (enabled seamless communication among machines such as robots, vehicles, drones, and other autonomous machines), Multi-Sensory Extended Reality (XR) (future research is on expanding the multi-sensory capability beyond visual and auditory, including smell and tactile sensation in XR), Distributed Sensing and Communications (this use case handles precision monitoring, smart infrastructure, and environmental sensing based on real-time data across distributed nodes to react dynamically around surroundings), and Personalized User Experiences (mainly oversee individual preferences and behavior patterns via AI-driven customization of services).

3.5 UPSIDES OF O-RAN ARCHITECTURE

The O-RAN architecture offers several benefits that are revolutionizing the telecoms sector. One of the primary revolutionary advantages is the interoperability among vendors. Contrary to conventional RAN systems, where operators tend to be locked into the hardware and software of a single vendor, O-RAN enables modules from multiple vendors to integrate seamlessly via standardized interfaces. This enables creativity to support innovation in the uptake of next-generation solutions and limits reliance on a single vendor, thereby conveniently promoting technological acceleration. Another significance of the O-RAN is the reduced COPEX/OPEX. As discussed above, using COTS infrastructure and open-source software reduces costs involved in designing and maintaining infrastructure. Similarly, the ability to source components from a wider range of producers leads to increased competition, thereby reducing prices. Here the benefits of O-RAN with 6G and 5G are discussed below:

O-RAN Deployment Optimization: O-RAN, openness, and AI-driven intelligence are consistent with the 5G and 6G vision of ultra-high performance and flexible networks, thus facilitating more agile and effective connectivity. It makes wireless cellular networks more flexible and better responsive to the broad range of future applications from XR to smart cities and more. Moreover, O-RAN's modularity allows operators to build networks more efficiently and add improvements without replacing the entire system, reducing long-term maintenance costs. The most persuasive of O-RAN's features is its ability to support AI-driven optimization. Operators can install RICS rApps and xApps that can use ML in real time to monitor and operate the network conditions in real time [47, 48]. This RIC intelligence in O-RAN assures optimality as well as enhances network efficiency.

Spectrum Efficiency and Optimization: The most valuable and scarce resource in cellular networks is the spectrum. In 5G, spectrum efficiency—defined as the maximum amount of data that can be carried over a bandwidth—is achieved with several O-RAN technologies, such as MIMO, dynamic spectrum sharing, beaming, and higher-order modulation. For 6G, this spectrum is used by multiple frequencies, including sub-THz bands, employed by various players from the telecom ecosystem. The limited nature of this resource necessitates its effective utilization enhanced by O-RAN's open architecture, which supports better spectrum sharing and dynamic spectrum allocation. 6G networks can achieve higher spectrum efficiency by using AI-driven spectrum management technologies in the O-RAN architecture. This will boost data speeds, reduce interference, and improve the overall performance of the network [49].

Global Connectivity and Inclusivity: The collaboration between O-RAN and 6G enables the democratization of cutting-edge wireless technology. In underserved and remote areas, O-RAN's goals of openness and interoperability may reduce costs, making 6G networks more widely available. O-RAN enables a more diverse ecosystem of suppliers and service providers, allowing for broader 6G deployment, which aids in bridging the digital gap and promoting worldwide connectivity. This factor is distinctive from technical harmony since it is based on a shared aim to make the globe better linked and inclusive.

Adaptive Infrastructure for Next-Generation Connectivity: O-RAN's adaptability and flexibility play a critical role in attaining more resilient future networks. O-RAN modular architecture ensures future wireless technology easily integrates and receives ongoing updates without requiring major redesigns. It is essential in ensuring 6G networks continue to function well and remain relevant as new use cases and challenges emerge. Moreover, integrating O-RAN with 6G enhances the capabilities of each technology and opens new frontiers of a faster, more reliable, more intelligent, and more inclusive future of connectivity.

Accelerating Innovation: O-RAN's open, modular architecture allows for a competitive vendor and solution provider environment, crucial to advancing the next-generation wireless network. O-RAN eliminates entry barriers and offers more chance for experimentation and cooperation to achieve this. O-RAN creates an atmosphere in which new ideas and concepts may be rapidly tested and implemented. This openness is required for 5G and next-generation wireless network 6G, which will continually require fresh concepts to meet the future's app requirements. Additionally, O-RAN and 6G combined allow for much faster innovation and keep the network in line with technical advancements, whether through the creation of new AI algorithms, better hardware, or unique applications.

Enhanced Network Intelligence: The AI-driven intelligence in O-RAN can forecast traffic, recognize anomalies, and automatically respond when a network is about to become congested or in case of breakdowns. In short, networks become more self-healing, resilient, and responsive to achieving high performance regardless of the circumstances. In addition to offering improved user experience, the intelligence automates network operations and thus reduces human intervention, which enhances operational efficiency. In other words, O-RAN architecture provides more flexibility, cost-effectiveness, and intelligence in network management, making it innovative for 5G and the future [34].

3.6 CHALLENGES AND FUTURE DIRECTIONS

The O-RAN architecture represents a considerable emphasis in the advance and operation of the cell network, focused on openness and disaggregation with the potential to allow greater adaptability and economies. It is an advantageous innovation, but there are some critical challenges to get it to its entire potential. An important challenge to address is serving the unproblematic interoperability of components from distinct vendors. Standard interface specifications often do not contain sufficiently detailed implementation instructions, which complicates interoperability and implementation. Moreover, security is another challenge since O-RAN open and disaggregated architecture expands the attack surface and makes it more likely to become a target of cyberattacks; effective security features are required to safeguard the network data and ensure availability. Additionally, the operationalizing of O-RAN components is challenging as the multiple vendors collaborate on particular components, which complicates things since coordinating and validating hardware and software from various vendors may delay deployment timelines and increase operational expenses [50, 51]. A few of the challenges are outlined here:

Interoperability and Standardization: The vision of openness in the O-RAN architecture implies a significant challenge in the form of interoperability of various vendors and various network elements. By enforcing that all vendors will indeed implement interfaces according to 3GPP and O-RAN standards. Moreover, deploying xApps and rApps which work together over the different near-RT and non-RT RICs will be prioritized in such vendor-independent environment. The main obstacle in expanding O-RAN deployments will be an FH interface and network that are efficient, scalable, and high-quality. This can be handled by rigorous testing and standardized protocols, so that integration of different vendors will be seamless and nonproblematic. Therefore, in the future work, it is crucial to strengthen universal standards (3GPP) and protocols, which are vendor-agnostic and support different vendor implementations, to be a main objective of the future initiatives.

Ultra-Low Latency Requirements: The disaggregated architecture of O-RAN can incur some latency on the component interfaces, perhaps considering the flexibility and cost-effectiveness. Thus, enhancing the real-time processing possibilities and optimization of the interfaces is needed to develop high-speed, low-jitter transport for O-RAN FH to fulfill 6G requirements. Ultimately, sophisticated applications require URLLC; hence, future research should target network architecture enhancements, such as cross-layer optimization and zero-touch automation, and the development of new technologies that reduce latency, like hardware acceleration.

Scalability: As networks grow larger and more intricate, scalability becomes more critical. To permit O-RAN solutions to expand properly, it is necessary to ensure the technical scalability of RAN components and manage large-scale installations efficiently. Thus, future work must include designing scalable architectures and management frameworks and meeting 6G technology's growing demands. Moreover, scaling does not lead to increasing power consumption; we use green technologies to assure the continuous growth attitude.

Effective AI/ML Algorithm Design, Testing, and Deployment: For the development of effective O-RAN AI/ML model (xApp/rApp), it required access to real datasets. Despite the attempt to create a standardized workflow, there were numerous limitations to develop process throughput in this manner. The three limitations that are encapsulated there are the following: (i) requiring thorough training and testing datasets for a massive deployment, (ii) accommodating online training of data-driven solutions without bottlenecking the operational performance of the RAN, and (iii) developing AI/ML algorithms that can be deployed in various different deployment scenarios in order to function through another unpredictable dataset.

Network Management Complexity: The flexibility of O-RAN makes it harder to manage a 6G network because it has to deal with parts from many different vendors. This requires the use of advanced automation technologies that do not need any assistance from humans. AI and ML will be important for making network administration easier and finding problems before they affect performance. In the future, we should create more advanced AI/ML algorithms to automate network maintenance and optimization. This will make sure that networks work reliably and effectively.

O-RAN-Enabled Spectrum Sharing Solutions: O-RAN specifications for programmable RAN nodes are opening new opportunities to develop spectrum-sharing systems. Although O-RAN requirements facilitate LTE/NR dynamic spectrum sharing, the underlying algorithms have several open problems to be addressed. Further research can look into the practical application of O-RAN-based real-time sensing, proactive, and reactive spectrum adaptation solutions, with a perspective on 3GPP and non-3GPP systems by following the specification of the O-RAN architecture.

O-RAN should be compatible with multi-operator spectrum sharing in the NR bands without degrading overall network performance, making O-RAN technology suitable for private network and enterprise solutions. Therefore, proper attention to spectrum allocation and management policies is needed once O-RAN is considered for use in 6G core networks. On the other hand, dynamic spectrum sharing is required to maximize spectrum utilization, but it should be supported by complementary legal frameworks allowing for fair and flexible access to spectrum resources.

Security in O-RAN: The RAN's open nature offers new enticing ways to network security, while expanding its potential threat space. Exploring new approaches to secure wireless networks, to make them more robust and resilient is critical. O-RAN approaches will become a practical and long-term replacement for traditional RAN implementations only if research is conducted and security strategies are well formulated. The strategy should benefit from O-RAN standards/specifications and improve the availability, resilience, and integrity of its deployments.

Moreover, in a multi-vendor environment, ensuring robust access control, guaranteeing data integrity, and protecting the network from cyberattacks remain major challenges. Future work on securing 6G deployments in O-RAN architecture must focus on developing robust security frameworks and include AI-centered security systems. Research in the field of highly secure encryption methods and anomaly detection systems adapted for open architecture of RAN should improve the overall security of networks.

Multi-Time Scale Control: In the O-RAN architecture, many control loops work parallelly but at different time scales. This makes it hard to keep the system stable, efficient, and able to resolve conflicts. It is essential to explore the advancement of multi-scale algorithms for uninterrupted operational efficiency. To reach high-level goals, it is very important to deal with possible instability, resolve conflicts between control loops, and automate the process of selecting the optimal control loops.

Energy Efficiency: Virtualization and closed-loop control provide real building blocks for optimal energy efficiency through dynamic network function allocation. As networks evolve toward 6G, optimizing energy consumption while sustaining performance is another significant challenge. As network functions move toward cloud operators, the efficiencies of cloud data centers will help to offset some of the costs. More research needs to be done for AI energy management systems to control xApps and rApps to take action or manage policies. These apps include energy efficiency targets and non-RT RIC/SMO orchestration that pivot toward energy efficiency for dynamic orchestration and the optimization goal.

Data Privacy and Sovereignty: Data privacy and sovereignty are critical for 6G and O-RAN optimizations. One of the challenges will be maximizing the use of AI-driven networks while still operating under the constraints of local legislation. Future work must address the needs of rapidly changing digital environments with privacy-preserving AI that comply with regional laws and regulations for data management frameworks. Collaboration between technology developers, data regulators, and regional lawmakers will be critical to creating balanced frameworks that allow ethically driven innovation.

In summary, integrating AI into O-RAN will become a cornerstone of network optimization by enabling intelligent, automated decision-making on resource allocation, fault detection, and traffic management. However, the application of real-time mission-critical AI in the outer loop for RANs depends on accuracy, reliability, and transparency, which are persistent challenges [51, 52]. Further, resource allocation techniques remain a concern of the future since networks will need to respond in real-time and adapt to the needs of users, which requires real-time algorithms to control resources at different layers and time scales. Ultimately, the success of O-RAN's evolution as a core technology of future 5G and 6G networks relies on overcoming these challenges through innovation, standardization, and collaboration efforts. This section covers the potential future challenges and developments relating to 5G and 6G since the integration of 6G with O-RAN presents several operational, technological, and regulatory challenges. Yet, overcoming them is crucial for unlocking all the opportunities to reshape the future of wireless communication.

3.7 CONCLUSION

O-RAN architecture opens the RAN from a closed vendor environment, allowing third parties to deploy innovative services. This makes RAN architecture more flexible, which is achieved by embedding an AI-powered hierarchical control structure. Openness and intelligence are the cornerstones of O-RAN's commitment to developing radio access networks. It seeks to drive the mobile sector toward an ecosystem of cutting-edge, multi-vendor, autonomous, interoperable RANs that are more agile, affordable, and perform better. This chapter looked closely at the O-RAN architecture, interfaces, upsides, and challenges with future prospects. We examined the openness, intelligence, flexibility, and cost-efficiency aspects characterizing O-RAN and found that they can facilitate innovation, cost reduction, and customization in 6G networks. In the future, there is plenty of promise for improving wireless communication by fusing O-RAN and 6G. AI-driven networks will grow more intelligent as they develop, enhancing security in real time, anticipating maintenance requirements, and streamlining operations. Finally, the continued research on the THz spectrum and quantum communications also helps attain a sufficiently large network to serve future demands. The O-RAN and 6G progressions are deeply dependent on edge and fog computing improvements, network slicing, and international collaboration, to deliver new-age, open, secure, and sustainable systems. The integration of O-RAN and 6G idealizes the future of wireless networks. Through this merge, future digital environments become more connected, agile, and productive as they address issues and conform to future trends.

REFERENCES

1. Frauendorf, José Luiz, and Érika Almeida de Souza. "The Evolution of RAN (Radio Access Network), D-RAN, C-RAN, V-RAN, and O-RAN." In *The Architectural and Technological Revolution of 5G*, 139–54. Cham: Springer International Publishing, 2022.
2. Penttinen, Jyrki T. J., Michele Zarri, and Dongwook Kim. *Open RAN Explained: The New Era of Radio Networks*. Hoboken, NJ: Wiley, 2024.
3. Wong, Ian C., Aditya Chopra, Sridhar Rajagopal, and Rittwik Jana. *Open RAN: The Definitive Guide*. Hoboken, NJ: Wiley–IEEE Press, 2023.
4. Polese, Michele, Leonardo Bonati, Salvatore D'Oro, Stefano Basagni, and Tommaso Melodia. "Understanding O-RAN: Architecture, Interfaces, Algorithms, Security, and Research Challenges." *IEEE Communications Surveys & Tutorials* 25, no. 2 (2023): 1376–1411.
5. Singh, Sameer Kumar, Rohit Singh, and Brijesh Kumbhani. "The Evolution of Radio Access Network towards Open-RAN: Challenges and Opportunities." In *2020 IEEE Wireless Communications and Networking Conference Workshops (WCNCW)*, 1–6. IEEE, 2020.
6. O-RAN ALLIANCE. *O-RAN ALLIANCE Overview*. Presentation, March 2025. https://mediastorage.o-ran.org/overview/O-RAN.Overview-of-the-O-RAN-ALLIANCE-presentation.pdf.
7. Wani, Mohamad, Mathias Kretschmer, Bernd Schröder, Andreas Grebe, and Michael Rademacher. "Open RAN: A Concise Overview." *IEEE Open Journal of the Communications Society* 6 (2024): 13–28.
8. Bonati, Leonardo, Michele Polese, Salvatore D'Oro, Stefano Basagni, and Tommaso Melodia. "Open, Programmable, and Virtualized 5G Networks." *Computer Networks* 182 (2020): 1–28.
9. Murti, Fahri. Wisnu, Jose. A. Ayala-Romero, A. Garcia-Saavedra, X. Costa-Pérez, and G. Iosifidis. "An Optimal Deployment Framework for Multi-Cloud Virtualized RANs." *IEEE Transactions on Wireless Communications* 20, no. 4 (2021): 2251–65.
10. ETSI. *ETSI TS 103 982 V8.0.0 (2024-01): Publicly Available Specification (PAS); O-RAN Architecture Description (O-RAN.WG1.OAD-R003-v08.00)*. Sophia Antipolis, France: ETSI, January 2024.
11. 3GPP. "NG-RAN; Architecture Description." *Technical Specification* (TS) 38.401, version 17.0.0, April 2022. http://www.3gpp.org/DynaReport/38401.htm.
12. Bonati, Leonardo, Salvatore D'Oro, Michele Polese, Stefano Basagni, and Tommaso Melodia. "Intelligence and Learning in O-RAN for Data-Driven NextG Cellular Networks." *IEEE Communications Magazine* 59, no. 10 (2021): 21–27.
13. O-RAN Alliance. Working Group 6. *O-RAN Cloud Architecture and Deployment Scenarios for O-RAN Virtualized RAN 2.02*. O-RAN.WG6.CAD-v02.02, 2021.
14. Jain, Raj, and Subir Paul. "Network Virtualization and Software Defined Networking for Cloud Computing: A Survey." *IEEE Communications Magazine* 51, no. 11 (2013): 24–31.
15. Luo, Jia, Qianbin Chen, and Lun Tang. "Reducing Power Consumption by Joint Sleeping Strategy and Power Control in Delay-Aware C-RAN." *IEEE Access* 6 (2018): 14655–67.
16. López-Pérez, David, Antonio De Domenico, Nicola Piovesan, Geng Xinli, Harvey Bao, Song Qitao, and Mérouane Debbah. "A Survey on 5G Radio Access Network Energy Efficiency: Massive MIMO, Lean Carrier Design, Sleep Modes, and Machine Learning." *IEEE Communications Surveys & Tutorials* 24, no. 1 (2022): 653–97.
17. 3GPP. *NR and NG-RAN Overall Description; Stage*-2. TS 38.300, version 17.1.0, 2022. http://www.3gpp.org/DynaReport/38300.htm.
18. O-RAN Alliance. Working Group 2. *O-RAN AI/ML Workflow Description and Requirements 1.03*. 2021.
19. 3GPP. *Radio Link Control (RLC) Protocol Specification*. TS 38.322, version 15.0.0, 2018. http://www.3gpp.org/DynaReport/38322.htm.
20. 3GPP. *Medium Access Control (MAC) Protocol Specification*. TS 38.321, version 15.0.0, 2018. http://www.3gpp.org/DynaReport/38321.htm.
21. 3GPP. *Physical Layer General Description*. TS 38.201, version 15.0.0, 2018. http://www.3gpp.org/DynaReport/38322.htm.
22. 3GPP. *Radio Resource Control (RRC) Protocol Specification*. TS 38.331, version 15.0.0, 2018. http://www.3gpp.org/DynaReport/38331.htm.
23. Abdalla, Aly S., Pratheek S. Upadhyaya, Vijay K. Shah, and Vuk Marojevic. "Toward Next Generation Open Radio Access Networks: What O-RAN Can and Cannot Do!" *IEEE Network* 36, no. 6 (2022): 206–13.
24. 3GPP. *E-UTRA and NR; Service Data Adaptation Protocol (SDAP) Specification*. TS 37.324, version 17.0.0, April 2022. http://www.3gpp.org/DynaReport/37324.htm.

25. O-RAN Alliance. Working Group 2. *O-RAN Non-RT RIC & A1 Interface: Use Cases and Requirements*. Version 4.00, July 2022.
26. O-RAN Alliance. Working Group 3. *O-RAN.WG3.E2GAP—Technical Specification Version 02.01*. July 2021.
27. O-RAN Alliance. Working Group 3. *O-RAN Near-Real-Time RAN Intelligent Controller, E2 Application Protocol 2.00*. O-RAN.WG3.E2AP—Version 02.00, July 2022.
28. O-RAN Alliance. Working Group 3. *O-RAN Near-Real-Time RAN Intelligent Controller E2 Service Model 2.00*. ORAN-WG3. E2SM—Version 02.00, July 2021.
29. O-RAN Alliance. Working Group 3. *O-RAN Near-Real-Time RAN Intelligent Controller E2 Service Model (E2SM) KPM 2.0*. ORAN-WG3. E2SM-KPM—Version 02.00, July 2021.
30. O-RAN Alliance. Working Group 1. *O-RAN Operations and Maintenance Interface 4.0*. O-RAN.WG1.O1-Interface.0—Version 04.00, November 2020.
31. Rob Enns, Martin Bjorklund, Juergen Schoenwaelder and Andy Bierman. "Network Configuration Protocol (NETCONF)." *RFC 6241*, June 2011. http://www.rfc-editor.org/rfc/rfc6241.txt.
32. 3GPP. *Management and Orchestration; Generic Management Services*. TS 28.532, version 17.0.0, March 2022. http://www.3gpp.org/DynaReport/28532.htm.
33. 3GPP. *Telecommunication Management; Generic Network Resource Model (NRM) Integration Reference Point (IRP); Solution Set (SS) Definitions*. TS 28.623, version 17.1.0, March 2022. http://www.3gpp.org/DynaReport/28623.htm.
34. 3GPP. *Management and Orchestration; Performance Assurance*. TS 28.550, version 17.0.0, March 2022. http://www.3gpp.org/DynaReport/28550.htm.
35. O-RAN Alliance. Working Group 6. *O-RAN O2 Interface General Aspects and Principles*. O2-GA&P-R003—Version 05.00, October 2025.
36. O-RAN Software Community. "O2 Interface API Documentation." 2025. https://docs.o-ran-sc.org/projects/o-ran-sc-pti-o2.
37. 3GPP. *System Architecture for the 5G System (5GS)*. TS 23.501, version 16.4.0, 2020.
38. O-RAN Alliance. Working Group 2. *A1 Interface: Application Protocol 3.01*. 2021.
39. 3GPP. *NG-RAN; F1 Application Protocol (F1AP)*. TS 38.473, version 15.0.0, January 2018. Accessed November 10, 2025. http://www.3gpp.org/DynaReport/38473.htm.
40. O-RAN Alliance. Working Group 4. *O-RAN Fronthaul Control, User and Synchronization Plane Specification 7.0*. ORAN-WG4.CUS.0-v07.00, July 2021.
41. O-RAN Alliance. Working Group 4. *O-RAN Management Plane Specification 7.0*. ORAN.WG4.MP.0-v07.00, July 2021.
42. 3GPP. *Study on NR Access Architecture and Interfaces*. TR 38.801, 2017. http://www.3gpp.org/DynaReport/38801.htm.
43. O-RAN Alliance. Working Group 1. *O-RAN Use Cases Analysis Report 13.0*. O-RAN.WG1.Use-Cases-Analysis-Report-R003-v13.00, February 2024.
44. O-RAN Alliance. Working Group 1. *O-RAN Use Cases Detailed Specification 13.0*.Alfter, Germany: O-RAN Alliance, 2024.
45. Khan, Imran, Krishna Kishor Tirupati, Pronoy Chopra, Er Aman Shrivastav, and Shalu Jain. "Challenges in Interworking of Multi-Vendor 5G O-RAN Networks." *International Journal of Computer Science and Engineering (IJCSE) ISSN (P): 2278–9960;* 2024.
46. Agarwal, Bharat, Ralf Irmer, David Lister, and Gabriel-Miro Muntean. "Open RAN for 6G Networks: Architecture, Use Cases and Open Issues." *IEEE Communications Surveys & Tutorials* 28, (2025): 2881–2924, doi: 10.1109/COMST.2025.3562429.
47. Habibi, Mohammad Asif, Bin Han, Merve Saimler, Ignacio Labrador Pavón, and Hans D. Schotten. "Towards an AI/ML-Driven SMO Framework in O-RAN: Scenarios, Solutions, and Challenges." In *2024 IEEE Future Networks World Forum (FNWF)*, 7–14. IEEE, 2024.
48. Kord, Aziz, Jason B. Coder, and Vu Le. "Evolving Open RAN Interoperability: A Large-Scale Definition." In *2024 IEEE International Conference on Communications Workshops (ICC Workshops)*, 2095–2100. IEEE, 2024.
49. Lotfi, Fatemeh, and Fatemeh Afghah. "Meta Reinforcement Learning Approach for Adaptive Resource Optimization in O-RAN." In *2025 IEEE Wireless Communications and Networking Conference (WCNC)*, 1–6. IEEE, 2025.
50. Liyanage, Madhusanka, An Braeken, Shahriar Shahabuddin, and Pasika Ranaweera. "Open RAN Security: Challenges and Opportunities." *Journal of Network and Computer Applications* 214 (2023): 103621.
51. Junejo, Yahya Sameen, et al. "Converging Towards Open RAN." *Journal of Mobile Multimedia* 20, no. 1 (2024): 49–86, doi: 10.13052/jmm1550-4646.2013.
52. Brik, Bouziane, et al. "Explainable AI in 6G O-RAN." *IEEE Communications Surveys & Tutorials* 27, no. 5 (2025): 2826–2859, doi: 10.1109/COMST.2024.3510543.

Chapter 4

Antenna technologies for 6G wireless communications

Sasmita Dash

4.1 INTRODUCTION

Wireless communication systems have evolved through generations to enhance data rate, latency, connectivity, and quality of service. Second-generation (2G) systems added digital modulation and circuit-switched voice communications, which were much more reliable than the first-generation (1G) analog systems. Third-generation (3G) systems increased network capacity and allowed basic mobile internet and multimedia services. Fourth-generation (4G) systems represented a major leap to broadband IP-based communications with increased spectral efficiency and data rates that support mobile broadband and video streaming through technologies like orthogonal frequency-division multiplexing (OFDM) and advanced modulation schemes. Continuing with these developments, fifth-generation (5G) systems include ultra-reliable low-latency communications (URLLC), massive machine-type connectivity for the internet of things (IoT), and enhanced mobile broadband (eMBB) using millimeter-wave (mm-wave) frequencies and massive multiple-input multiple-output (MIMO) architectures [1]. Future generations of wireless systems, sixth-generation (6G), are envisioned to expand network capacity, provide terabit-per-second (Tbps) data rates, and natively integrate sensing, imaging, and artificial intelligence (AI) into network operations [2].

Antennas have traditionally been a significant enabling technology for wireless systems, but with the demands of massive MIMO and beamforming in 5G and beyond (6G), antennas are being pushed to meet tighter and tighter performance requirements in a smaller form factor [3]. With severe propagation losses at mm-wave and terahertz (THz) frequencies, high-gain radiation patterns and precise beam control are required, making advanced antenna technologies, such as reconfigurable antenna arrays, reconfigurable intelligent surfaces (RIS), and metasurfaces (MSFs), active areas of research to dynamically adapt to changing propagation conditions, improve energy efficiency, and accommodate dense network deployments without corresponding increases in power consumption or hardware complexity [4–10]. Several performance criteria must be met by next-generation antenna systems, such as having wide operational bandwidths to accommodate the use of higher-frequency bands for multi-gigabit and terabit data rates [11], having high-gain and high-efficiency designs for mm-wave and THz frequencies to compensate for increased path losses, and maintaining stable radiation characteristics for reliable link quality and energy-efficient operation [3, 12–14].

In addition, the emergence of mobile and IoT devices requires the miniaturization of antennas without sacrificing performance, and new materials and design techniques are being explored to achieve this goal [11]. The need for antennas to reconfigure in real time in frequency, radiation pattern, and polarization to better adapt to dynamic wireless environments and various service requirements enhances system adaptability and spectral efficiency [4]. As new applications, such as XR, holographic communications, and dynamic digital

DOI: 10.1201/9781003637721-4

twins, are emerging and high-performance computing and AI are rapidly advancing, wireless communication technologies are experiencing unprecedented growth. While 5G mm-wave systems are still being deployed around the world, 3GPP has already released 3GPP Release 18, also known as 5G-Advanced, which is a major milestone on the path to future 6G networks as defined by the IMT-2030 framework [15, 16].

The transition from 5G to 6G is expected to introduce several key capabilities with significant implications for antenna technologies, as shown by comparing the IMT-2020 and IMT-2030 performance indicators [17]. The expected development of augmented and virtual reality applications and digital twins requires operation at higher frequency bands and deployment of extremely large-scale massive MIMO systems to achieve eMBB data rates of up to 1 Tbps. Although higher frequencies are not optimal for wide-area coverage, they will be necessary to support bandwidth-hungry 6G use cases. Introduced in 3GPP Release 15, massive MIMO uses large arrays of antennas to provide beamforming and spatial multiplexing, greatly enhancing coverage, throughput, and signal quality, but scaling to even larger array sizes in 6G will come with cost, complexity, and energy-efficiency issues.

Second, integrated sensing and communications (ISAC) is a key feature of 6G, which turns the traditional communication network into a dual-purpose system for both environmental sensing and connectivity [18]. ISAC relies on the interaction of radio signals with the environment and needs antenna systems that can generate both wide beams for coverage and steerable narrow beams for high-resolution sensing. Most current ISAC implementations use fully digital arrays, while analog and hybrid array architectures may provide low-cost and energy-efficient alternatives [19].

Third, side link communications supporting vehicle-to-everything (V2X), device-to-device (D2D), and machine-type communications are expected to play important roles in industrial automation and intelligent transportation systems [20]; these applications require dynamic multibeam antenna technologies that support efficient one-to-many communications under strict energy and cost constraints. Likewise, unmanned aerial vehicles (UAVs) are expected to become an essential part of future wireless infrastructures, which will require flexible antenna solutions for aerial and ground communications.

Fourth, to address the limited coverage of existing terrestrial networks, 6G systems are expected to combine terrestrial and non-terrestrial networks (NTN), including satellites, high-altitude platform stations (HAPS), and UAVs, to provide ubiquitous global connectivity and resilient communication services in emergencies [21]. These platforms bring special antenna challenges, including high-gain transmitarrays and reflectarrays for LEO satellites, adaptive multibeam antennas for HAPS, and cost-effective high-gain terminals for mobile users. Direct-to-device satellite communications are a particularly challenging but promising application that demands large-scale multibeam satellite antennas to support moderate data rates.

Although the 6G standardization efforts will mainly focus on system-level specifications, many emerging system architectures fundamentally rely on advances in antenna design to realize their performance gains, and therefore, recent progress in antenna technologies for 6G systems is summarized in this chapter. This chapter provides a comprehensive review of fundamental antenna principles, covering the fundamentals of 6G wireless systems with a focus on emerging frequency bands in the mm-wave and THz regimes with their associated propagation characteristics, opportunities, and limitations. This chapter systematically discusses major antenna technologies that enable 6G communications, such as mm-wave and THz antennas, massive MIMO systems, RIS, reconfigurable and multibeam antennas, antenna collocation strategies, and in-band full-duplex architectures. It also outlines the increasing role of AI and machine learning (ML) for antenna modeling, optimization, and real-time adaptation. Finally, it identifies critical technical challenges and emerging research directions and opportunities for future antenna solutions.

4.2 FUNDAMENTAL ANTENNA PRINCIPLES FOR 6G WIRELESS SYSTEM

Antennas are the heart of any wireless communication system and are the interface between guided electromagnetic waves and free-space propagation, making antenna fundamentals even more critical for the next generation of wireless networks as they move toward 6G, with ever-increasing demands on bandwidth, gain, directivity, energy efficiency, and beamforming capabilities. Operating frequency bands and propagation challenges provide a foundation for designing next-generation wireless systems.

4.2.1 Frequency bands

Modern wireless communication systems use a variety of frequency bands, each with its own unique advantages and design challenges that affect antenna architecture and network performance. The sub-6 GHz frequency bands (700 MHz–6 GHz) still serve as the foundational bands for macro-cellular networks because of their advantageous propagation characteristics, such as strong diffraction, good penetration, and wide-area coverage, for legacy and modern technologies such as 4G, Wi-Fi, and much of 5G New Radio (NR) deployments [22]. Table 4.1 presents frequency bands and propagation characteristics for advanced wireless systems. The evolution to 5G and beyond has increased the need for advanced antennas that can handle ultra-high data rates, ultra-low latency, massive connectivity, and seamless operation across sub-6 GHz, mm-wave, and THz frequency bands. For 5G networks, antennas must support massive MIMO, high-efficiency beamforming, and multi-band operation to provide eMBB, URLLC, and massive IoT services [23]. Massive MIMO arrays, which have tens to hundreds of antenna elements, are an important component of 5G NR deployments and are critical for increasing spectral efficiency and spatial multiplexing.

Large contiguous bandwidths of multi-gigabit-per-second data rates are available at mm-wave frequencies in the 24–100 GHz range [24], but mm-wave propagation involves high free-space path loss, sensitivity to blockage, and atmospheric absorption [25], which require high-gain, directional antennas and dense small-cell deployments to maintain reliable links, thus necessitating narrow-beam radiation patterns and adaptive beam-steering capabilities for the antenna architectures operating in this band. Phased arrays, leaky-wave antennas, lens antennas, and reflectarrays have become a must for mm-wave communication systems, and compact and low-profile solutions such as MSF-based phased arrays and

Table 4.1 Frequency bands and propagation characteristics for advanced wireless systems

Frequency band	*Typical range*	*Applications*	*Propagation challenges*	*Opportunities*
Sub-6 GHz	0.5–6 GHz	5G, LTE, Wi-Fi	Moderate path loss, penetration limited in dense urban areas	Wide coverage, good penetration, reliable links
Millimeter-wave (mm-wave)	30–300 GHz	5G eMBB, backhaul	High free-space path loss, blockage sensitivity, limited diffraction	Large bandwidth for multi-Gbps links, directional beams, massive MIMO
Terahertz (THz)	0.1–10 THz	6G ultra-high speed, sensing, imaging	Severe atmospheric absorption, short-range coverage, material penetration limited	Ultra-wide bandwidth for Tbps links, compact high-gain antennas, integrated communication and sensing

substrate-integrated waveguide (SIW) antennas are promising candidates for integration into handheld devices, small cells, and vehicular communication platforms [26, 27].

In addition to mm-wave, the THz frequency range (0.1–10 THz) provides vast spectral resources for ultra–high capacity communications, real-time holographic data transmission, wireless chip-to-chip interconnects, and nanoscale sensing applications, which are all expected to be achieved in wireless systems to support data rates beyond 1 Tbps as wireless systems evolve to 6G [25]. THz antenna design is particularly challenging because of severe propagation losses and size constraints; however, the electrically tunable plasmonic properties of graphene and other 2D materials have been identified as a potential solution for highly compact, reconfigurable, and broadband THz antennas suitable for integrated and adaptive communication front ends [28–30].

4.2.2 Propagation challenges and opportunities

Modern wireless communication systems face both significant challenges and unique opportunities associated with propagation at higher frequencies. Free-space path loss increases with frequency, making mm-wave and THz links highly sensitive to distance and requiring antennas with high directivity to maintain acceptable link budgets [31]. Furthermore, atmospheric and molecular absorption attenuate high-frequency signals and limits long-range propagation in THz frequency bands [32]. Poor diffraction is another propagation characteristic of high-frequency electromagnetic waves, and such waves are easily blocked and scattered by obstacles such as buildings, foliage, vehicles, and even human bodies. This requires stringent requirements on antenna systems to use narrow-beam radiation patterns, adaptive beam steering, and intelligent propagation control techniques, such as RIS, to maintain reliable and resilient connectivity [33].

However, the short wavelengths of mm-wave and THz frequency bands can offer ultra-directional beamforming, massive and ultra-massive MIMO architectures, compact on-chip and in-package antenna solutions, and advanced reconfigurable antenna platforms, including graphene and two-dimensional-material-based THz antennas, enabling extremely high data rates, ultra-low latency, and enhanced spatial multiplexing that shows the great potential of THz communications for 6G and beyond [34–37]. With the rapid development of wireless technologies from the sub-6 GHz to mm-wave and THz regimes, the antenna design methodologies should evolve to meet the propagation challenges and maximize the advantages brought by high-frequency operation [38–40].

4.3 ANTENNA TECHNOLOGIES FOR 6G COMMUNICATION SYSTEMS

5G wireless systems require large bandwidths and high directional gain, and as such, mm-wave architectures have become the dominant architecture for the antennas used in 5G systems [1]. Massive MIMO antenna arrays that enable spatial multiplexing and adaptive beam steering have been studied extensively for next-generation wireless systems [23]. In addition to conventional antenna paradigms, the passive manipulation of electromagnetic waves and programmable control of the propagation environment have been realized by the emerging RIS technology [10, 41, 42], which is considered a key enabling technology for 6G. Meanwhile, the THz frequency band is considered one of the critical frontiers for 6G communications and has drawn significant attention to the design of THz antennas and arrays using plasmonic effects and graphene-based materials to realize compact, tunable, and efficient radiators operating at frequencies above 100 GHz [43, 44].

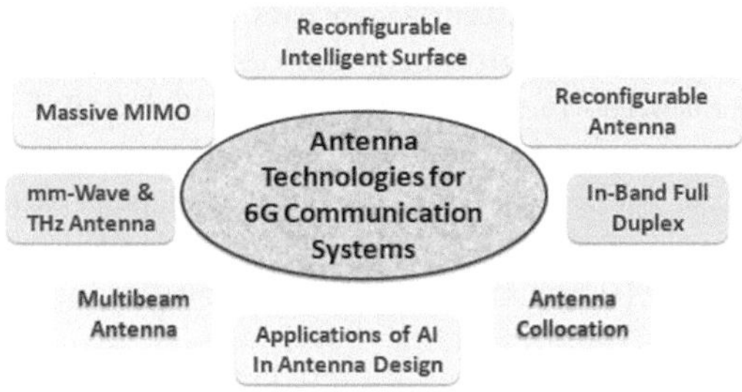

Figure 4.1 Antenna technologies for 6G communication systems.

The programmable RIS platform is anticipated to be a pivotal platform for 6G networks to achieve passive beamforming, dynamic coverage adaptation, and low-power link optimization, which can significantly improve network capacity, robustness, and energy efficiency when combined with massive MIMO and mm-wave/THz transceivers [45]. In addition, AI will be deeply embedded into antenna design and control, allowing real-time beam adaptation, environment-aware operation, and digital twin–based optimization for autonomous and self-configuring wireless networks [45].

The collective advances of 5G and 6G antenna technologies, including massive MIMO, mm-wave phased arrays, graphene-based THz antennas, RIS-enabled propagation control, and intelligent electromagnetic interfaces, are foundational for next-generation applications like autonomous mobility, immersive XR, smart healthcare, and space–air–ground-integrated networks. Compared with previous generations of wireless systems, which mainly focused on physical-layer performance improvement by means of novel modulation, multiplexing, and signal-processing techniques, future 6G systems need to consider not only spectral efficiency but also other metrics, such as total energy consumption and deployment cost. Consequently, 6G antenna designs are required to deliver not only high capacity and flexibility but also energy- and cost-efficient operation. In this context, the following section outlines several major antenna technologies that are expected to play a central role in realizing the performance targets of 6G wireless systems, as illustrated in Figure 4.1.

4.3.1 mm-wave and THz antennas

Australia has joined the US in the rollout of 5G mm-wave technology. It has been demonstrated that 5G mm-wave can deliver Gbps data rates, making it a desirable option for small cells that cover data-hungry hotspots. However, mm-wave propagation's quasi-optical nature restricts coverage to mostly outdoor settings, and NLoS coverage is particularly difficult in open spaces devoid of building clusters. RIS or other relay devices may be required for NLoS coverage. Meanwhile, 5G mm-wave has proven to be an excellent alternative to fixed wireless, offering broadband services to residences and businesses without wired connections. It is anticipated that some airborne and spaceborne systems will use THz and mm-wave as they transition to 6G. Low-cost, low-energy, high-gain, multibeam mm-wave antennas are essential due to the large number of airborne systems. Recently, several mm-wave antennas and THz antennas have been explored to meet the requirement of 6G communication networks [12, 35, 46, 47]. However, it is important to note that THz adoption in the 6G timeframe is much more difficult than mm-wave adoption. For consumer devices, the existing THz power sources and receivers are not developed enough. THz systems are challenging to use as an outdoor coverage solution for 6G due to their propagation characteristics, which include high attenuation loss and weather susceptibility. Near-field communications (NFC) is an intriguing application for sub-THz systems. The near-field region can reach several hundred meters when a very electrically large THz antenna or RIS is used. Beyond Tbps data rates, significantly higher user densities, with greatly enhanced security and dependability, precise 3D positioning, and other innovative applications not feasible in the far-field, will be made possible by this [48, 49]. It is a difficult task for the antennas, mm-wave, and THz research communities to fully utilize this potential with affordable sub-THz transceivers.

4.3.2 Graphene-based THz antennas

The advent of 6G wireless communication, exploiting mm-wave to THz frequency bands, demands antennas capable of high directivity, tunability, and compact form factors. Conventional metallic and planar THz antennas face limitations such as high material losses, low radiation efficiency, and stringent fabrication tolerances at these frequencies. Recent advances in graphene and nanomaterial-based antennas provide promising solutions to overcome these challenges while enabling novel functionalities [50–53]. Recent efforts in THz antenna research have demonstrated the potential of graphene and nanomaterial-based radiators to overcome propagation limitations and design challenges at mm-wave/THz bands. Graphene nanoantennas have shown improved miniaturization, directivity, and frequency tunability owing to high carrier mobility and surface conductivity control, making them promising candidates for THz links in 6G systems [8]. Graphene nanoantennas exploit the high carrier mobility and electro-optically tunable surface conductivity of graphene, which allows for miniaturized antenna designs with enhanced directivity and radiation efficiency [8]. By dynamically adjusting graphene's chemical potential, these nanoantennas support frequency reconfigurability, impedance stability, and adaptive beam steering, which are crucial for maintaining reliable THz links in dense network environments [42].

Graphene MSFs extend these capabilities by actively manipulating the propagation of THz waves. Engineered meta-atoms embedded in these MSFs can achieve perfect absorption, polarization insensitivity, and beam shaping, thereby improving coverage, energy efficiency, and link robustness in obstructed scenarios [41]. Integration of MSFs with graphene-based antenna arrays enables dynamic control of reflection and scattering, providing an unprecedented level of programmability in the wireless environment.

Ultra-wideband graphene antennas have been demonstrated with orthogonal beam patterns and directional radiation, offering high capacity and robustness for 6G THz links [12]. These designs not only support multi-Gbps to Tbps data rates but also facilitate adaptive multibeam operation, essential for massive MIMO configurations. Furthermore, AI-assisted design and optimization frameworks applied to graphene antennas allow for real-time performance prediction, parameter tuning, and interference mitigation, making complex THz antenna design tractable [46].

4.3.3 Massive MIMO

All digital massive MIMO will unavoidably result in extremely high energy consumption and high manufacturing costs, as was already realized in 5G technology development [54]. As a result, major vendors introduced hybrid antenna arrays to 5G mm-wave base stations. Both analog and digital beamformers are used in a hybrid antenna array [55]. In particular, digital beamformers are created by combining analog beamformers that are used as subarrays. The digital beamformer offers a way to process complex signals like beam acquisition and tracking, while the analog subarrays help reduce expenses and energy usage. In this manner, the digital beamformer complexity is decreased by a factor equivalent to the subarray's element count. The hybrid array architecture is a desirable option for some massive MIMO implementations in 6G due to its trade-off between beamforming performance, energy consumption, and cost. The productions of identical phase shifters in wide mm-wave bands, which can result in beam squint that compromises system reliability and efficiency, and the losses in phase shifters and the feed network, which tend to worsen at high frequencies, are two of the antenna community's biggest problems. Additionally, from a system perspective, the literature has not adequately addressed the efficient support of multibeams in the subarray.

4.3.4 Reconfigurable intelligent surface (RIS)

The metasurface (MSF) is a major advancement in antenna technology over the past 20 years [56]. MSFs are usually created by arranging electrically tiny scatterers or holes into a two-dimensional pattern at a surface or interface, giving them properties not typically found in natural materials. MSFs have been used in antenna design to create multibeams, shape the beam pattern, and increase bandwidth [57]. Active MSFs were suggested as reconfigurable antennas with sophisticated analog signal-processing capabilities for transceivers of the future [58].

RIS is the reconfigurable MSF in 5G and 6G systems. Since mm-wave and THz waves are quasi-optical, NLoS propagation would become a significant barrier to communications link establishment. Consequently, both academia and industry have shown a great deal of interest in RIS [59]. An RIS can typically reroute electromagnetic waves toward specific areas in the form of an MSF. They offer significant advantages for localization and mapping in intricate NLOS environments, in addition to communications. As an alternative to traditional relays, which typically have complicated hardware and high power consumption, RISs can also extend range in integrated terrestrial and non-terrestrial wireless networks. Despite the fact that massive MIMO in RISs has been the subject of many publications in the wireless communication community, it is important to remember that RISs must be inexpensive and energy-efficient for the majority of applications. Therefore, unless there are significant advancements in smart materials that allow for low-cost and low-energy beamforming, reconfigurable RISs and hybrid RISs with both passive and active elements may be more practical. Additionally, optically transparent RISs will lessen the physical and visual impact of extensive deployment.

4.3.5 Reconfigurable antennas

For more than 20 years, reconfigurable antennas have been the subject of extensive research [12, 60]. Differing from the conventional antennas that are physically fixed once deployed, reconfigurable antennas can change their bandwidth, radiation patterns, and polarizations on the fly to deliver optimal system performance, be it data rates, communications range, or information interceptibility. The dependability and losses of switches and tunable devices are major obstacles to the industry's adoption of reconfigurable antennas. Since the 1990s, liquid antennas have been investigated as an alternative low-loss technology. Reconfiguration is accomplished in such antennas by reversibly altering their geometry, which subsequently modifies the beam patterns and operating frequency [7, 14, 20]. To make the solution workable, a number of issues still need to be resolved. These consist of system complexity and compactness, reconfiguration speed, and cost.

Reconfigurable antennas have been a popular choice for cognitive wireless systems since their inception. A number of reconfigurable antenna types now have an ideal application space thanks to the development of intelligent 6G networks. These include reconfigurable frequency and beam pattern antennas for moving platform applications that need a moderate reconfiguration speed, like inexpensive LEO satellite terminal antennas and UAV antennas, for example. It is anticipated that the reconfigurability feature will manifest in a number of ways, including bandwidth adaptation to facilitate dynamic spectrum allocation and possibly more for analog beamforming in both analog beamformers and RISs. A reconfigurable mm-wave dual-polarized antenna array system for base station antennas of the future was recently described. In contrast to traditional switching techniques, a polarization mixing method was used to achieve continuous beam width control over a wide range. Wider beams were produced by simultaneously stimulating two polarizations; the system's beam width could be continuously adjusted by simply adjusting a few phase shifters [47]. A reconfigurable layer and wide-angle impedance matching (WAIM) structure allow the design to achieve high radiation efficiency and wide beam-scan angles. In particular, it is possible to dynamically alter the antenna positions at the transmitter and/or receiver to improve the channel conditions and enhance communications performance. However, there are inherent drawbacks to movable antennas, such as slow response times and feeding challenges.

4.3.6 Multibeam antennas

Sectorization has been an important strategy to boost system capacity at base stations. As a result, base station antennas are usually made to cover specific fixed sectors. Supporting dynamic multiple beamforming will become standard with 5G-Advanced and 6G. Analog multibeam antennas can be used alone, as a feed for a larger antenna, or in subarrays within a hybrid antenna system. These antennas can be made to be inexpensive and consume little energy. Quasi-optical multibeam antennas have advanced quickly in recent years [61]. Unfortunately, these antennas are nearly exclusively used for fixed multibeams, with very few exceptions. Multibeam analog antennas that are individually steerable continue to be a significant challenge.

4.3.7 Antenna collocation

The cellular industry has always faced significant challenges in acquiring new base station/antenna locations. Consequently, antenna collocation has become a standard technique in earlier mobile wireless network generations. There are two possible negative effects on system performance when antenna arrays operating in different bands coexist. Mutual coupling is the first, and cross-scattering between antenna elements in different frequency bands is the second. The primary cause of distorted radiation patterns is usually the cross-scattering from lower frequency, larger antennas to higher frequency, smaller antennas. Such distortion in the antenna patterns must be suppressed in order to guarantee quality coverage because base station antennas are usually built to provide specific fixed or dynamic angular coverage; this process is known as de-scattering. De-scattering becomes more challenging for the more bands a base station must support. In the early stages of the 6G era, base stations will have to accommodate the collocation of 6G, 5G, and probably 4G antennas. Thus, it is necessary to develop new wideband and multibeam de-scattering techniques. Effectively serving UAVs and LEO satellites while assisting mobile users on the ground is another difficult problem for base station antennas. This necessitates a paradigm change in antenna design.

4.3.8 In-band full-duplex

There has been a lack of frequency resources to support new systems due to the exponential growth in connectivity and data rates, and this spectrum scarcity issue is getting much worse for 6G. One of the main solutions to the issue of spectrum scarcity is in-band full-duplex (IBFD) technology, which enables transceivers to transmit and receive in the same frequency band simultaneously. Simultaneously transmit and receive (STAR) systems are another name for IBFD. Additionally, it is very helpful for joint sensing and communications because it is easy to sense the environment by directly analyzing the signals that are scattered throughout the surrounding area. Self-interference (SI), or the signal that a receiver receives from a collocated transmitter, is a basic problem for IBFD systems [62]. Compared to the signal received from an intended distant transmitter, SI is substantially stronger. In an IBFD system, the SI must be canceled before any receiver can operate because it uses the same frequency band but is much more powerful than the intended signal. Ideally, the SI would need to be sufficiently lowered to below the noise floor in order to allow an IBFD operation. The SI power is typically 70–120 dB higher than the signal of interest. Even worse, the SI is highly susceptible to phase noise, nonlinear distortion, time variance, and frequency instability. When cost-sensitive and form factor–constrained mobile devices are taken into account, these difficulties are made worse. One must use a comprehensive strategy to address the issue concurrently in the digital, RF, and propagation/antenna domains in order to effectively cancel the SI. A cutting-edge prototype that achieved 75 dB SI cancellation over a 50 MHz band recently proved the viability of IBFD systems [62]. It used two decoupled patch antennas for SI mitigation in the propagation domain, a joint analog and digital SI cancellation scheme, and an adaptive least mean square (ALMS) RF filter with a track/hold function for analog domain SI cancellation. It is simple to use decoupling techniques to isolate transmitting and receiving antennas if they are used separately. Orthogonal polarizations, orthogonal antenna modes, near-field cancellation, and feed network isolation are additional techniques. When the transmitting and receiving antennas are positioned closely together in dual-polarized arrays, the problem becomes considerably more difficult. When a single dual-polarized antenna array is used for both transmission and reception, the situation becomes even more difficult. Future research can greatly benefit from finding solutions to these useful antenna configurations.

4.3.9 Applications of AI in antenna design

AI, and more especially ML, has developed quickly over the past ten years and has affected many different industries. AI is being used more and more in the antenna industry to design intricate electromagnetic structures, increase design accuracy, and shorten development times. In recent years, significant efforts have been made to apply AI to the design of antenna elements, antenna arrays, electromagnetic MSFs, and other structures. AI will become more important due to the difficulties with 6G antenna designs, and the combination of AI and reconfigurable antennas to make the entire wireless system cognitive may be even more intriguing. Some of the most common antenna types will be covered in this section, including the application of AI to antennas. It is debatable whether the majority of current antenna research is relevant given that 6G will bring about a revolution in information and communications systems. As a result, the topics are selected based on current industry trends and research interests.

With the increasing complexity of 6G antenna systems and the rapid advancement of AI, it is expected to play a crucial role in future antenna design and operation. Traditional antenna design methods rely on computationally expensive, time-consuming iterative full-wave

Table 4.2 6G antenna technologies and their potential

Key antenna technologies	*Potential*
Hybrid antenna arrays for MIMO	Peak data rate, massive connectivity, capacity, cost reduction, energy saving
Multibeam antennas	Sensing, coverage, connection density, security, resilience, energy saving
Antenna collocation	Interoperability, coverage, sustainability
RIS	Coverage, cost, and energy saving
Reconfigurable antennas	Network performance, spectrum sharing, interoperability
In-band full-duplex	Spectral efficiency
mm-wave and THz antennas	Peak data rate, capacity
Wearable antennas	Sensing, connectivity, digital twin
Artificial intelligence as a design tool	Energy efficiency, spectrum efficiency, reliability, user experience

electromagnetic simulations using tools like HFSS and CST, whose large design space precludes fast performance prediction. Therefore, data-driven surrogate models based on ML techniques have been widely used to predict antenna performance with significantly reduced computational cost. With well-trained ML models, fast performance prediction, automated optimization, and even inverse antenna design (i.e., desired antenna characteristics directly yield optimal geometrical parameters) can be realized. In addition, as AI is expected to become an integral part of 6G systems, AI can be used to assist in situ and adaptive antenna reconfiguration to achieve the optimal performance in dynamic radio environments.

Recent studies on AI-based antenna design primarily focus on surrogate modeling, performance optimization, and inverse design. Neural network–based models, such as convolutional neural networks (CNNs) and deep regression networks, have demonstrated high accuracy in predicting antenna responses from EM-simulated datasets [62]. AI techniques have also been extensively applied to antenna array synthesis, significantly accelerating traditional iterative optimization approaches. ML-based models enable rapid prediction of excitation amplitudes, phases, and element positions to achieve desired radiation patterns, including beam steering, null placement, and sidelobe suppression. Neural network–based methods have been reported for phase-only and amplitude-only array synthesis, as well as for pattern-reconfigurable and nonuniform array designs [63, 64]. Given the dynamic nature of future wireless networks, reinforcement learning has emerged as a promising approach for realizing cognitive antenna systems capable of adapting radiation patterns and bandwidth in real time according to system and environmental requirements.

AI has further demonstrated strong potential in MSF and reflectarray design, where conventional full-wave optimization becomes particularly challenging due to high-dimensional design spaces. Moreover, deep-learning-enabled generative approaches, including conditional GANs and variational autoencoders, have been employed for the efficient synthesis of single- and multi-layer MSFs, significantly reducing design complexity while enabling high-performance and multifunctional MSF antennas [65]. Table 4.2 summarizes 6G antenna technologies and their potential.

4.4 CHALLENGES AND EMERGING OPPORTUNITIES IN 6G ANTENNA TECHNOLOGIES

The transition to 6G wireless communication represents a fundamental paradigm shift that requires radical innovations in antenna technology and system design to meet significantly higher data rate, latency, and reliability demands. While 5G largely uses sub-6 GHz and mm-wave bands, 6G aims to utilize mm-wave and THz frequency bands to provide multi-Gbps to Tbps data rates and to facilitate new applications such as holographic telepresence, ultra–high resolution sensing, autonomous systems, and the tactile internet. These extreme frequency bands, however, pose significant propagation and hardware challenges that require innovative antenna concepts and design methodologies to maintain link reliability and network performance [66]. However, operating at THz bands poses major challenges, because high free-space path loss, atmospheric absorption, and diffraction effects in the THz regime severely limit the communication range and link reliability, which necessitates dense small-cell deployments, distributed access points, and sophisticated beamforming and beam-tracking techniques to maintain coverage and throughput. In addition, efficient and compact THz antennas are difficult to develop because of inherent material losses, low radiation efficiency, and stringent fabrication tolerances for plasmonic and on-chip designs; and accurate channel models for these bands are still under active research [2].

Severe propagation impairments such as high free-space path loss, atmospheric absorption, limited diffraction, and blockage effects are inherent in mm-wave and THz bands, which constrain the communication range and coverage and consequently require ultra-directional antennas, dense deployment of small cells or distributed access points, and sophisticated beamforming and beam-tracking techniques to maintain link robustness in dynamic environments. In order to address these problems, THz-frequency graphene-based nanoantenna designs with high directivity and tunability have been studied in an attempt to compensate for propagation losses by manipulating plasmonic waves through control over material conductivity. Compared with conventional metallic nanoantennas, the relatively higher miniaturization, directivity, and radiation efficiency of a single graphene patch are achieved because the surface conductivity of graphene can be electro-optically tuned at THz frequencies [8]. Design strategies have been proposed to dynamically change the surface conductivity via the chemical potential of graphene, allowing for reconfigurability of frequency and impedance for adaptive beam steering and broadband operation at 0.1–10 THz to support the bandwidth demands of next-generation wireless links with mitigation of the high path losses in the development of THz communication systems.

Besides the propagation challenges, massive and ultra-massive MIMO antenna arrays required for spatial multiplexing and high spectral efficiency add more complexity due to mutual coupling, element correlation, and calibration issues and can degrade the beamforming performance if not properly addressed by advanced isolation strategies and calibration techniques [11]. Additionally, full-duplex systems in the same band that offer high spectral efficiency by allowing simultaneous transmission and reception suffer from inherent SI that requires effective cancellation circuits built into the antenna front ends and sophisticated analog/digital interference suppression methods. Real-world performance is further limited by hardware impairments, including phase noise and nonlinear RF behavior, particularly at high frequencies. However, 6G presents a wealth of new opportunities facilitated by advanced antenna concepts, intelligent surfaces, and AI-enabled design methods. Antennas based on nanomaterials (e.g., graphene and metamaterials) are tunable, highly compact radiators with the potential to be frequency-agile at THz frequencies and can be made to perform better with the right decoupling structures to mitigate mutual coupling in dense MIMO configurations.

RIS can go even further by programming the propagation environment itself, dynamically controlling reflections and scatterings to enhance coverage, energy efficiency, and link robustness in both line-of-sight and obstructed conditions [67]. Graphene-based antennas and MSFs offer tunable, compact radiators for THz communication, which can be frequency reconfigurable, dynamically beam-shaping, and multifunctional, all of which are important features for high-capacity, energy-efficient 6G networks [12]. For instance, research

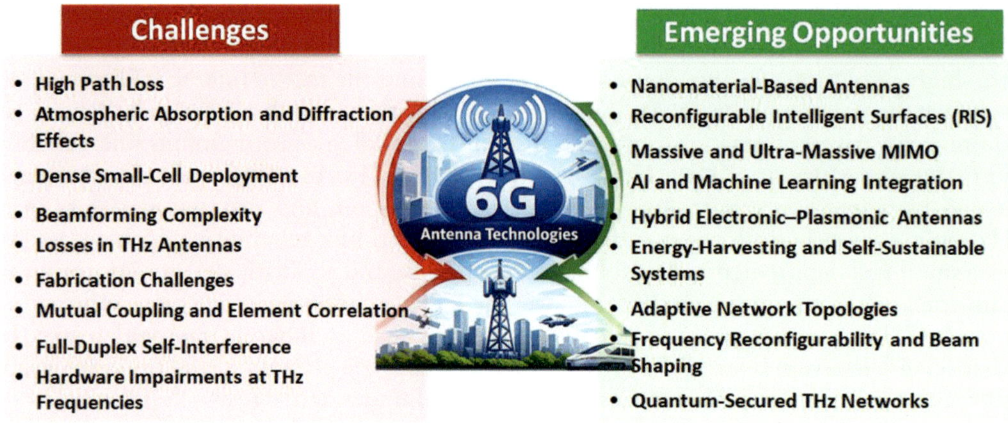

Figure 4.2 Challenges and opportunities in 6G antenna technologies.

has shown ultra-wideband graphene-based directional antennas that can work over broad THz spectra with reconfigurable beam patterns to improve coverage and link quality. More advanced designs include graphene RIS and MSFs that manipulate THz wave propagation via engineered meta-atoms to achieve perfect absorption, frequency tuning, and polarization insensitivity and can be integrated with antenna elements to program the wireless environment to optimize propagation characteristics and blockage mitigation in complex scenarios [42]. Figure 4.2 presents challenges and opportunities in 6G antenna technologies.

6G enhancements continue to rely on massive MIMO and multibeam antenna architectures to enable precise beam steering, spatial multiplexing, and increased spectral efficiency, and when combined with AI and ML algorithms, these systems can intelligently adapt to environmental and network dynamics in real time, optimizing beamforming, parameter tuning, and interference management. AI-driven frameworks have shown potential in optimizing parameters and predicting performance, such as learning-based regression models for THz MIMO antenna design that can achieve wide bandwidth, high radiation efficiency, and high isolation through data-driven optimization [68].

These intelligent frameworks not only improve the prediction of performance but also automate tasks that would be impractical to execute through classical electromagnetic design techniques, such as AI-based adaptive beamforming strategies that dynamically update antenna patterns and resource allocation based on real-time channel state information to counteract the dynamic network conditions and high user mobility, as well as RIS integrated with AI for secure and resilient communication paradigms, such as quantum-secured THz networks, which point to the deep synergies between emerging paradigms and 6G security considerations [69, 70].

Antenna design and network resource management can benefit from the integration of AI and ML, which brings considerable potential for automating the optimization of parameters and for adapting to real-time conditions. AI-driven frameworks have been demonstrated to outperform traditional design methods in estimating performance metrics, optimizing beamforming strategies, and dynamically adapting radiation patterns according to heterogeneous channel conditions, especially in THz MIMO systems with a high-dimensional design space and analytical modeling intractability, which can thus yield new levels of antenna performance that cannot be achieved through traditional methods. The future of 6G antennas and communication systems will be co-designed and integrated with multiple technologies, including hybrid electronic–plasmonic antennas; energy-harvesting/self-sustainable systems;

MSF-integrated architectures; and ultra-massive MIMO arrays with scalable hardware.; In addition, adaptive network topologies like cell-free architectures for more uniform coverage and lower latency; and wide-scale application of AI-assisted design and optimization will support context-aware and autonomous radio resource management. Further standardization of the THz spectrum, and development of reliable fabrication methods for transitioning these concepts from research labs to robust infrastructure supporting the next generation of ubiquitous, intelligent, and immersive wireless systems.

4.5 CONCLUSION

The move to 6G wireless communication systems marks a paradigm shift in the thinking about antenna technologies, with 6G operating across sub-6 GHz, mm-wave, and THz frequency bands, where antennas are pivotal elements of the air interface that can affect system capacity, latency, reliability, and energy efficiency. In this chapter, we have discussed the frequency bands and propagation characteristics that are relevant to 6G, the severe path loss, blockage sensitivity, and hardware constraints at higher frequencies, and the opportunities made possible by highly directional and adaptive antenna solutions.

This chapter has presented a comprehensive review of key antenna technologies for 6G communication systems, including massive and ultra-massive MIMO, multibeam antennas, antenna collocation, reconfigurable antennas, in-band full-duplex systems, and MSF-based architectures such as RIS. In particular, special emphasis has been given to mm-wave and THz antennas, where emerging material platforms, especially graphene-based nanoantennas, provide promising avenues for miniaturization, tunability, and dynamic beam control to meet the stringent requirements of spectral efficiency, spatial multiplexing, and coverage enhancement in future dense and heterogeneous networks. Additionally, 6G antenna systems are viewed as a transformative enabler that can benefit from the integration of AI and ML to provide adaptive, context-aware operation in highly dynamic environments through AI-assisted design, optimization, and real-time beam management, which is particularly important for managing the complexity of ultra-massive MIMO arrays, RIS-assisted propagation, and THz communication links.

Although great strides have been made, there are still challenges in the areas of fabrication scalability, hardware impairments, mutual coupling, SI in full-duplex systems, and the standardization of THz spectrum usage, which will need to be addressed using holistic co-design approaches that jointly consider antennas, materials, RF front ends, and network architectures, along with hybrid electronic–plasmonic antennas, energy-efficient and self-sustainable designs, and AI-native antenna systems. Future 6G antenna technologies are expected to integrate materials, intelligent surfaces, massive array architectures, and data-driven design methodologies, and continued interdisciplinary research and close collaboration between academia, industry, and standardization bodies will be necessary to translate these advances from laboratory demonstrations to robust, scalable, and ubiquitous 6G wireless infrastructure.

REFERENCES

1. Kaur, K., Kumar, S. & Baliyan, A. (2020) 5G: A new era of wireless communication. *International Journal of Information Technology*, 12(3), 619–624.
2. Alraih, S., Shayea, I., Behjati, M., Nordin, R., Abdullah, N.F. & Nandi, D. (2022) Revolution or evolution? Technical requirements and considerations towards 6G mobile communications. *Sensors*, 22(3), 762.

3. Basherlou, H.J., Parchin, N.O. & See, C.H. (2025) Antenna design and optimization for 5G, 6G, and IoT. *Sensors*, 25(5), 1494.
4. Liu, M., Li, M., Liu, R., Liu, Q. & Swindlehurst, A.L. (2025) Reconfigurable antenna arrays: Bridging electromagnetics and signal processing. *arXiv preprint*, arXiv:2510.17113.
5. Dash, S. (2025) *Reconfigurable dual-band terahertz antenna using bilayer graphene for 6G wireless communication*. Indian Patent IN 202531101908.
6. Dash, S. & Patnaik, A. (2024) Advancements in terahertz antenna design. *arXiv preprint*, arXiv:2412.19156.
7. Dash, S., Psomas, C. & Krikidis, I. (2025) Graphene-based beam-reconfigurable liquid antenna for 5G mmWave wireless systems. *Frontiers in Communications and Networks*, 6, 1560311.
8. Dash, S., Liaskos, C., Akyildiz, I.F. & Pitsillides, A. (2020) Nanoantenna design for THz communication: Material selection and performance enhancement. In: *Proceedings of the 7th ACM International Conference on Nanoscale Computing and Communication (NanoCom)*.
9. Dash, S., Soni, G., Patnaik, A., Liaskos, C.K., Pitsillides, A. & Akyildiz, I.F. (2021) Switched-beam graphene plasmonic nanoantenna in the terahertz wave region. *Plasmonics*, 16(1), 1–10.
10. Dash, S., Psomas, C., Krikidis, I., Akyildiz, I.F. & Pitsillides, A. (2022) Active control of THz waves in wireless environments using graphene-based reconfigurable intelligent surfaces. *IEEE Transactions on Antennas and Propagation*, 70(10), 8785–8797.
11. Rahman, A.R. & Kumar, P.M. (2024) Future prospects of 6G wireless communication: Review of antenna technologies. *Discover Electronics*, 1, 36.
12. Dash, S., Psomas, C., Patnaik, A. & Krikidis, I. (2022) An ultra-wideband orthogonal-beam directional graphene-based antenna for THz wireless systems. *Scientific Reports*, 12, 22145.
13. Dash, S. & Patnaik, A. (2018) Graphene plasmonic bowtie antenna for UWB THz applications. In: *Proceedings of the 24th National Conference on Communications (NCC)*, pp. 1–4.
14. Dash, S., Patnaik, A. & Kaushik, B.K. (2019) Performance enhancement of graphene plasmonic nanoantennas for THz communication. *IET Microwaves, Antennas & Propagation*, 13(1), 71–75.
15. Chen, W. et al. (2023) 5G-Advanced toward 6G: Past, present, and future. *IEEE Journal on Selected Areas in Communications*, 41(6), 1592–1619.
16. Gong, T. et al. (2024) Holographic MIMO communications: Theoretical foundations, enabling technologies, and future directions. *IEEE Communications Surveys & Tutorials*, 26(1), 196–257.
17. Guo, Y.J. & Ziolkowski, R.W. (2021) *Advanced antenna array engineering for 6G and beyond wireless communications*. Hoboken, NJ: Wiley.
18. Lu, S. et al. (2024) Integrated sensing and communications: Recent advances and ten open challenges. *IEEE Internet of Things Journal*, 11(11), 19094–19120.
19. Wu, K., Zhang, J.A., Huang, X., Heath, R.W. & Guo, Y.J. (2023) Green joint communications and sensing employing analog multi-beam antenna arrays. *IEEE Communications Magazine*, 61(7), 172–178.
20. Shrivastava, R., Hegde, S. & Blume, O. (2023) Sidelink evolution toward 5GA/6G: Future considerations for group communications. *IEEE Communications Standards Magazine*, 7(1), 24–30.
21. Wigard, J. et al. (2023) Ubiquitous 6G service through non-terrestrial networks. *IEEE Wireless Communications*, 30(6), 12–18.
22. Rappaport, T.S. et al. (2017) Wireless communications: Past, present, and future. *IEEE Communications Magazine*, 55(5), 16–26.
23. Larsson, E.G., Edfors, O., Tufvesson, F. & Marzetta, T.L. (2014) Massive MIMO for next-generation wireless systems. *IEEE Communications Magazine*, 52(2), 186–195.
24. Rappaport, T.S., Sun, S., Mayzus, R., Zhao, H., Azar, Y., Wang, K., Wong, G.N., Schulz, J.K., Samimi, M. & Gutierrez, F. (2013) Millimeter wave mobile communications for 5G cellular: It will work! *IEEE Access*, 1, 335–349.
25. Nagatsuma, T. et al. (2016) Advances in terahertz communications. *Nature Photonics*, 10(6), 371–379.
26. Balanis, C.A. (2016) *Antenna theory: Analysis and design*, 4th ed. Hoboken, NJ: Wiley.
27. Prakash, S., Dash, S. & Patnaik, A. (2019) Reconfigurable circular patch THz antenna using graphene stack-based SIW technique. In: *IEEE Indian Conference on Antennas and Propagation (InCAP)*, pp. 1–3.
28. Low, T. & Avouris, P. (2014) Graphene plasmonics for terahertz to mid-infrared applications. *ACS Nano*, 8(2), 1086–1101.
29. Dash, S. & Patnaik, A. (2017) Sub-wavelength graphene circular patch nanoantenna for UWB THz applications. In: *International Conference on Nanotechnology: Idea, Innovations & Initiatives (ICN:3I)*, India.

30. Dash, S. & Patnaik, A. (2019) Reconfigurability in THz plasmonic antennas using monolayer and bilayer graphene. In: *Graphene Week 2019*, Helsinki, Finland.
31. Akyildiz, I.F., Jornet, J.M. & Han, C. (2014) Terahertz band: Next frontier for wireless communications. *Physical Communication*, 12, 16–32.
32. Han, C. & Akyildiz, I.F. (2015) Distance-aware PHY and MAC protocols for terahertz communication systems. *IEEE Transactions on Communications*, 63(3), 727–742.
33. Liaskos, C. et al. (2019) A novel communication paradigm for programmable indoor wireless environments. *Ad Hoc Networks*, 87, 1–16.
34. Zhang, H. et al. (2019) 6G wireless networks: Vision, requirements, architecture, and key technologies. *IEEE Vehicular Technology Magazine*, 14(3), 28–41.
35. Dash, S. & Patnaik, A. (2021) Advancement in terahertz antenna design and performance. In: *Next Generation Wireless Terahertz Communication Networks*. CRC Press, Taylor & Francis Group, pp. 193–212.
36. Hosseininejad, S.E. et al. (2018) MAC-oriented programmable terahertz PHY via graphene-based Yagi–Uda antennas. In: *IEEE WCNC*, Barcelona, Spain.
37. Dash, S. & Patnaik, A. (2017) A comparative analysis of graphene, CNT, and copper dipole antennas. In: *URSI Regional Conference on Radio Science (URSI-RCRS)*.
38. Dash, S. (2025) *Graphene-based beam-reconfigurable liquid antenna for 6G mmWave wireless systems*. Indian Patent IN 202531005958.
39. Dash, S. (2020) Performance analysis of graphene plasmonic nanoantenna using different substrates at the terahertz band. *ITEE Journal of Information Technology and Electrical Engineering*, 9(2), 1–5.
40. Dash, S., Psomas, C. & Krikidis, I. (2023) Selection of metallic liquid in sub-6 GHz antenna design for 6G networks. *Scientific Reports*, 13, 20551.
41. Dash, S., Liaskos, C., Akyildiz, I.F. & Pitsillides, A. (2020) Graphene hypersurface for manipulation of THz waves. *Materials Science Forum*, 1009, 63–68.
42. Dash, S., Liaskos, C., Akyildiz, I.F. & Pitsillides, A. (2019) Wideband polarization-insensitive reconfigurable graphene metasurface for THz wireless environments. In: *IEEE MTT-S International Microwave Workshop Series*.
43. Dash, S. & Patnaik, A. (2017) Graphene-loaded frequency reconfigurable metal antenna. In: *IEEE International Conference on Antenna Innovations & Modern Technologies*.
44. Dash, S. & Patnaik, A. (2019) Sub-wavelength graphene planar nanoantenna for THz applications. *Materials Today: Proceedings*, 18, 1336–1341.
45. Wu, Q., Zhang, S., Zheng, B., You, C. & Zhang, R. (2021) Intelligent reflecting surface-aided wireless communications: A tutorial. *IEEE Transactions on Communications*, 69(5), 3313–3351.
46. Haque, M.A. et al. (2024) Machine learning-based novel-shaped THz MIMO antenna for future 6G applications. *Scientific Reports*, 14, 32162.
47. Zeng, F., Ding, C. & Guo, Y.J. (2024) Reconfigurable mmWave antenna array with continuous beamwidth control. *IEEE Transactions on Antennas and Propagation*, 72(4), 3092–3103.
48. Monemi, M. et al. (2025) A study on Characterization of near-field sub-regions for phased-array antennas. *IEEE Transactions on Communications*, 73(5), 2964–2979.
49. An, J. et al. (2024) Near-field communications: Research advances, potential, and challenges. *IEEE Wireless Communications*, 31(3), 100–107.
50. Dash, S. & Patnaik, A. (2018) Graphene plasmonic THz antenna for large near-field enhancement. In: *OPTCT 2018*, India.
51. Dash, S., Liaskos, C., Akyildiz, I.F. & Pitsillides, A. (2020) Impact of graphene thickness on electromagnetic modelling of antennas. *Materials Science Forum*, 1009, 75–80.
52. Dash, S. & Patnaik, A. (2017) Dual-band reconfigurable plasmonic antenna using bilayer graphene. In: *IEEE APS/URSI*.
53. Dash, S. & Patnaik, A. (2018) Performance of graphene plasmonic antennas in comparison with counterparts for low-terahertz applications. *Plasmonics*, 13(6), 2353–2360.
54. Roh, W. et al. (2014) Millimeter-wave beamforming as an enabling technology for 5G cellular communications. *IEEE Communications Magazine*, 52(2), 106–113.
55. Huang, X., Guo, Y.J. & Bunton, J. (2010) A hybrid adaptive antenna array. *IEEE Transactions on Wireless Communications*, 9(5), 1770–1779.
56. Holloway, C.L. et al. (2012) An overview of metasurfaces: Physics and applications. *IEEE Antennas and Propagation Magazine*, 54(2), 10–35.
57. Bukhari, S.S., Vardaxoglou, J.C. & Whittow, W.G. (2019) A metasurfaces review: Definitions and applications. *Applied Sciences*, 9(13), 2727.
58. Shlezinger, N. et al. (2021) Dynamic metasurface antennas for 6G extreme massive MIMO communications. *IEEE Wireless Communications*, 28(2), 106–113.

59. Di Renzo, M. et al. (2020) Smart radio environments empowered by reconfigurable intelligent surfaces. *IEEE Journal on Selected Areas in Communications*, 38(11), 2450–2525.
60. Qi, L. et al. (2022) Low-cost beam-reconfigurable directional antennas. In: *Antenna and Array Technologies for Future Wireless Ecosystems*. Wiley.
61. Guo, Y.J. et al. (2021) Quasi-optical multi-beam antenna technologies for B5G and 6G. *IEEE Open Journal of Antennas and Propagation*, 2, 807–830.
62. Wu, Q., Wang, H. & Hong, W. (2020) Multistage collaborative machine learning for antenna modelling. *IEEE Transactions on Antennas and Propagation*, 68(5), 3397–3409.
63. Le, A.T. et al. (2024) In-band full-duplex prototype with joint self-interference cancellation. *IEEE Transactions on Microwave Theory and Techniques*, 72(9), 5540–5549.
64. Yang, X. et al. (2023) Real-time pattern synthesis for large-scale conformal arrays. *IEEE Transactions on Antennas and Propagation*, 71(12), 9559–9570.
65. Cui, C. et al. (2022) Pattern-reconfigurable nonuniform linear arrays using neural networks. *IEEE Transactions on Antennas and Propagation*, 70(6), 4355–4368.
66. Liu, P., Chen, L. & Chen, Z.N. (2022) Deep-learning-enabled synthesis for broadband metacells. *IEEE Transactions on Antennas and Propagation*, 70(7), 5024–5034.
67. Qamar, F. et al. (2024) Massive MIMO for B5G/6G wireless communications. *Information*, 15(8), 442.
68. Ogenyi, F.C. et al. (2025) AI-native 6G communications. *Frontiers in Communications and Networks*, 6, 1655410.
69. Thomas, S. et al. (2025) A Survey on advancements in terahertz technology for 6G: Systems, Circuits, Antennas, and Experiments. *IEEE Open Journal of the Communications Society*, 6, 1998–2016.
70. Raza, A. et al. (2025) TeraRIS NOMA-MIMO communications for 6G and beyond. *arXiv preprint*, arXiv:2508.05130.

Chapter 5

Massive MIMO and beamforming

Neetu Joshi and Pimmy Gandotra

5.1 INTRODUCTION

Wireless communication plays a major role in the growth of today's high-demanding internet and phone services. The development of various generations in wireless communication has led to increased throughput with reduced latencies. Some latest techniques, such as beamforming, massive multiple-input multiple-output (MIMO), and reconfigurable intelligent surfaces (RIS), have become quintessential in this development, particularly in 5G networks and beyond. There are a few critical factors associated with privacy and security that have come along with the implementation of these techniques [1–5]. The signals propagating through space suffer degradation due to multipath fading and interference. There are security concerns in massive MIMO and beamforming that can cause errors in channel estimation. The attacking interface rapidly enhances with increasing network density, executing billions of devices simultaneously. It is required to put in place security measures designed for large MIMO, as these vulnerabilities can affect the confidentiality, availability, and integrity of wireless services [6–8]. Massive MIMO systems allow the widespread acquisition of location and other personal data. This makes the user's privacy a significant problem in addition to security. MIMO networks produce fine-grained user data as they may serve multiple users at one time. It might expose profiles and patterns of behavior if improperly encrypted. Data privacy laws become more stringent globally, resulting in ignorance of user privacy and loss of confidence. Protecting privacy is the top priority in the design of massive MIMO systems [7–10]. As wireless networks continue to develop, maintaining user privacy and guaranteeing strong security have become urgent requirements. Wireless network threats include network attacks, malicious data breaches, and unauthorized access. Proactive measures adapted to new technologies and attack vectors are necessary due to the expansion of cybersecurity vulnerabilities in order to improve the security of upcoming wireless communication systems. The widespread use of wireless networks and extensive data collection has led to worries about the privacy of personal data.

Various privacy-enhancing technologies (PETs) and protocols are investigated to serve multiple users. The comprehensive integration of security and privacy is required to create reliable communication environments in 6G networks and beyond. This integration requires careful consideration of the potential and difficulties attached to it. This chapter discusses the status of massive MIMO systems, their advantages, and potential in detail. Various beamforming techniques along with RIS surfaces are explained. In the last section, precoding techniques are described [11–14]. The main advantages and characteristics of massive MIMO, its current uses and implementations, and an analysis of the risks and weaknesses inherent in massive MIMO systems are covered in this section. Massive MIMO, a state-of-the-art technology in wireless communication networks, is used to increase spectral efficiency and capacity. It uses several antennas at the base station (BS). Standard MIMO

DOI: 10.1201/9781003637721-5

systems use a few antennas, whereas massive MIMO uses dozens or even hundreds of antennas to serve several customers at once. It effectively regulates interference, beam form, and spatial multiplexing, which raises data rates, enhances signal quality, and expands network capacity. Since the advent of massive MIMO technology, the performance and capacity of wireless communication networks have significantly increased. The wireless networks developed from 3G to the upcoming 6G with these innovations [15–17]. Wireless communication systems have advanced significantly from 3G to the next generation of 6G networks [18–20]. New features, technologies, and capabilities have been introduced by each and every iteration to satisfy the increasing demands of mobile communication.

5.2 MASSIVE MIMO FUNDAMENTALS

Massive MIMO is a major force behind the development of wireless communications. It is a crucial part of pervasive 5G networks, and helps meet the needs for highly reliable, minimal latency, and expanded mobile broadband communication. Step-change increases in network capacity, connection dependability, and spectrum efficiency are made possible by the technology. Thomas Marzetta, the innovator of massive MIMO, claims that by expanding the number of service antennas without restrictions, massive MIMO systems might easily and power-efficiently approach theoretical capacity limits [21–25].

Several other important 5G technologies can function at scale by virtue of massive MIMO. This covers ultra-dense small cell deployments, large machine-type connections, and millimeter wave communication. Massive MIMO is a key enabler for 5G and beyond because of its sophisticated spatial multiplexing and interference management capabilities. Massive MIMO has undoubtedly played a crucial role in the development of the fifth generation of wireless networks. New incentives, however, are anticipated to spur additional development and progress in large MIMO capabilities for 6G and beyond [26, 27].

5.2.1 Advantages of massive MIMO technology

Massive MIMO has several advantages. The following are some of the main reasons for anticipating huge MIMO technology in upcoming wireless systems:

i. *Pervasive connectivity and improved coverage:* The technology of 6G is expected to create a connection between rural and urban locations by offering extensive coverage. Massive distributed MIMO has the potential to expand coverage to the most remote regions that aren't yet accessible by conventional terrestrial networks by combining space-based and terrestrial technologies.
ii. *Exuberant spike in data requirements:* New technologies such as holographic communication, streaming HD video, and sophisticated virtual and augmented reality applications are expected to drive an exponential rise in data demand, which 6G is expected to support. To meet these high data rate needs, massive MIMO can effectively use the available spectrum and manage a large number of spatial streams.
iii. *Extremely minimal latencies and ultra-high credibility:* The technology of 6G seeks to provide these characteristics in order to facilitate crucial uses such as industrial automation, remote surgery, and driverless cars.
iv. *Blended communication and sensing*: In the 6G era, communication and sensing could be combined to create wireless networks that are more intelligent and contextually aware. Real-time environmental monitoring and other intelligent wireless applications can be supported by advanced massive MIMO techniques that allow for combined

communication and sensing. Additionally, it also provides sustainable development, energy savings, broad bandwidths, IoT extensibility, and link distribution. It allows the usage of vertical commercial enterprises with artificial intelligence (AI)-aided massive MIMO optimization methods. There is a wide range of antennas used in networks for communication systems, like reconfigurable antenna, massive MIMO antennas, beam-steering antennas, meta-surface antennas, antennas for IoT applications, and RIS.

For 6G, massive MIMO technology is essential, particularly in the 28 GHz FR-2 band. The drawback of having a narrow antenna beam is that it necessitates the use of several antennas, such as array antennas. Many antenna elements are frequently seen in arrays. This results in the antenna being proportionately larger. Additionally, digital signal processing is necessary for both tracking a user with a narrow beam and using MIMO [6] methods. This, along with the need to communicate extra channel information (to enable channel adoption), negates some of the power consumption advantages that narrow beams provide. By increasing the throughput according to the number of antennas, MIMO systems aid in the utilization of spatial diversity.

5.2.2 Potential of massive MIMO technology

Massive MIMO technology holds great potential for applications in future wireless communication systems. A few of these potential applications of massive MIMO include the following: massive MIMO improving signal quality, massive MIMO improving spectral efficiency, massive MIMO supporting multiple users and dense devices, and massive MIMO enabling a wider coverage. By using numerous antennas, the designed large-scale MIMO systems are able to mitigate the impacts of fading and interference, which enhances signal quality. Advanced signal processing methods like beamforming and spatial multiplexing are supported by massive MIMO systems, which improve network performance overall by lowering interference and improving signal quality. Moreover, it enhances spectral efficiency through massive MIMO as there are multiple antennas at the BS, allowing the simultaneous transmission and reception of multiple data streams.

Massive MIMO provides higher data rates and increased network capacity with a notable increase in spectral efficiency. The users can be served simultaneously by massive MIMO systems, which improve user experience, increase network capacity, and allow for more effective resource allocation. It provides wider coverage. It precisely shapes the radiation pattern to focus beams. Massive MIMO can improve wireless network coverage in remote locations. It increases signal strength and coverage by focusing the signal on particular users or areas, enabling dependable communications. It can be used for non-terrestrial wireless communication systems. The deployment of very high-throughput satellites in Low Earth Orbit (LEO) and Geostationary Earth Orbit (GEO) satellite constellations relates to a paradigm shift in global connectivity. The low-latency, high-speed data transfer capabilities are promised by future wireless communication networks like 6G and beyond, bringing in a new era of applications ranging from cutting-edge remote sensing to ubiquitous internet access. A wide range of applications, including huge machine-type communications, will be supported by satellites, which are positioned as essential enablers for the upcoming generation of wireless networks. All of these applications are emphasized by the general requirement for secure connectivity. Standardized and interoperable communication protocols are becoming more and more necessary as the requirement for connectivity outside of Earth's borders increases. It should be highlighted, nonetheless, that maintaining a reliable and secure communications system while coordinating the complex interactions of hundreds or perhaps thousands of satellites would present significant hurdles.

Massive MIMO makes the assumption that the number of transmit antennas (m) on the non-terrestrial platform will always exceed that of single receive antennas (k) that are terrestrial, maritime, or aeronautical. Non-terrestrial networks will serve user equipments (UEs) in remote locations with low user density, a sparse user population, and relatively high latency in the fully converged future wireless communication systems that are envisioned. Significant propagation delay and comparatively high and unpredictable Doppler shifts and spreads are anticipated in the channel propagation characteristics of non-terrestrial networks [20–25]. It is necessary to use suitable signal processing to sufficiently compensate for these channel obstructions.

5.3 PRINCIPLES, TYPES, AND TECHNIQUES OF BEAMFORMING IN MASSIVE MIMO

Power-hungry digital beamforming with somewhat less functionality is the current research direction in hybrid beamforming techniques. Some studies attempt to combine massive MIMO and hybrid beamforming. Antennas serve as both spatial and frequency filters in order to accommodate multiple users, and they may also help achieve the frequency selectivity that cognitive radio systems require. These systems, like television white spaces (TVWS), are permitted to function in a relatively broad frequency range. Current generations, such as 5G cellular and Wi-Fi, have begun to use multiple antennas at mm waves, which are typically designated for frequencies between 30 and 300 GHz. Currently, attempts are being made to use full-duplex (FD) techniques in millimeter-wave communications.

Wireless communication has begun to emerge as one of the essentials of the modern world. It has resulted in several new technologies that now control our daily lives, including deep space research, telemedicine, connected and driverless cars, the Internet of Things (IoT), and so forth. Since the advent of these new technologies, there is a growing demand for high bandwidth and fast data rates because of the quickly increasing number of devices connected through public wireless networks. There is a lot of interest in employing sub-millimeter and millimeter-wave bands (3–300 GHz) since wireless networks have hit their limit and there is not enough bandwidth for future advancements [26, 27]. The mm-wave spectrum that isn't being used can help with bandwidth shortages and accommodate the increasing user base. However, the mm-wave band has a short wavelength (1–100 mm) and faces a number of problems, such as multipath delay, vegetation loss, high path loss, low diffraction, atmospheric losses, and rain attenuation. For instance, research found that mm-wave bands can exhibit attenuation of up to 6 dB/km with a strong downpour of 25 mm/h at 30 GHz, which has a major effect on the world's tropical regions [27].

A signal processing technique called beamforming creates a highly directional beam by concentrating the power emitted from a source in a single direction [28, 29], shown in Figure 5.1 (a) and (b). The problems of path loss, fading, and multipath delay can be solved by this highly directional beam. Antenna arrays work in coherence to provide constructive interference with high signal strength only in a specific direction, termed a beam. Other direction patterns deliver radiation nulls. Beamforming-enabled BSs with a 200 m radius (thus forming picocells) can be used for and beyond 5G and 6G networks, according to a number of field studies [30–33]. Such concentrated beams can have their direction controlled and guided by a reconfigurable antenna system. Beam steering antennas are the name given to such antennas. In the realm of antennas and propagation, beam steering antennas have emerged as a very appealing research topic for use in satellite communications, aerospace, defense, and not only 5G/6G networks.

Beamforming and beam steering can be accomplished most easily with phased-array antennas. To comprehend a phased-array antenna, consider an N-slit diffraction experiment. In this experiment, the total beam from the array is the consequence of coherent

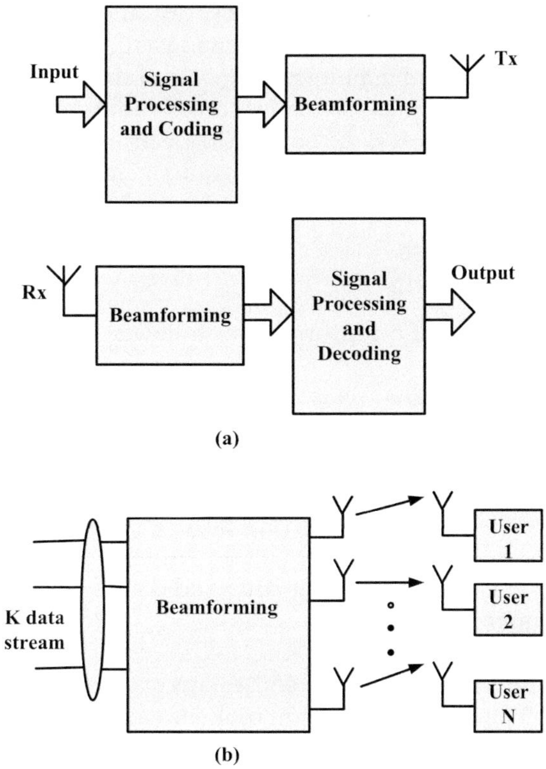

Figure 5.1 Beamforming in (a) MIMO and (b) massive MIMO.

constructive interference from N sources. The array introduces a temporal delay in the radiated beam by feeding each antenna element with a phase shift. Time-delayed beams from each radiating unit interfere constructively, allowing for directional control. The array factor can be used to estimate the tilt angle and radiation pattern for such antennas. There are numerous examples of these phased-array antennas in the literature. The stacked array antenna is among the most widely used designs. Although it can produce extremely directional, skewed beams, reconfiguring it is challenging. Various methods for gain and directivity enhancements are commonly found in the literature [34–37].

A number of novel methods were investigated in order to accomplish a reconfigurable phase shift among the radiating elements, as phased arrays have the potential to produce a beam tilt. Offering a separate phase shifter to every radiating element is a straightforward technique, but the result is a very large and expensive system. Various feeding networks can also produce these phase shifts in order to implement a beam steering (or beam switching) design. The Butler [38] and Nolen [39] matrices are two examples of such well-liked structures. Through various phase distributions, an N-elemental array can be controlled by a standard N × N Butler matrix, which can also produce N orthogonal beams. Shifters of 45° phase and a 30-dB quadrature coupler are used to create the most basic version of a Butler matrix [40, 41]. Wide beam steering is possible with such matrices since they may be tuned for considerably bigger antenna arrays. A phased-array Butler matrix antenna is employed with eight programmable phase shifters [38].

Nolen matrices, as opposed to Butler matrices, employ external phase shifters in addition to couplers with various coefficients [35]. They have proven to be able to create 2D beamforming and multibeam antenna arrays [28, 36] and [37]. Such matrices can offer considerable flexibility and are simple to design and install, but they require a big form factor, which raises the total size and manufacturing cost of the antenna system. Over the past few decades, phased-array antennas have advanced significantly and are currently widely employed for a variety of purposes. Although they can offer wide beam guiding with a tiny form factor, their high power consumption requirements and high manufacturing costs make them unsuitable for system-scale deployment.

5.3.1 Types of beamforming

There are basically three types of beamforming as follows:

Analog beamforming: The signal phase is adjusted before transmission. It uses a single beam at a time. It allows the usage of phase shifters and RF components and has applications in 4G/5G BSs.

Digital beamforming: Signal phase is adjusted in baseband. It uses multiple beams. It utilizes signal processing algorithms that have applications in applications: massive MIMO systems

Hybrid beamforming: Combination of analog and digital beamforming and application in mm-wave communication

The beamforming structures' spectrum and energy efficiency were further examined in studies in [26]. Additionally, this investigation took into account the link between the energy efficiency and spectrum efficiency curves with the highest energy efficiency point. Additionally, examined were the effects on energy efficiency and the performance of energy efficiency for a specific spectrum efficiency rating. In [26], the authors also provided a reference signal design for the hybrid beamforming structure. When it comes to channel estimation, the beamforming structure outperforms the method that just relies on an analog beamformer.

When a delay-sensitive broadcast is needed, the authors in [27] based their plan on the results of comparing the delays of a radio's access points (AP). Due to the utilization of a spectrum broker, this strategy provides an ideal TVWS. In a 5G communication network, the spectrum broker is used to control energy consumption while accounting for the evaluation of queue delays and the delay-aware method. The investigations [22] concentrated on minimizing the effects of interference on the signal while also sending a powerful beamforming signal in the direction of the intended users. An antenna that can send equal beamforming signals in all directions is needed for the suggested architecture. To provide precise beamforming, the antenna must also have an increased gain and be omnidirectional.

5.3.2 Techniques of beamforming

There are various techniques for beamforming, broadly classified as follows:

Fixed beamforming: The beam direction is fixed, and it uses phase antenna arrays.

Adaptive beamforming: The beam direction changes dynamically with user requirements. It uses smart antenna systems.

Switched beamforming: The beam switching takes place depending on signal strength.

Adaptive beamforming has shown clear industrial importance recently, as seen in the use of contemporary wireless communication systems. An overview of the use of adaptive beamforming in cellular systems was given by studies in [28]. This study went on to say that adaptive beamforming, which takes into account the Nuevo 5G technology, is a breakthrough in cutting-edge technologies like long-term evolution (LTE). The importance of large and adaptive beamforming in 5G was covered in more detail in [28], which also claimed that 5G could support ever-increasing network services, including mobile broadband and the IoT [42–48].

5.4 CHANNEL ESTIMATION AND PRECODING TECHNIQUES

The process of determining the channel conditions between the transmitter and receiver is defined as Channel Estimation. It is a critical function in a wireless communication system, as a wireless communication signal introduces effects like multipath fading, interference, delay, and noise. These need to be critically addressed for efficient and reliable communication [49].

5.4.1 Importance of channel estimation

The phenomenon of channel estimation is essential in a wireless communication transmission for the following reasons:

- For the correct decoding of the signals
- In order to support adaptive modulation and coding (AMC)
- To inculcate handover abilities and mobility management
- For performing beamforming (especially in massive MIMO systems) and precoding

Methods of channel estimation, adopted in 4G, 5G, include the use of demodulation reference signals (DRS) and cell-specific reference signals (CRS). The user here utilizes the known pilot symbols for channel estimation and equalization of the received signal. While 4G systems mostly use Frequency Division Multiplexing (FDD) and Channel State Information (CSI) is fed back to the BS, 5G systems utilize TDD, where the channel reciprocity allows estimation of the DL channel by the BS via the UL SRS.

5.4.2 Precoding techniques

Precoding techniques are applied in wireless communication networks to shape the signals being transmitted, acquiring knowledge of the wireless channel. The process of precoding occurs at the transmitter end as shown in Figure 5.2. It is an effective measure to reduce interference. The primary objectives are to precode the signals, focusing on interference reduction and maximization of the signal quality at the receiver, while improving throughput, reliability, and spectral efficiency of the system. These techniques are vital in massive MIMO systems, since the complexity of the massive MIMO system is expressed through the same techniques.

Once the channel characteristics of the channel are exploited, precoding algorithms directly mitigate of anomalies like co-channel interference and multipath fading, while enhancing the spectral efficiency. The schemes used for precoding massive MIMO are zero forcing (ZF), maximum ratio combining (MRC), matched filtering, peak to average power precoding, and Minimum Mean Square Error (MMSE). Although linear precoders at times endure performance degradation, they are reasonably simple, making them a competent choice for designing transmitters. However, non-linear precoders possess high

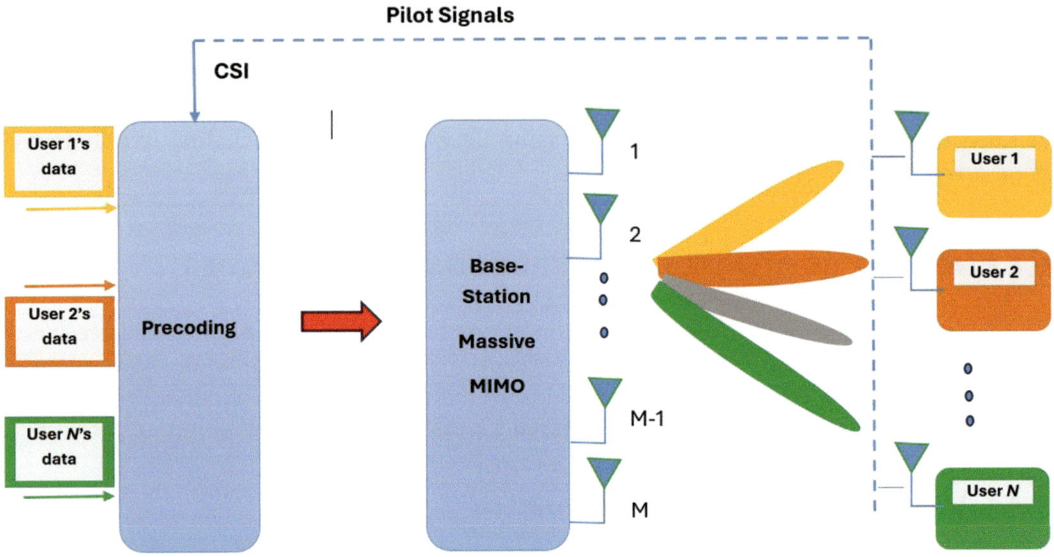

Figure 5.2 Fundamental precoder system.

computational complexity while achieving an acceptable performance. Precoding in wireless communication networks comprises a sequence of steps, as stated below:

i. *Estimation of CSI:* Preceding the precoding, the transmitter necessitates accurate information on the state of the wireless channel. By sending pilot signals or via training sequences from receiver to transmitter, CSI is acquired. Channel response is estimated by the transmitter on the basis of the received pilot signals.
ii. *Calculation of the precoding matrix:* Upon CSI estimation, the transmitter calculates the precoding matrix. This matrix is designed for the transformation of data symbols to specific spatial patterns.

5.4.2.1 Linear precoding schemes

The simplest precoding scheme is single-user precoding, where the signal is focused on the intended user, while knowing the channel conditions [50]. This section discusses the various linear precoding schemes.

5.4.2.1.1 Zero-Forcing (ZF) precoding

A spatial signal processing approach for multi-user interference elimination is ZF or null steering. If a channel's current condition is known to a transmitter prior to broadcasting, pseudo-inverse of the channel matrix is used by it. For an access point with 'N' transmitters, and 'M' users having a single receive antenna, forming a multi-antenna downlink system, for $M \leq N$, a user 'u' will have received signal given as

$$y_u = h_u^T x + n_u \tag{5.1}$$

where $u = 1,2,3,\ldots,M$, and transmitted symbols 'x' are represented as

$$x = \sum_{i=1}^{M} \sqrt{P_i} S_i W_i \tag{5.2}$$

n_u represents the noise signal, h_u^T represents the channel vector, and W_i represents the linear precoding vector, P_i represents the transmit power, S_i represents the message signal with mean '0' and variance '1'. As a concise representation, the signal model can be represented in vector form.

In case of a large user count in a network, with a transmitter having widespread knowledge of downlink CSI, ZF-precoding can aid the system to operate at almost the full capacity. In case the transmitter's CSI is inadequate/limited, the performance for ZF precoding decreases. This decrease in performance is proportional to the CSI at the transmitter. ZF requires a large amount of feedback overhead, which helps it achieve the maximum multiplexing gain (in terms of SNR). A considerable amount of multi-user interferences in the network result in throughput degradation. With a poor CSI at the transmitter side, the interference persists in the network and cannot be eliminated. These challenges are overcome with the use of ZF precoding, to some extent.

5.4.2.1.2 Maximum ratio combining (MRC) precoding

MRC precoding combines multiple received signals for achieving better performance. For a given channel, the gains of channels are combined such that they are inversely proportional to the mean square noise level for it. Also, a unique set of proportionality constants are assigned to individual channels. When numerous Gaussian noise channels are combined, MRC turns out to be the most effective approach. Suppose the receiver has M antennas. When small-scale fading occurs, a channel having an impulse response with variance ∂^2, the instantaneous power, exponentially dispersed, is expressed as

$$P^2 = \frac{1}{\partial^2}\exp-\left(\partial^2/h^2\right) \tag{5.3}$$

MRC beamformer is expressed as

$$w = \frac{h}{\|h\|} \tag{4.4}$$

5.4.2.1.3 MMSE precoding

In a multi-cell massive MIMO system, it is at times tedious to allocate the training sequence. This must be critically addressed in the case of multi-cell massive MIMO systems. MMSE precoding works efficiently in such systems by reducing the pilot contamination. It aims to strike a balance between amplification of noise and interference cancellation. The mean square error minimization is practiced between the desired signal and the signal received, using the information from the channel. This results in a good trade-off between mitigation of interference and enhancement of noise.

5.4.2.1.4 Matched filter precoding (maximum ratio transmission (MRT))

Modern MIMO systems deploy matched filter detectors, the simplest fundamental linear detector. It maximizes the power of the received signal by aligning the transmitted signal in the direction of the channel vector. The channel precoding vector is described as

$$w = \frac{h^*}{\|h\|} \tag{5.5}$$

This ensures a constructive addition at the receiver, thereby enhancing SNR. In the case of massive MIMO systems, these systems scale well and perform close to optimally due to channel hardening. MRT, however, does not have the capability to suppress inter-user interference.

5.4.2.1.5 Peak-to-average power ratio (PAPR) precoding

Use of linear power amplifiers in mMIMO systems, incorporating multiple antennas in the system, increases the hardware costs along with higher energy consumption. As a result, this precoding technique aims at reducing the impact of power amplifier non-linearity. For the same reason, the PAPR is kept minimum, as high PAPR values cause distortion in the in-band and spectral spreading. PAPR issues are resolved using techniques like filtering, amplitude clipping, coding, and transmitting sequences partially.

5.4.2.1.6 Continuous envelope (CE) precoding

Another scheme, which reduces the PAPR for broadcasting signals, is the continuous envelope (CE) precoding. To meet the objectives of this precoding scheme, it uses low-cost amplifiers that are readily available and possess a high efficiency. Under certain channel conditions, there is, however, a probability of the CE precoding scheme increasing the array power.

5.4.2.1.7 Quantized precoding

Weighing the advantages and disadvantages of a system with multiple antennas is critical. There is a requirement for restricting the system, so as to avoid any inefficacies. Due to the inclusion of multiple RF chains, the BSs nowadays have become extensively power hungry, complicated, and costly. To reduce the complexity of the hardware, power consumption, and the requirements for data transfer, low-resolution digital representations (quantized values) are proposed to be used in quantized precoding. This type of precoding is digitally applied, and the output is first quantized and then sent to the RF chains. This is an important type of precoding for enabling practical implementations and supporting features like beamforming and spatial multiplexing.

5.4.2.2 Non-linear precoding

These methods are deployed for actively cancelling interference in the network, with the use of advanced signal processing methods. The non-linear precoding is useful in cases of a smaller number of users and a smaller number of antennas in the system. Dirty paper coding (DPC), vector perturbation (VP) precoding, Tomlinson–Harashima precoding (THP), along with auxiliary network approaches are commonly used methods that are categorized under non-linear precoding. Other techniques include sphere decoding (SD), which incorporates the use of a maximum likelihood detector. These methods are complex, in comparison to linear methods; however, they improve the system performance significantly. They have limited use due to the complexity involved in their usage and scalability challenges. These may find a significant place in the AI-RAN and future 6G networks.

5.5 RECONFIGURABLE INTELLIGENT SURFACES IN MIMO

Downlink transmissions are supported by RIS [51–53] by utilizing the RIS structures along with the BS and the UE, as is depicted in Figure 5.3.

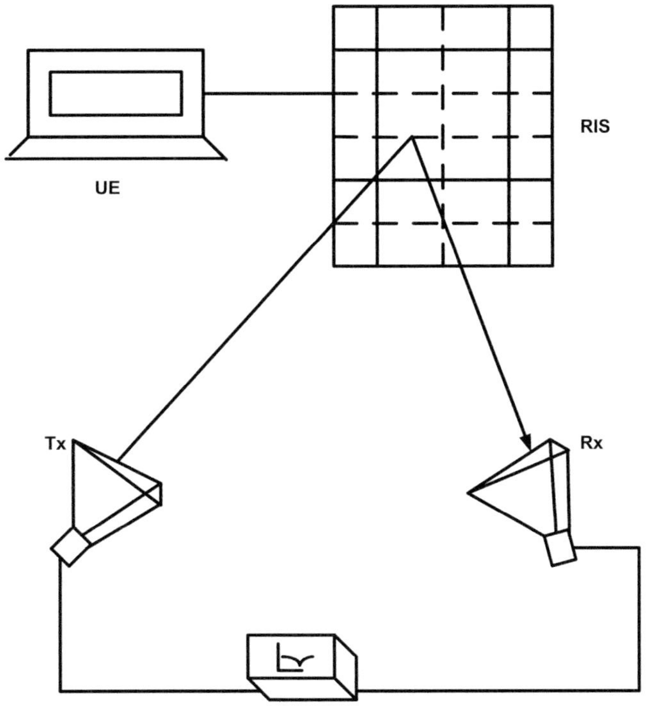

Figure 5.3 Reconfigurable intelligent surface (RIS).

A 2D (two-dimensional) surface, comprising reconfigurable elements (mostly small meta-surfaces), comprises the RIS. These can be instructed to transmit, reflect, or diffract waves on the basis of the communication system requirements. It also exhibits the capability to optimize the environment for signal propagation. The designs for RIS can be simulated by using High Frequency Structure Simulator (HFSS). This also assists in high-frequency simulations (including THz, millimeter waves). Unit cells comprise the RIS surface, performing as miniature reconfigurable elements for optimization, as per the precise requirements of the network communication. These interact with the EM waves, which are also critical for designing efficient meta-surfaces. Further to optimize the RIS performance, various parameters can be assessed in a real-world scenario, including radiation patterns, S-parameters, and signal propagation [54–56].

5.5.1 RIS antenna

The depiction of RIS in Figure 5.4 aims to improve the overall antenna performance with respect to the parameters listed below:

1. *Coverage extension*: Reflection of signals near/around the obstacles extends wireless network coverage. This is specifically effective in urban environments with dense building structures.
2. *Beamforming*: Highly directional beams can be generated by RIS, capable of focusing signals for a desirable receiver, thereby improving the strength of the signals received along with interference reduction.
3. *Diversity gain*: RIS can create multiple signal paths and improve the reliability of the connection, particularly in fading environments.

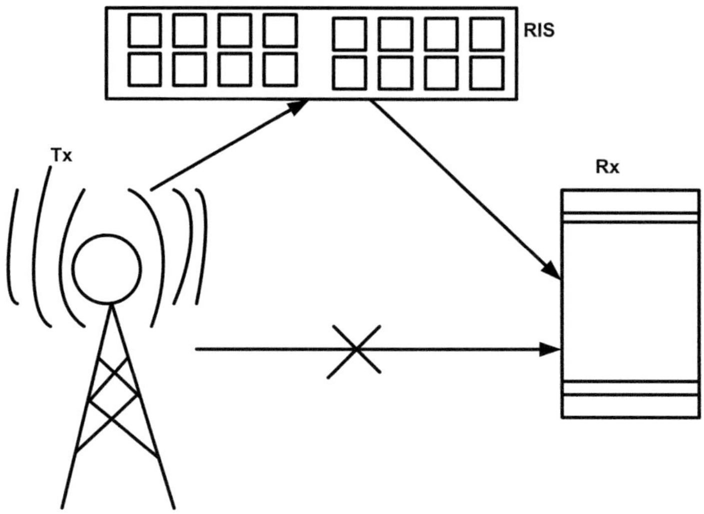

Figure 5.4 Development cycle of MIMO techniques.

4. *Interference mitigation*: RIS can be used to null out interfering signals, improving the signal-to-interference and noise ratio.

RIS improves signal quality, coverage, and energy efficiency. The incoming signal is directed toward the receiver by tuning the phase of the reflecting elements. The reflecting elements can be passive, which only perform phase tuning, or active, which can be used for beam steering and modulation.

5.5.2 Components and mechanism of RIS

Meta-material elements (unit cells), typically small reconfigurable antennas or scatterers.
Control mechanisms (e.g., phase shifters) to modify the wave's behavior.

1. *Incoming signal*: A signal is transmitted from the source (e.g., a BS).
2. *RIS interaction*: The signal impinges upon the RIS, which is composed of numerous small, programmable reflecting elements.
3. *Signal manipulation*: Each element in the RIS can independently adjust the phase and amplitude of the reflected signal.
4. *Reflected signal*: The RIS reflects the manipulated signal toward the receiver.
5. *Constructive interference*: The phase shifts of the reflecting elements result in constructive interference at the receiver, resulting in a stronger and more reliable signal.

Here, the phase changes can be modeled to interact with RIS, aiding in beam-forming and directionality. The unit cell in RIS is a reconfigurable meta-surface, which could be a microstrip patch antenna or a reconfigurable meta-surface with controllable phases. A large number of unit cells form the RIS surface. Its size can be defined to cover a specific coverage area. Next, the designs can be analyzed to obtain the desired radiation pattern and beam-forming gain.

5.5.3 RIS for modulation

RIS-based systems can suffer from multiplicative fading. Use of active elements helps in mitigating multiplicative fading. An active element amplifies signals in addition to phase tuning. Two types of modulation are possible.

Hybrid reflection modulation (HRM): It uses a combination of active and passive elements. It is not only used to mitigate multiplicative fading but also to encode information through indexing.

Media-based modulation (MBM): It uses passive elements as RF mirrors that are controlled by information bits. It exploits the reflection characteristics of RF mirrors to create a modulation alphabet.

RIS can be used for beamforming. This beamforming happens in the medium, not at the transmitter or receiver. HRM modulation utilizes the principle of index modulation. N elements in the space are partitioned into G tiles, each tile having S reflecting elements. Each tile can be either active or passive. The number of active tiles varies depending on the information bits. Each bit sequence corresponds to a unique number of active tiles. For example, consider an interspace having N = 32 elements symmetrically partitioned into G = 4 tiles, where $W_{hrm} = \{w_1, w_2, \ldots w_G\}$ constitute the HRM signal set. MBM modulation offers MIMO benefits at reduced RF hardware complexity and improved performance due to the good distance properties of the MBM alphabet. RF mirrors are switched ON/OFF based on the mirror activation pattern MAP/MBM bits to create a different propagation environment. It can be used for Radar sensing to transmit a radar sensing signal. It uses receiver echo to estimate the delay and Doppler of targets in the scene. The RADAR scene is completely described by the DD spreading function of 'echo'. The DD spreading function has peaks at the target locations in the DD domain. A good radar waveform must be localized well in the DD domain. It can be used for sensing the vehicular environment. TD pulses can resolve well along delay (not along Doppler). FD pulses can resolve well along Doppler (not along delay). DD pulses can resolve well along delay and Doppler.

5.6 MASSIVE MIMO IN 5G AND BEYOND

For 5G networks, the evolution of massive MIMO has marked a pivotal advancement. Deploying several antennas that transmit and receive simultaneously through MIMO technology increases spectral efficiency and data throughput. Massive MIMO was created in response to the unquenchable need for greater capacity and faster data speeds. With the installation of an exceptional number of antennas at the BSs, this pioneering method supports multiple parallel connections, along with superior spatial multiplexing. A critical part of 5G networks, the 'Large MIMO', is a transformative technology aiding in satisfying the demand and enhancing connectivity [15]. The past two decades have witnessed a pioneering phase in the evolution of antenna technology. MIMO techniques adoption is an integral driver in the evolution of the next generation of wireless communication networks. While employing two to four antenna elements in the initial phase, MIMO has evolved to large-scale MIMO arrays, deploying hundreds of antennas. MIMO has been supporting improvements in link reliability, spectral efficiency, and capacity of networks.

A major leap toward MIMO development is massive MIMO. Radical expansion in the count of antennas for multi-user MIMO systems aids in unlocking substantial gains in terms of throughput, spectral efficiency, and energy efficiency, compared to the traditional small-scale MIMO systems. The 'massive' scaling of arrays of antennas distinguishes this recent

MIMO development and supports unmatched capabilities in terms of spatial multiplexing. Initial research in the field of MIMO mostly focused on point-to-point applications. These used two to four antennas on the radio while servicing a small number of users and achieving two to three times the gains over the antenna systems with a single antenna. Capacity gains were, however, restricted as a result of inter-user interference in Multi-user MIMO, which has emerged as a novel research area in the 2000s. These paved the way for massive MIMO in the 2010s, intending 10× or higher capacities over the conventional MU-MIMO. This is achieved by augmenting the antenna arrays to simultaneously serve tens of users. Ground-breaking research by Marzetta has shown that adding to the service antenna count without any bounds aids the massive MIMO approach to the theoretical limits of capacity in an elementary and power-effective way. This has established massive MIMO in meeting the required demands for 4G and 5G networks [31]. Massive MIMO has been a fundamental technology for enabling the full-scale roll-out of 5G networks over the past few years. The capabilities unlocked by the massive MIMO technology have been critical for achieving the 5G performance targets like data rates of 1–10 Gbps, 100 times improvement in traffic capacity, >100 Mbps user throughput, along with 5–10 times gains in spectral efficiency.

With the exploitation of the spatial domain in massive MIMO, large antenna arrays with more than 100–200 elements simultaneously serve many tens of users using the same resource (in terms of time and frequency). Multiplexing in the spatial domain improves the network capacity, spectral efficiency, and per-user throughput. Additionally, there is an enhancement in the energy efficiency and reliability in the links, for concentrating signals into narrow beams, using precoding schemes and beamforming. Massive MIMO also plays a pivotal role in enabling several key 5G technologies for a comprehensive 5G operation. These technologies include ultra-dense small cell deployments, massive machine type connectivity, millimeter wave communication, and others. The advanced capabilities for interference management and spatial multiplexing accomplish massive MIMO as a universal enabler for 5G wireless communication networks. While being a foundational technology for 5G wireless communication networks, massive MIMO also drives further innovation and advances 6G and beyond capabilities. The timelines of evolution of the MIMO technology are depicted in Figure 5.5.

The key motivations for advancing toward massive MIMO, with special focus on the 6G timeframe, are listed here.

1. *Exponential advancement in demand for data*: The technology of 6G is expected to accommodate the exponential surge in the data demand, as a result of the emerging technologies like holographic communication, 8K/16K video streaming, and advanced virtual and augmented reality applications. The competence of the massive MIMO technology to manage the substantial number of spatial streams and effectively harness the available spectrum will be critical in achieving the requirements for high data rates.
2. *Higher-frequency bands, wider bandwidths*: The technology of 6G intends to investigate higher frequency bands, like sub-terahertz and terahertz, leveraging wider bandwidths. These can achieve even higher data rates with massive bandwidths.

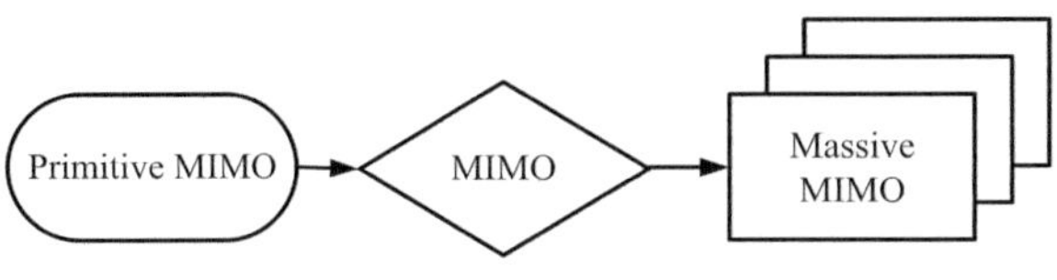

Figure 5.5 Illustration of RIS-assisted communication system.

Large antenna arrays in the massive MIMO system can benefit from the intensified bandwidths available and address the hurdles introduced due to higher frequency propagation.

3. *Density in connections and IoT scalability*: The technology of 6G envisages a massive count of connected devices, including IoT devices and sensors. A scalable solution is offered by massive MIMO in managing the unprecedented density of connections. These enable seamless connectivity, along with resource allocation for the vast device count, per unit area.
4. *Diversity in applications for various verticals*: The technology of 6G is expected to perform exceptionally in a wide range of vertical industries, each having a distinctive communication requirement. Applications like wireless sensing, robotics, and industrial automation shall be demanding substantially reliable, low-latency communication. The adaptive beamforming and interference management capabilities of massive MIMO can cater to these diverse application needs.
5. *Integrated communication and sensing*: The technology of 6G pursues integration of communication and sensing capabilities, thereby enabling additionally intelligent and context-aware networks. Advanced massive MIMO techniques can accelerate joint communication and sensing, providing real-time environmental awareness.
6. *Ultra-high reliability and low latency*: The ability of massive MIMO technology to decrease the latency through advanced beamforming and interference management can help meet the rigorous requirements for the diverse applications.
7. *Ubiquitous coverage with connectivity*: The technology of 6G seeks to provide ubiquitous coverage, even in remote and challenging terrains/areas. Massive-distributed MIMO, integrating space-based and terrestrial systems, can expand coverage to areas presently underserved by the conventional terrestrial networks.
8. *Energy efficiency and sustainability*: With the scaling of communication networks, energy efficiency in the networks becomes paramount. Advanced massive MIMO techniques can help in improving energy efficiency either by optimizing the

Table 5.1 Features, technologies, and benefits of massive MIMO

Description	*Technologies*	*Benefits*
Exponential increase in data-driven technologies	Holographic communication	Massive MIMO utilize large number of streams and utilize spectrum efficiently
High frequency and huge bandwidth	Millimeter wave and terahertz wave communication	Use of arrays of massive MIMO
Increased connections with scalability	IoT and sensors	Massive MIMO enables scalability to allow large number of connections
Adaptive beamforming and interference mitigation	Industrial IoT	Massive MIMO provides diverse vertical applications
Communication with sensing	Radar sensing and high frequency communication	Massive MIMO allows integrated communication and sensing
Low latency and high reliability	Remote healthcare, Autonomous vehicles	Massive MIMO provides low latency along with high reliability
Large coverage	LEO orbit satellite communication	Massive MIMO provide extensive and 360° coverage
Power optimization and interference reduction	Green communication	Massive MIMO improves system efficiency
Use of AI/ML	Wireless communication with AI/ML	Massive MIMO adapts communication with AI

transmission power or reducing interference levels, and/or enabling spectrum allocation dynamically.

9. *AI-driven optimization in MIMO*: The MIMO technology has the capability to optimize allocated resources, achieve gains with beamforming, and manage interference in a dynamic way. AI-driven MIMO optimization can bring about further gains in performance, capacity, and adaptability in 6G networks. Table 5.1 shows the benefits and underlying technologies with distinctive features of massive MIMO.

5.7 CONCLUSION

This chapter talks about the massive MIMO systems with beamforming techniques. The fundamentals of MIMO systems are discussed, with a focus on the potential and advantages of MIMO technology. Massive MIMO features, specific to 6G and beyond, have been highlighted to gain insight into the technology. Channel estimation has been discussed, along with various precoding schemes to get a better insight into the overall system performance in MIMO and massive MIMO systems. Although these technologies provide several advantages for the present and next-gen networks, certain challenges still persist in the networks, which can be addressed to enhance the potential of the technology.

REFERENCES

1. Shi, E., Björnson, E., Sanguinetti, L. and Debbah, M. (2022) 'Wireless energy transfer in RIS-aided cell-free massive MIMO systems: opportunities and challenges', *IEEE Communications Magazine*, 60(3), pp. 26–32. https://doi.org/10.1109/MCOM.001.2100671.
2. Daryoush, A., Bontzos, K. and Herczfeld, P. (1986) 'Optically tuned patch antenna for phased array applications', *Proceedings of the IEEE AP-S International Symposium Digest*, January, pp. 361–364. https://doi.org/10.1109/APS.1986.1149762.
3. Van Blaricum, M.L. (2000) 'A brief history of photonic antenna reconfiguration', *Proceedings of the International Topical Meeting on Microwave Photonics*, November, pp. 9–12. https://doi.org/10.1109/MWP.2000.889772.
4. Mohammadi, M., Mobini, Z., Ngo, H.Q. and Matthaiou, M. (2025) 'Ten years of research advances in full-duplex massive MIMO', *IEEE Transactions on Communications*, 73(3), pp. 1756–1786. https://doi.org/10.1109/TCOMM.2024.3464414.
5. Sadiq, U., Hayat, S., Umar, A., Ali, U., Tahir, F.A. and Flint, J.A. (2017) 'Design, fabrication and measurement of triple band frequency reconfigurable antennas for portable wireless communications', *AEU – International Journal of Electronics and Communications*, 81, pp. 236–242. https://doi.org/10.1016/j.aeue.2017.07.028.
6. Tewari, N., Joshi, N. and Srivastava, S. (2023) 'A compact quad-band circular cavity-backed substrate integrated waveguide antenna for millimetre-wave applications', *Microwave and Optical Technology Letters*, 65(9). https://doi.org/10.1002/mop.33766.
7. Tewari, N., Joshi, N. and Srivastava, S. (2023) 'A novel reconfigurable H-plane horn leaky-wave substrate integrated waveguide MIMO antenna for K-band', *AEU – International Journal of Electronics and Communications*, 170, p. 154832. https://doi.org/10.1016/j.aeue.2023.154832.
8. Tewari, N. and Joshi, N. (2022) 'A circularly polarized cross-slot cavity-backed substrate integrated waveguide antenna for 60 GHz applications', *Proceedings of the 8th International Conference on Signal Processing and Communication (ICSC)*, pp. 1–5. https://doi.org/10.1109/ICSC56524.2022.10009409.
9. Tewari, N., Joshi, N. and Srivastava, S. (2022) 'A circularly polarized cylindrical cavity-backed substrate integrated waveguide antenna for millimetre-wave applications', *Proceedings of IEEE MAPCON*, pp. 1562–1567. https://doi.org/10.1109/MAPCON56011.2022.10047690.
10. Prakash, S., Dash, S. and Patnaik, A. (2018) 'Reconfigurable circular patch THz antenna using graphene stack-based SIW technique', *Proceedings of IEEE InCAP*, pp. 1–3. https://doi.org/10.1109/INCAP.2018.8770855.

11. Majumder, B., Krishnamoorthy, K., Mukherjee, J. and Ray, K.P. (2016) 'Frequency-reconfigurable slot antenna enabled by thin anisotropic double-layer metasurfaces', *IEEE Transactions on Antennas and Propagation*, 64(4), pp. 1218–1225. https://doi.org/10.1109/TAP.2016.2526081.
12. Majumder, B., Kandasamy, K., Mukherjee, J. and Ray, K.P. (2015) 'Dual-band dual-polarized frequency reconfigurable antenna using metasurface', *Proceedings of the Asia-Pacific Microwave Conference (APMC)*, pp. 1–3. https://doi.org/10.1109/APMC.2015.7411790.
13. Casu, G., Catalin, M. and Andrei, K. (2014) 'Design and implementation of microstrip patch antenna array', *Proceedings of the IEEE 10th International Conference on Communications*, pp. 1–4. https://doi.org/10.1109/ICCOMM.2014.6866738.
14. Dwivedi, R.P., Kommuri, U.K. and Veeramani, N. (2015) 'Design and simulation of wide-band patch antenna for wireless application', *Proceedings of IEEE SPIN*, pp. 1–4. https://doi.org/10.1109/SPIN.2015.7095300.
15. Rappaport, T.S., Xing, Y., MacCartney, G.R., Molisch, A.F., Mellios, E. and Zhang, J. (2017) 'Overview of millimeter-wave communications for fifth-generation (5G) wireless networks—with a focus on propagation models', *IEEE Transactions on Antennas and Propagation*, 65(12), pp. 6213–6230. https://doi.org/10.1109/TAP.2017.2734243.
16. Pi, Z. and Khan, F. (2011) 'An introduction to millimeter-wave mobile broadband systems', *IEEE Communications Magazine*, 49(6), pp. 101–107. https://doi.org/10.1109/MCOM.2011.5783993.
17. Rappaport, T.S. *et al.* (2013) 'Millimeter wave mobile communications for 5G cellular: it will work!', *IEEE Access*, 1, pp. 335–349. https://doi.org/10.1109/ACCESS.2013.2260813.
18. Geng, S., Kivinen, J., Zhao, X. and Vainikainen, P. (2009) 'Millimeter-wave propagation channel characterization for short-range wireless communications', *IEEE Transactions on Vehicular Technology*, 58(1), pp. 3–13. https://doi.org/10.1109/TVT.2008.924990.
19. Narekar, N.P. and Bhalerao, D.M. (2015) 'A survey on obstacles for 5G communication', *Proceedings of ICCSP*, pp. 831–835. https://doi.org/10.1109/ICCSP.2015.7322610.
20. Shamsan, Z.A. (2020) 'Rainfall and diffraction modeling for millimeter-wave wireless fixed systems', *IEEE Access*, 8, pp. 212961–212978. https://doi.org/10.1109/ACCESS.2020.3040624.
21. MacCartney, G.R., Zhang, J., Nie, S. and Rappaport, T.S. (2013) 'Path loss models for 5G millimeter wave propagation channels in urban microcells', *Proceedings of IEEE GLOBECOM*. https://doi.org/10.1109/GLOCOM.2013.6831690.
22. Kim, K.-W. *et al.* (2020) 'Millimeter-wave diffraction-loss model based on over-rooftop propagation measurements', *ETRI Journal*, 42(6), pp. 827–836. https://doi.org/10.4218/ETRIJ.2019-0411.
23. Rappaport, T.S. *et al.* (2017) 'Small-scale, local area, and transitional millimeter-wave propagation for 5G communications', *IEEE Transactions on Antennas and Propagation*, 65(12), pp. 6474–6490. https://doi.org/10.1109/TAP.2017.2734159.
24. Zhang, P. *et al.* (2020) 'Measurement-based 5G millimeter-wave propagation characterization in vegetated suburban macrocell environments', *IEEE Transactions on Antennas and Propagation*, 68(7), pp. 5556–5567. https://doi.org/10.1109/TAP.2020.2975365.
25. Samimi, M.K. and Rappaport, T.S. (2016) 'Local multipath model parameters for generating 5G millimeter-wave 3GPP-like channel impulse response', *Proceedings of EuCAP*, pp. 1–5. https://doi.org/10.1109/EUCAP.2016.7481410.
26. Rahim, H.M. *et al.* (2017) 'Foliage attenuation measurement at millimeter wave frequencies in tropical vegetation', *Proceedings of IEEE MICC*, pp. 241–246. https://doi.org/10.1109/MICC.2017.8311766.
27. Zhao, Q. and Li, J. (2006) 'Rain attenuation in millimeter wave ranges', *Proceedings of ISAPE*, pp. 1–4. https://doi.org/10.1109/ISAPE.2006.353538.
28. Roh, W. *et al.* (2014) 'Millimeter-wave beamforming as an enabling technology for 5G cellular communications', *IEEE Communications Magazine*, 52(2), pp. 106–113. https://doi.org/10.1109/MCOM.2014.6736750.
29. Li, Y., Gao, B., Zhang, X. and Huang, K. (2020) 'Beam management in millimeter-wave communications for 5G and beyond', *IEEE Access*, 8, pp. 13282–13293. https://doi.org/10.1109/ACCESS.2019.2963514.
30. Azar, Y. *et al.* (2013) '28 GHz propagation measurements for outdoor cellular communications using steerable beam antennas in New York City', *Proceedings of IEEE ICC*. https://doi.org/10.1109/ICC.2013.6655399.

31. Larsson, E.G., Edfors, O., Tufvesson, F. and Marzetta, T.L. (2014) 'Massive MIMO for next generation wireless systems', *IEEE Communications Magazine*, 52(2), pp. 186–195. https://doi.org/10.1109/MCOM.2014.6736761.
32. Murdock, J.N. *et al.* (2012) 'A 38 GHz cellular outage study for an urban outdoor campus environment', *Proceedings of IEEE WCNC*, pp. 3085–3090. https://doi.org/10.1109/WCNC.2012.6214335.
33. Alobaidi, O.R. *et al.* (2022) 'Transparent antenna for green communication feature: a systematic review', *IEEE Access*, 10, pp. 12286–12321. https://doi.org/10.1109/ACCESS.2020.3044435.
34. Oh, S. *et al.* (2001) 'A broadband microstrip antenna array for LMDS applications', *Microwave and Optical Technology Letters*, 32(1), pp. 35–37. https://doi.org/10.1002/mop.10084.
35. Wincza, K. and Gruszczynski, S. (2011) 'Microstrip antenna arrays fed by a series-parallel slot-coupled feeding network', *IEEE Antennas and Wireless Propagation Letters*, 10, pp. 991–994. https://doi.org/10.1109/LAWP.2011.2167491.
36. Dzagbletey, P.A. *et al.* (2016) 'Stacked microstrip linear array with highly suppressed side-lobe levels', *IET Microwaves, Antennas & Propagation*, 11(1), pp. 17–22. https://doi.org/10.1049/iet-map.2016.0161.
37. Dzagbletey, P.A. and Jung, Y.-B. (2018) 'Stacked microstrip linear array for millimeter-wave 5G baseband communication', *IEEE Antennas and Wireless Propagation Letters*, 17(5), pp. 780–783. https://doi.org/10.1109/LAWP.2018.2816258.
38. Khalily, M. *et al.* (2015) 'Design of phased arrays of series-fed patch antennas for 28-GHz applications', *IEEE Antennas and Wireless Propagation Letters*, 15, pp. 1305–1308. https://doi.org/10.1109/LAWP.2015.2505781.
39. Chang, C.-C., Lee, R.-H. and Shih, T.-Y. (2009) 'Design of a beam switching/steering Butler matrix', *IEEE Transactions on Antennas and Propagation*, 58(2), pp. 367–374. https://doi.org/10.1109/TAP.2009.2037693.
40. Cao, J.-Z. and Chen, F.-C. (2022) 'Design of continuously steerable Nolen matrix-based beamforming networks', *Proceedings of ICMMT*. https://doi.org/10.1109/ICMMT55580.2022.10023059.
41. Ibrahim, S.Z. and Bialkowski, M.E. (2009) 'Wideband Butler matrix in microstrip-slot technology', *Proceedings of APMC*, pp. 2104–2107. https://doi.org/10.1109/APMC.2009.5385262.
42. Breed, G. (2009) 'Transmission line and lumped element quadrature couplers', *High Frequency Electronics*, pp. 44–48. Available at: http://www.highfrequencyelectronics.com.
43. Wang, X. *et al.* (2019) 'Compact 2-D multibeam array antenna fed by planar cascaded Butler matrix', *IEEE Antennas and Wireless Propagation Letters*, 18(10), pp. 2056–2060. https://doi.org/10.1109/LAWP.2019.2937254.
44. Wang, D. *et al.* (2022) 'Switched and steered beam end-fire antenna array fed by tunable Butler matrix', *IEEE Transactions on Antennas and Propagation*, 70(7), pp. 5383–5392. https://doi.org/10.1109/TAP.2022.3142334.
45. So, H. *et al.* (2023) 'Staircase array antenna with stacked Butler matrix for concurrent multi-beams', *IEEE Access*, 11, pp. 76638–76646. https://doi.org/10.1109/ACCESS.2023.3297659.
46. Abhishek, A. *et al.* (2020) 'Design of beam steering antenna for 5G at 28 GHz using Butler matrix', *Proceedings of ICCCS*. https://doi.org/10.1109/ICCCS49678.2020.9276492.
47. Ding, K. and Kishk, A. (2018) 'Two-dimensional Butler matrix concept for planar array', *Proceedings of IEEE IMS*, pp. 632–635. https://doi.org/10.1109/MWSYM.2018.8439616.
48. Gu, C. *et al.* (2017) 'Dual-band electronically beam-switched antenna using slot active FSS', *IEEE Transactions on Antennas and Propagation*, 65(3), pp. 1393–1398. https://doi.org/10.1109/TAP.2016.2647578.
49. Tarafder, P. *et al.* (2025) 'Channel estimation in 5G-and-beyond wireless communication', *Electronics*, 14(4), p. 750. https://doi.org/10.3390/electronics14040750.
50. Jacobsson, S. *et al.* (2017) 'Quantized precoding for massive MU-MIMO', *IEEE Transactions on Communications*, 65(11), pp. 4670–4684. https://doi.org/10.1109/TCOMM.2017.2723000.
51. Gros, J.-B. *et al.* (2021) 'A reconfigurable intelligent surface at mmWave based on a binary phase tunable metasurface', *IEEE Open Journal of the Communications Society*, 2, pp. 1055–1064. https://doi.org/10.1109/OJCOMS.2021.3076271.
52. Zhang, Y. *et al.* (2022) 'A highly pattern-reconfigurable planar antenna with 360° beam steering', *IEEE Transactions on Antennas and Propagation*, 70(8), pp. 6490–6504. https://doi.org/10.1109/TAP.2022.3161514.
53. Wan, X. *et al.* (2021) 'Reconfigurable sum and difference beams based on a binary programmable metasurface', *IEEE Antennas and Wireless Propagation Letters*, 20(3), pp. 381–385. https://doi.org/10.1109/LAWP.2021.3050808.

54. Araghi, A. *et al.* (2022) 'Reconfigurable intelligent surface (RIS) in the sub-6 GHz band', *IEEE Access*, 10, pp. 2646–2655. https://doi.org/10.1109/ACCESS.2022.3140278.
55. Fong, B.H. *et al.* (2010) 'Scalar and tensor holographic artificial impedance surfaces', *IEEE Transactions on Antennas and Propagation*, 58(10), pp. 3212–3221. https://doi.org/10.1109/TAP.2010.2055812.
56. Rana, B., Cho, S.-S. and Hong, I.-P. (2023) 'Review paper on hardware of reconfigurable intelligent surfaces', *IEEE Access*, 11, pp. 29614–29634. https://doi.org/10.1109/ACCESS.2023.3261547.

Chapter 6

AI and machine learning in wireless networks

Pritika Kujur and Priyanka Kujur

6.1 INTRODUCTION

The fast development of wireless communication systems has resulted in an increase in the amount of complexity that is associated with the administration of networks, the distribution of resources, and the provision of services. Large, dynamic, and diverse wireless networks are a feature of today's modern wireless networks. In addition to this, they are not within the purview of the conventional rule-based and deterministic techniques. It is important to keep this in mind since artificial intelligence (AI) and machine learning (ML) are rapidly becoming prominent complementary techniques that enable intelligent and adaptive decision-making throughout the gradient of wireless network stack layers. By utilizing time-sensitive real-time data in addition to previous patterns and characteristics, AI and ML algorithms are able to successfully optimize a wide variety of operational processes. As a result, the whole capabilities of AI and ML are important and required for the future of wireless networks in next-generation networks such as fifth generation (5G) and the aspirational sixth-generation (6G) networking standards [1]. A substantial movement in the landscape of wireless communications is represented by the incorporation of AI into wireless networks, which not only promotes integration but also marks a significant transformation. Cognitive and self-organizing networks are being brought into existence as a result of this development. These networks are able to learn from their surroundings, adapt to their surroundings, and adjust sample input based on real-time user observations.

It is possible for AI to process vast volumes of information pertaining to users, devices, and infrastructure, which is one of the most important advantages of using AI in wireless networks. ML models are able to create judgments that are actionable and focused on increasing the quality of service (QoS) and response latency. These decisions are made by recognizing patterns and forecasting future states. Methods for achieving this include supervised learning, reinforcement learning (RL), and deep learning, all of which provide suitable and usable solutions for any layer in the network stack, including physical layer signal processing, medium access (MAC) layer, and the network layer. From the perspective of next-generation wireless technologies, the role AI plays in supporting network slicing, beamforming, massive MIMO optimization, and intelligent handover is indispensable, with growing applications that require reliability and ultra-low latency, such as remote surgeries, autonomous vehicles, smart cities, and the Internet-of-Things (IoT) [2]. Despite the potentially beneficial applications of AI/ML, there are many challenges to integrating them into wireless networks. This could include, but is not limited to, obtaining training data of proper quality, computational capability limitations of on-device edge devices, model interpretability, and data privacy and security. Addressing these challenges will require interdisciplinary research to develop sound and comprehensive frameworks for model validation and collaborations between industry and academia.

 DOI: 10.1201/9781003637721-6

In this chapter, a view of AI and ML in wireless networks is presented. The organization of this chapter is carried out as follows. Section 6.2 discusses the AI-driven applications used in wireless networks. Section 6.3 provides details of ML techniques for wireless networks. Section 6.4 discusses AI and ML in next-generation wireless networks, while case studies and real-world implementations are presented in Section 6.5. Section 6.6 elaborates on challenges and future research directions. In the last section, we provide the concluding remarks followed by the future scope.

6.1.1 Fundamentals of AI and machine learning in wireless networks

AI can be defined as the ability for machines or software systems to emulate human cognition while completing tasks that include reasoning, learning, and making decisions. In wireless networks, AI will be used to automate complex processes, aid in the adaptation of variable tasks, and provide intelligent decisions in real time. The greater the heterogeneity of networks and user applications/needs, the more reliance AI can provide on wireless systems to be operationally independent. ML is a subfield of AI that can help wireless systems learn through the use of history and real-time data without being inherently programmed. Instead of defining relationships between input and output with code, ML uses data-driven models to determine patterns, predict future network states, and optimize network behavior and performance. An advantage of using ML in wireless networks is the quantity of variable conditions that can include sample size of movement, signal quality, bandwidth use, or interference levels in real time [3, 4].

Supervised learning is one of the more common ML techniques that is currently being used in wireless networks. This model uses labeled datasets where the input–output pairs of data are known and is ideal for predicting user behavior, classifying modulation, or identifying network traffic type. By learning the relationships between the input values and the desired output, the supervised learning models can predict new samples against the training data in real time. Unsupervised learning, conversely, does not use labeled data like supervised approaches but instead seeks out hidden patterns and structures in an unlabeled dataset. Unsupervised learning typically includes clustering users based on their behaviors, anomaly detection in the network traffic, and segmentation of usage profiles. Unsupervised learning allows wireless networks to identify non-obvious insights that can lead to greater flexibility and security. RL is another significant ML technique that works very well in dynamic environments like wireless networks. On RL, the agent will learn on the fly by interacting with the environment, as it will receive rewards or penalties as feedback. This trial-and-error learning is an optimal solution for tasks such as dynamic spectrum access, power control, and handover optimization, even when making real-time decisions under uncertainty [5, 6].

Data is a prerequisite for any AI and ML applications in wireless and other systems. In wireless networks, user devices, the access points, and even the core of the networks generate a lot of data, including signal-to-noise ratios, throughput, latency, and mobility traces. For learning algorithms to be successful and to generalize advantageously under new conditions, learning algorithms should involve an initial pre-processing phase, including feature engineering, labeling the output, and/or labeling the data responsibly. AI and ML can be used across all layers of wireless network architecture. The physical layer can benefit from ML according to optimized signal detection, channel estimation, and interference avoidance techniques. The MAC layer can benefit from ML in terms of adaptive scheduling and resource allocation [7]. The network layer encompasses the use of AI for smart or autonomous routing as well as moving management. The network is able to incorporate all of these layers in either layer 0, 1, 2, or 3. Together, these can form a single holistic intelligent network edge that can adjust and self-optimize. With strong consumer demand, modern wireless architectures, like 5G and beyond, have required wireless networks to become more AI-native. The use of edge computing will enable the AI model to deploy edge AI, while the core of the advanced wireless network implements centralized training and coordination of resource optimization while adjusting to multiple contexts with great response time. The variety of tools to support the development and deployment of AI in wireless networks is considerable. For instance, TensorFlow and PyTorch are popular open-source tools for developing ML models. Other tools in RL, such as OpenAI Gym and RLlib, further empower researchers to simulate and test various dynamic wireless situations. Beyond the research domain, industry-wide initiatives, including O-RAN (Open Radio Access Network) architecture, promote collaboration in defining standards that utilize AI as a necessity in future wireless infrastructure [8].

6.1.2 Role of artificial intelligence in wireless networks

Wireless networks are becoming more intelligent and flexible as a result of the use of AI and ML. In order to maintain a connection, the use of algorithms powered by AI allows for the optimization of network traffic and the dynamic distribution of network resources. The construction of a more accurate prediction of network congestion is the means by which this objective is attained. ML also improves security by allowing a rapid reaction to potential cyberthreats and recognizing abnormalities in the behavior of traffic. This is accomplished via the identification of traffic patterns. The use of AI helps to increase the efficiency of spectrum management by reducing interference and increasing the quantity of data that can be sent. When it comes to the development of wireless networks that are headed toward 5G and beyond, AI and ML play a vital role in the enhancement of performance as shown in Figure 6.1 [9].

AI enhances the security of wireless networks by identifying weaknesses and stopping cyber attacks. To block unauthorized access, use patterns of ML algorithms to identify atypical behavior in tracking network activity. As IoT devices, smart cities, and self-driving cars become more prevalent, these skills are becoming more and more necessary as networks become bigger and more intricate. AI is becoming more and more crucial in next-generation

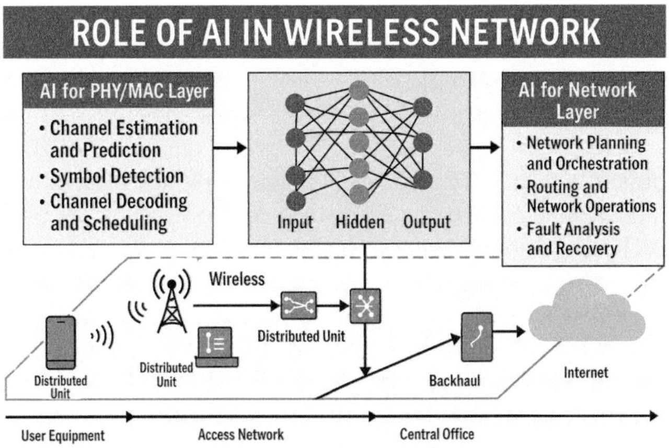

Figure 6.1 Role of AI in different layers of wireless network.

networks, especially 5G and 6G networks, where it helps make intelligent network slicing and edge computing possible. AI makes the service more reliable and responsive by giving critical apps priority access to the right and enough network resources. AI also makes wireless networks more stable and uses less energy. This will lead to a smarter and more connected future as a result [10, 11].

6.1.3 Role of ML in wireless network

It is impossible to overstate the importance of ML algorithms in terms of enhancing the functioning and productivity of wireless networks. Adaptive algorithms are being used to address the difficulties of resource and system parameters performance and reliability. This is due to the fact that wireless networks have become more complicated and dynamic in nature. Various ML techniques, with their specific strengths, are implemented across different levels and the various parts of the architecture of wireless communication systems as shown in Figure 6.2. Supervised learning is one of the more commonly used techniques when there is access to labeled data. In supervised learning, an algorithm learns through input–output pairs to perform predictions or classifications. Various supervised learning models can help wireless networks by addressing tasks such as signal classification, recognition of modulation types, and predicting user activity. The algorithms used in supervised learning include decision trees, logistic regression, support vector machines (SVM), and random forests to train models designed to effectively detect and classify network events [8, 12].

Unsupervised learning occurs when there are no labels defined for the data at hand, and is useful for identifying hidden patterns, relationships, or clusters within a large dataset of unlabeled data. Wireless systems apply unsupervised learning techniques for anomaly detection, traffic profiling, and clustering users who have similar behaviors. For example, some efficient methods for clustering are k-means clustering, DBSCAN, and hierarchical clustering. All of these algorithms have been developed to help networks process unlabeled data, therefore helping the networks adapt to unknown, or changing, scenarios. RL is particularly useful

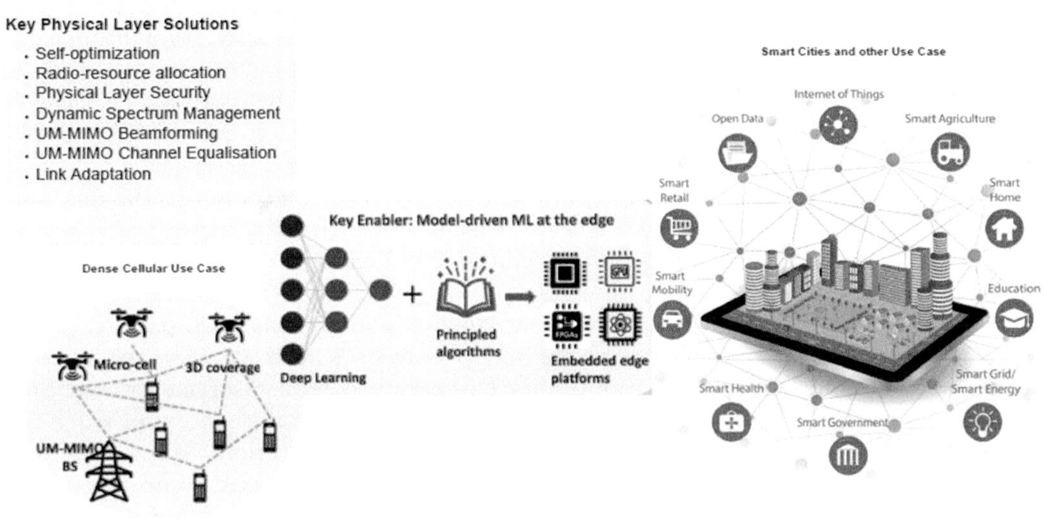

Figure 6.2 Role of ML in wireless network.

when decisions are enforced through trial-and-error, on whatever wireless tasks. In RL, an agent interacts with the network environment, receives feedback (rewards or penalties) based on its decisions, and iteratively learns over time optimal actions. Given the dynamic nature of wireless tasks, as you can see, the RL framework is very well suited to tasks such as channel selection, power control, and handover decisions. Some applied RL techniques that attempt to optimize network decision-making in real time are Q-learning and Deep Q-Networks (DQN) [13].

Deep learning builds upon conventional ML using multi-layered neural networks to deal with complex or high-dimensional data. Areas of application for deep learning in wireless communications include spectrum sensing, signal detection, and predictive maintenance. Examples of commonly used architectures include Convolutional Neural Networks (CNNs) for extracting spatial features from radio signals, recurrent neural networks (RNNs) or long-short-term memory (LSTM) networks for modeling time-dependent behavior such as user mobility or rapid changes in traffic profiles. One of the most recent developments in ML is the concept of federated learning, which takes a "collaborative learning" perspective in which many devices collaboratively train a shared model without directly exchanging sources of raw data. In doing this, federated learning maintains privacy and reduces the communication overhead, which makes it relevant to distributed wireless systems such as IoT systems, and/or edge networks [14].

6.2 AI-DRIVEN APPLICATIONS IN WIRELESS COMMUNICATIONS

AI is rapidly becoming a key enabler in improving wireless communication systems. As networks become more extensive, more intricate, and more user-driven, traditional model-based approaches to optimization are not sufficient. AI enables adaptive, data-driven solutions that will significantly improve the performance, reliability, and intelligence of the many-layered wireless communications typical of next-generation wireless networks. Dynamic resource allocation is a key application of AI in wireless networks. Where bandwidth and power are limited, AI models, particularly RL, can act in real time to allocate resources in a manner that maximizes usage in a methodical, efficient way. By learning from usage patterns and environmental conditions, the utilization of the spectrum can be optimized, and users can access frequencies in a fair way. Managing interference is also a valuable place for AI use in communication networks. In highly populated areas of operation, such as urban areas or large IoT systems, overlapping signals can contribute to the performance degradation of all communication sources. AI algorithms can be employed to isolate the sources of interference, to modify transmission parameters, and facilitate and recommend new channels or beam paths to lessen displacement and/or interference and improve the quality of wireless communications [15, 16].

AI is another way to improve mobility prediction and handover optimization. Wireless networks are required to provide seamless service while users move from one cell or access point to another. ML models can be utilized to identify structural patterns in user movements, allowing predictions about the handover to be generated before the handover occurs. A more holistic approach to mobility will certainly reduce service interruptions. Minimizing interruptions is critical for real-time services (i.e., in the context of video calls or online gaming). AI also fits into the category of threat detection. Many traditional intrusion detection and prevention systems either rely on signatures or known attacks persistently updated with the latest threat landscape. AI-based systems not only quickly identify malicious code for known attacks but also allow for learning from the behavioral state of the network to detect abnormal behaviors that indicate an intrusion attempt. These behaviors may include denial-of-service attacks or even spoofing. In short, these systems help solve significant security challenges facing wireless networks by supplying some level of early indication of suspicious activity or even automating prevention in some cases [17, 18].

Another useful component of wireless networks is related to traffic prediction and classification. Wireless networks with base stations often handle long-running services that run in the background, such as video, voice, and control packets. Deep learning models within AI systems can identify and predict specific types of traffic based on observed current and historical trends. This type of inquiry addresses both short- and long-term traffic planning. Improved planning of wireless systems can lead to the proactive dynamic optimization of the network resources and the wireless systems before congestion sets in. The energy efficiency of wireless devices and systems is becoming more and more critical, particularly for remote and battery-powered devices. In energy-aware communication, AI can also provide the learned power and sleep cycle patterns of various communication transmitters in the system. Device battery life can be extended, system energy consumption lowered, and network power used during transmission can be minimized, all in the name of the AI system determining the system communication power. The antenna systems and beamforming associated with massive AI systems also assist in the most recent wireless systems. Advanced wireless systems utilize focused radio energy that beams to track the user. AI can modify beam patterns based on user location, signal strength, and interference contexts to maximize coverage and system throughput. A network that is SON capable is driven by AI models that make decisions regarding AI capabilities, which allows the network to become more resilient, more efficient, and easier to manage [20, 21]. Various AI-driven applications in wireless communications are depicted in Figure 6.3.

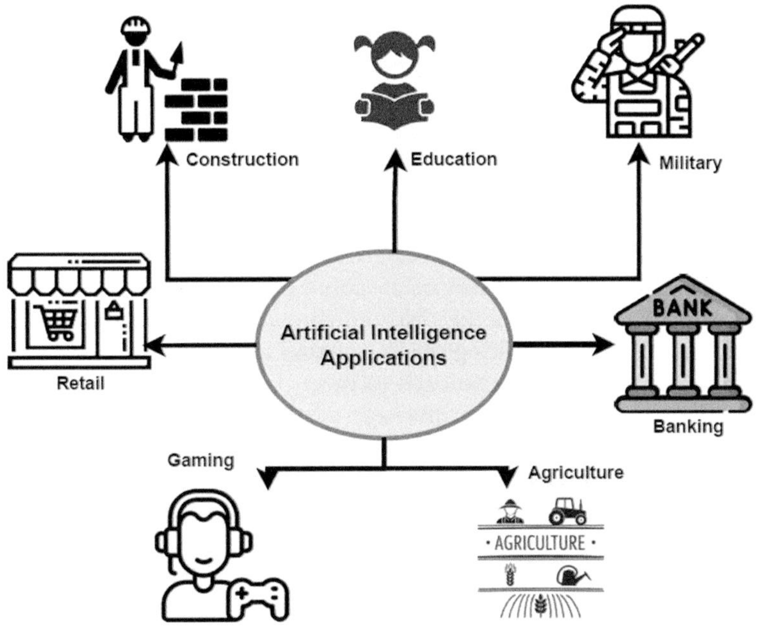

Figure 6.3 AI-driven applications in wireless communications.

6.3 AI AND MACHINE LEARNING IN NEXT-GENERATION WIRELESS NETWORKS

Wireless networks of 5G and the emerging 6G are expected to provide ultra-reliable, high-speed, and low-latency connectivity for new use cases spanning smart cities, autonomous vehicles, augmented reality, and massive IoT connectivity. To meet the increasing demands of next-generation networks, AI and ML are playing an increasingly important role in increasing the network's intelligence and operation. AI and ML are enabling next-generation, heterogeneous networks to manage the high complexity of real-time operation, including the automated management of device heterogeneity, dynamic traffic, and changing channel states. The capabilities of AI and ML enable networks to analyze massive amounts of historical and real-time data at scale, providing actionable intelligence to predict user behavior, maximize resource allocation, and adapt to changes in the environment. One area where AI and ML will be important is automated networks and self-optimizing capabilities. The network is expected to be self-configuring, self-healing, and self-optimizing with little to no human-in-the-loop interaction. ML algorithms will autonomously access network performance metrics to make recommendations that improve coverage, capacity, and energy efficiency, enabling the network to autonomously manage non-real-time network capabilities, reducing latency and increasing reliability [22].

Another significant example is in the application of dynamic spectrum management. Increasing wireless service demand and spectrum scarcity are significant challenges facing network managers; cognitive radio techniques driven by AI detail how well cognitive radio networks, as AI-enabled wireless devices, empower the network to learn how to sense available frequencies in the wireless environment, dynamically make spectrum manipulation decisions based upon overall demand, interference, and quality. This is a powerful AI/ML-based way to maximize spectrum use while minimizing interference. AI and ML also improve mobility management and handover in future next-generation networks. AI models can help predict user mobility and the state of the network, resulting in improvements in handover timing and cell selection, to ensure users' devices remain seamlessly connected while travelling at high speeds, especially in vehicular networks. The security of systems that include wireless technologies is getting increasingly difficult to achieve as these systems grow more interconnected. Smart systems that make use of AI techniques are able to protect themselves from sophisticated, persistent attacks. This is accomplished by analyzing patterns that are connected with the behavioral norms of a nearby network and identifying irregularities in network traffic and device utilization. A person cannot do this. These intelligent and adaptable technologies are able to protect your systems from spoofing, jamming, and distributed denial of service assaults. AI also makes it possible to implement enhanced network slicing, which is the distinguishing characteristic of 5G and subsequent technologies. It does this by generating several virtual networks on the same piece of physical infrastructure, each of which is tailored to meet the requirements of a particular application. AI-based systems dynamically manage and coordinate these slices by modifying parameters in real time. This helps to guarantee that interdependent slices do not impact the performance and latency of the other slices, which ultimately assists them in accomplishing their objectives [23, 24].

6.4 CASE STUDY AND REAL WORLD IMPLEMENTATION

AI and ML are currently being utilized in the operational and strategic enhancements of modern wireless networks, moving beyond the realm of theory. The wide variety of applications in the business indicates how the two technologies are resolving issues that were previously challenging and enhancing the overall performance of networks, as well as their dependability and the level of satisfaction experienced by users.

6.4.1 Case 1: AI-driven network optimization

One interesting example comes from the implementation of AI and autonomous network optimization using 5G in an operator's deployment. Here, a famous telecommunications service provider was able to analyze masses of datasets produced from network elements, user devices, and traffic flows using ML algorithms. Based on that, it used RL methods to autonomously leverage the network toward modifying a set of parameters, such as power levels, spectrum, and antenna beamforming/steering of the antennas, for maximum achievable coverage and throughput. The operator noticed that, with the use of AI-driven self-optimization as shown in Figure 6.4, there was a much-reduced level of calls dropped and improved data speeds for its customers during peak usage hours-a huge overall improvement, if anything [23].

6.4.2 Case 2: Predictive maintenance in wireless infrastructure

One example of a practical use is the use of predictive maintenance for wireless infrastructure is shown in Figure 6.5. Using previous sensor data from base stations and core network gear, service providers have developed ML models to estimate when equipment is likely to break. These models are used to predict when equipment will fail. These kinds of technologies are utilized in predictive maintenance in order to identify indications of imminent failure. This enables operators to repair equipment before it completely fails, which ultimately results in a reduction in downtime. This has the effect of lowering operational costs and increasing reliability. Reliability is critical in mission-critical applications, such as emergency services and industrial automation. ML applications can also be found in dynamic spectrum management in cognitive radio networks. In this regard, ML systems, through the

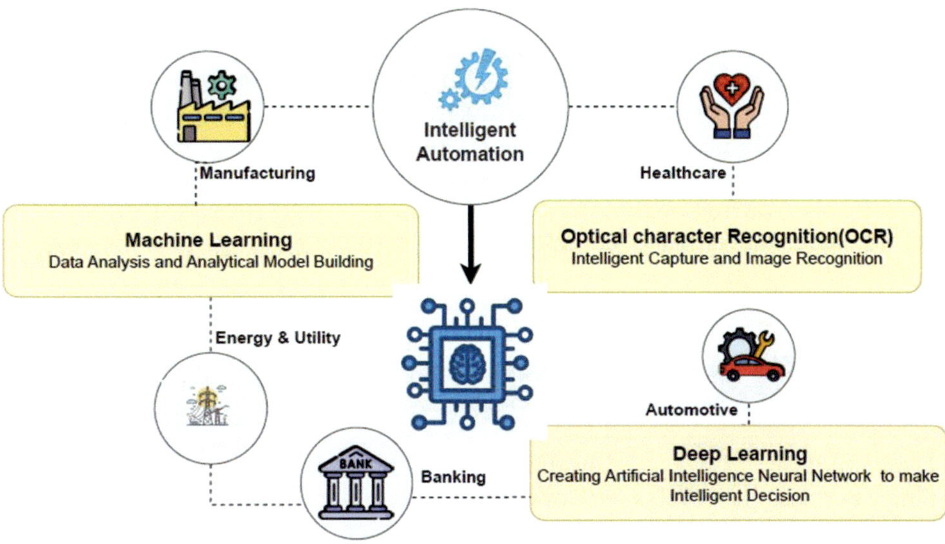

Figure 6.4 AI-driven network optimization.

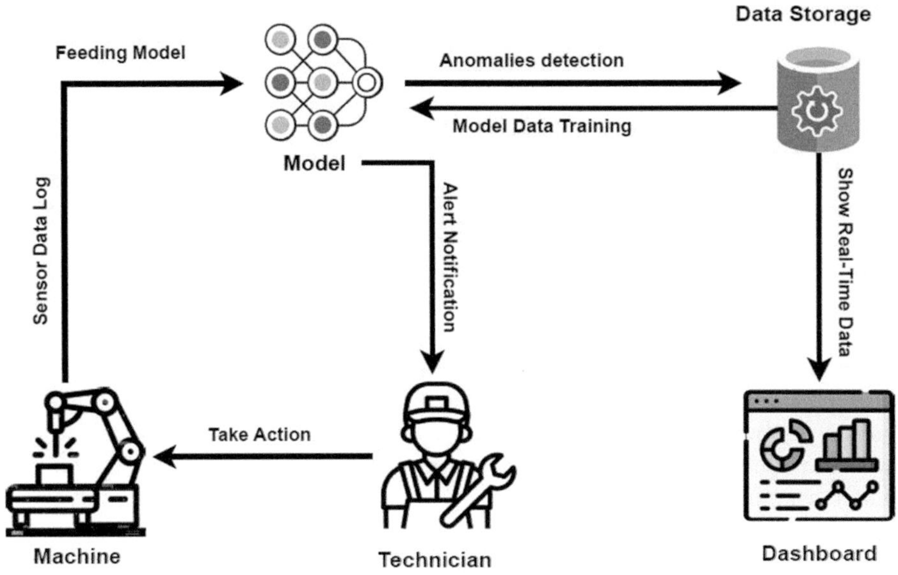

Figure 6.5 Predictive maintenance in wireless infrastructure.

use of ML algorithms, are capable of observing patterns in the usage and environment of the spectrum, thus offering an assignment of frequencies to users in flight and without interference, improving spectrum use. The use of ML should therefore improve the management of spectrum demand, particularly in urban environments where demands can change rapidly in dense settings. Using real-time spectrum sensing with AI will ultimately allow the operator to satisfy user demands without the need to acquire more spectrum [22].

6.4.3 Case 3: AI in next-generation wireless networks

A major telecommunications provider has adopted AI and ML technologies to optimize predictive maintenance for their 5G networks. With the help of RL, the system autonomously adjusted power levels, spectrum allocations, and antenna beamforming. Furthermore, the AI-powered predictive maintenance tools, which analyzed base station sensor data and detected unpredictable hardware failure risks, resulted in improved operational downtime predictive maintenance. This provided greater operational expense savings. Reliability improved 35%, data speeds increased 20%, and significantly fewer calls were dropped during peak hours. This demonstrates the successful AI-enabled self-learning adaptive and efficient communications system, while paving the way for the future of 6G [25].

6.5 CHALLENGES IN AI AND MACHINE LEARNING FOR WIRELESS NETWORKS

i. *Data Quality and Availability*: High-quality labeled data is necessary in large volumes, but data in wireless networks can be noisy, incomplete, and biased. Gathering enough labeled data for supervised learning is often a complex task [26, 27].
ii. *Computational Complexity and Resource Constraints*: A key challenge with AI and wireless devices is the limited processing capacity and the constrained battery life. In

these cases, the running of complex AI models locally becomes difficult. There is a need for efficient algorithms and model compression to make AI more manageable in these situations [26].

iii. *Security and Privacy Issues*: Adversarial assaults can change the inputs of AI models so that they give wrong outputs. Preserving user privacy during collaborative learning (e.g., federated learning) is a significant concern [27].

iv. *Dynamic and Heterogeneous Network Environments*: Wireless networks have characteristics that change quickly, such as mobility, interference, and traffic that varies. This means that AI models need to be able to react in real time. Challenges arise when attempting to design models that are capable of online learning and rapid adaptation [27].

v. *Interoperability and Lack of Standardization*: In general, AI solutions are proprietary and do not adhere to any common standards. As a result, it is difficult to link many providers and different kinds of networks. The fragmentation makes it more difficult to adopt on a widespread scale [24].

6.6 FUTURE RESEARCH DIRECTIONS

6.6.1 Development of lightweight and energy-efficient models

We need to make models that are both lightweight and energy-efficient in order to get the most out of our computers while using the least amount of resources. Lightweight versions are the greatest choice when it comes to things like memory and power. This is especially true for devices on the edge, the IoT, and mobile networks, all of which have big problems. Techniques such as model pruning and quantization reduce the complexity of deep learning architectures without loss of accuracy. Efficient neural network design methods, like TinyML and optimized transformers, are ways through which AI applications consume less energy. Lightweight and energy-efficient models contribute to green AI since they reduce the carbon footprint of large computations while keeping high performance [28].

6.6.2 Explainable AI (XAI)

Explainable AI (XAI) is a collection of techniques and methods applied to make AI models more transparent, interpretable, and understandable for human users. Because deep learning models and other traditional AI systems act like black boxes, it is often hard to figure out why they came to certain conclusions. XAI wants to close this information gap about how AI comes to particular conclusions so that AI-driven judgments are more trustworthy, fair, and responsible [29].

6.6.3 Integration with emerging technologies

It explores how AI can be combined with quantum computing, blockchain, and other technologies for enhanced security and privacy. Integration of various emerging technologies with AI and ML is transforming industries in general through enhanced automation and efficiency. AI-enabled solutions are being mixed with IoT, blockchain, edge computing, and AR to create intelligent, adaptive systems. AI-powered blockchain-based solutions automate and enhance the security and transparency of financial transactions and supply chain management. Edge computing also benefits from AI by processing data closer to the source and reducing latency and bandwidth use [30].

6.6.4 Multi-agent and distributed learning

In cooperative learning methodologies, the efforts in a network of nodes can be combined to enhance the efficiency of the network itself. Regarding AI and ML, some of the most important techniques are multi-agent and distributed learning for creating systems that can work together. One of the most important aspects of multi-agent learning is that it involves a large number of AI agents working together in the same environment. These agents are able to learn from one another and improve their decision-making abilities as a group. This technique is called multi-agent learning. Most autonomous systems and optimization of networks rely on this heavily because, in such contexts, agents need to coordinate their actions effectively. Distributed learning, on the other hand, allows multiple devices or servers to collaborate in order to train AI models. In addition to reducing the amount of computing load, it also promotes scalability [31].

6.6.5 AI-driven network automation and self-organization

It enhances hands-free network management by facilitating various automated tasks, such as defect detection and performance optimization. AI-powered wireless network automation and self-organization are causing a paradigm shift in modern radio networks. These technologies perform autonomous administration and optimization without human interference, enabling networks to self-configure and self-learn with utmost network efficiency [31].

6.6.6 Robust and privacy-preserving AI techniques

Construct models that are resistant to adversarial attacks while maintaining their confidentiality, while working together and sharing data. Data privacy strategies fend off unethical AI systems that are also insecure while maintaining the performance and resilience of AI systems against adversarial assaults. Both ethical and adversarially robust AI systems can gain from the data privacy AI approaches, which help them in striking a balance between privacy and performance. Advanced AI privacy protection methods and adversarial resilience are essential for healthcare, finance, and autonomous vehicle AI systems.

6.6.7 Standardization and interoperability frameworks

This movement wants to see more standardized protocols and frameworks established. By using these frameworks and protocols as a standard, intelligent systems can more readily keep working normally across different wireless networks. Reliable standards and interoperability frameworks are an important part of making this happen. It is possible to make it easier to move data between different systems, sectors, and technologies by creating clear standards and methods for interoperability at the same time. By specifying multiple data structures and formats, standardization helps maintain things the same across several platforms. This makes it easy to put things together. On the other hand, interoperability is all about making it possible for systems to talk to each other and work together well, with the goal of making it easier for people to work together and coordinate across a wide range of technology settings [32].

6.7 CONCLUSION

Adding AI and ML to wireless networks has made them more connected, efficient, and flexible, thus opening the door to further innovations. AI technologies such as autonomous network optimization, predictive maintenance, and dynamic spectrum management will make the networks more reliable, efficient in managing resources, and safe. With wireless networks inching closer to 6G technology, AI is predicted to enable intelligent communication among networks and their self-organization in the near future, thereby making wireless infrastructures flexible, energy-efficient, and strong. The continued growth of AI and ML will ensure that networks have the capacity to process massive volumes of data in real time. This will further facilitate seamless communication and enhance the user experience greatly. Wireless networking, coupled with global connectivity, will be driven in the future by higher levels of automation, security, and efficiency.

REFERENCES

1. Hemrajani, P., Dhaka, V. S., Bohra, M. K. and Gupta, A. K. (2022). Artificial Intelligence in Wireless Communication. In: Kanthavel, R., Ananthajothi, K., Balamurugan, S., Ganesh, R. K. (eds) Artificial Intelligent Techniques for Wireless Communication and Networking, pp. 317–334. John Wiley & Sons, Ltd. https://doi.org/10.1002/9781119821809.ch20
2. Du, J., Jiang, C., Wang, J., Ren, Y., & Debbah, M. (2020). Machine Learning for 6G Wireless Networks: Carrying Forward Enhanced Bandwidth, Massive Access, and Ultrareliable/Low-Latency Service. IEEE Vehicular Technology Magazine, 15(4), 122–134. https://doi.org/10.1109/MVT.2020.3019650
3. Syed, Q. S., Hussain, S., & Bashir, I. (2025). Artificial Intelligence and Machine Learning as Pioneers in Advancing 5G/6G Network Capabilities: Evolution, Introduction, Antecedents, and Consequences. In: Miryala, R., Lee, Y., Mazurek, M., Chendragiri, M. (eds) 5G/6G Advancements in Communication Technologies for Agile Management, pp. 215–232. IGI Global Scientific Publishing.
4. Kataria, D., Walid, A., Daneshmand, M., Dutta, A., Enright, M. A., Gu, R., Lackpour, A., Ramachandran, P., Wang, H., Chen, C.-M., Chng, B., Darema, F., Das, P., Lala, T. K., Chng, B., Ranpara, R., Kloza, B., Borst, M., & Polk, C. (2023). Artificial Intelligence and Machine Learning. https://doi.org/10.1109/fnwf58287.2023.10520629
5. El khediri, S., Benfradj, A., Thaljaoui, A., Moulahi, T., Ullah Khan, R., Alabdulatif, A., & Lorenz, P. (2024). Integration of Artificial Intelligence (AI) with Sensor Networks: Trends, Challenges, and Future Directions. Journal of King Saud University – Computer and Information Sciences, 36(1), 101892. ISSN 1319-1578. https://doi.org/10.1016/j.jksuci.2023.101892.
6. Mian, S. M., Shawez, M., & Kaur, A. (2024). Artificial Intelligence (AI), Machine Learning (ML) & Deep Learning (DL): A Comprehensive Overview on Techniques. Applications and Research Directions, 1404–1409. https://doi.org/10.1109/icscss60660.2024.10625198
7. Vadher, A., Patel, M., Thaker, J., & Bhise, A. (2024). "AI-Based Security System for 5G Enabled IoT," *Proceedings of the Second International Conference on Emerging Trends in Information Technology and Engineering (ICETITE)*, Vellore, India, pp. 1–7. https://doi.org/10.1109/ic-ETITE58242.2024.10493287.
8. Ahammadi, A., Hassan, W. H., & Shamsan, Z. A. (2022). "An Overview of Artificial Intelligence for 5G/6G Wireless Networks Security," *International Conference Control and Robots*, pp. 1–6. https://doi.org/10.1109/ICCR56254.2022.10024692
9. El-Hajj, M. (2025). Enhancing Communication Networks in the New Era with Artificial Intelligence: Techniques, Applications, and Future Directions. Network, 5(1), 1. https://doi.org/10.3390/network5010001
10. Xue, J., Qu, Z., Zhao, S., Liu, Y., & Lu, Z. (2023). "Data-Driven Next-Generation Wireless Networking: Embracing AI for Performance and Security," *32nd International Conference on Computer Communications and Networks (ICCCN)*, Honolulu, Hawaii, pp. 1–10. https://doi.org/10.48550/arXiv.2306.06178
11. Priya, C., Pemula, L., & Shaik, N. (2024). Beyond Conventional Methods: Machine Learning Approaches for Advancing Wireless Network Performance. In Futuristic Trends in Network & Communication Technologies, V3B1, pp. 23–36. Iterative International Publishers (IIP), Selfypage Developers Pvt Ltd. https://doi.org/10.58532/v3bbnc1p1ch3.
12. Gunturu, V., Singh, C., Patankar, N., & Praveena, S. (2024). 16 Exploring the Potential of Artificial Intelligence in Wireless Sensor Networks. In: de Albuquerque, V., Raj, P., Yadav, S. (eds) Toward Artificial General Intelligence: Deep Learning, Neural Networks, Generative AI, pp. 331–344. Berlin, Boston: De Gruyter. https://doi.org/10.1515/9783111323749-016
13. Hota, L., Nayak, B. P., & Kumar, A. (2024). Machine Learning Algorithms for Optimization and Intelligence in Wireless Networks. In: Bhowmick, A., Choukiker, Y. K., Singh, I., Nallanathan, A. (eds) 5G and Beyond Wireless Communications, pp. 306–332. CRC Press. https://doi.org/10.1201/9781032625034-16.

14. Kim, T., Vecchietti, L. F., Choi, K., Lee, S., & Har, D. (2021). Machine Learning for Advanced Wireless Sensor Networks: A Review. IEEE Sensors Journal, 21(11), 12379–12397. https://doi.org/10.1109/JSEN.2020.3035846
15. Delwar, T. S., Aras, U., Mukhopadhyay, S., Kumar, A., Kshirsagar, U., Lee, Y., Singh, M., & Ryu, J.-Y. (2024). The Intersection of Machine Learning and Wireless Sensor Network Security for Cyber-Attack Detection: A Detailed Analysis. Sensors, 24(19), 6377.
16. Sur, S. N., Vishwakarma, P., & Palit, S. (2024). Artificial Intelligence Revolutionizing Wireless Communication Systems. In: Sur, S. N., Imoize, A. L., Bhattacharya, A., Kandar, D., Banerjee, J. S. (eds). Artificial Intelligence for Wireless Communication Systems, pp. 1–24. CRC Press. https://doi.org/10.1201/9781003517689-1
17. Gundabatini, S. G., Kolluru, S. B., Ratnam, C. H. V., & Krupa, N. N. (2023). DAAM: WSN Data Aggregation Using Enhanced AI and ML Approaches. In: Biswas, A., Islam, A., Chaujar, R., Jaksic, O. (eds) Microelectronics, Circuits and Systems. Lecture Notes in Electrical Engineering, vol 976. Springer, Singapore. https://doi.org/10.1007/978-981-99-0412-9_49
18. Zhang, H., Xu, S., Xing, Z., Li, F., Xin, J., & Zhang, C. (2023). "Research on Applications of AI in Wireless Network." In *Proceedings of the 7th International Conference on Control Engineering and Artificial Intelligence (CCEAI '23)*, Association for Computing Machinery, New York, NY, USA, pp. 31–36. https://doi.org/10.1145/3580219.3580226
19. Chaudhary, A. (2025). Enhancing Wireless Connectivity Through AI-Driven Antenna Technologies. In: Hammouch, Z., Jamil, O. (eds) Convergence of Antenna Technologies, Electronics, and AI, pp. 335–346. IGI Global Scientific Publishing. https://doi.org/10.4018/979-8-3693-3775-2.ch013
20. Wang, J., Li, R., Jian, W., Ge, Y., Zhang, Q., & Shi, W. (2020). Artificial Intelligence and Wireless Communications. Journal of Zhejiang University Science C, 21(10), 1413–1425. https://doi.org/10.1631/FITEE.1900527
21. Ghadi, Y. Y., *et al.* (2024). Machine Learning Solutions for the Security of Wireless Sensor Networks: A Review. IEEE Access, 12, 12699–12719. https://doi.org/10.1109/ACCESS.2024.3355312
22. Raut, A. R. & Khandait, S. P. (2021). "Machine Learning Algorithms in WSNs and Its Applications," *2021 International Conference on Computational Intelligence and Computing Applications (ICCICA)*. Nagpur, India. pp. 1–5. https://doi.org/10.1109/ICCICA52458.2021.9697319
23. Jain, A., Jain, P., Verma, A., Gopal, L., & Chaudhary, P. (2024). "AI-Driven Dynamic Network Slicing for Resource Optimization in 5G Networks: Implementation and Performance Evaluation," *International Conference on Cybernation and Computation (CYBERCOM)*, Dehradun, India. pp. 720–724. https://doi.org/10.1109/cybercom63683.2024.10803129
24. Al-Khafaji, M. F. K. & Elwiya, L. (2022). "ML/AI Empowered 5G and beyond Networks," *International Congress on Human-Computer Interaction, Optimization and Robotic Applications (HORA)*, Ankara. Turkey. pp. 1–6. https://doi.org/10.1109/HORA55278.2022.9799813
25. Jurado-Lasso, F. F., Marchegiani, L., Jurado, J. F., Abu-Mahfouz, A. M., & Fafoutis, X. (2022). A Survey on Machine Learning Software-Defined Wireless Sensor Networks (ML-SDWSNs): Current Status and Major Challenges. IEEE Access, 10, 23560–23592. https://doi.org/10.1109/ACCESS.2022.3153521
26. SK, W. H., N. R, W. B., Chaubey, S. K., Mehmood, Z., J, K. S., & G, S. S. (2025). "AI-Integrated Sensor Data Analytics for Real-Time Decision-Making in Wireless Sensor Networks," *Proceedings of the International Conference on Machine Learning and Autonomous Systems (ICMLAS)*, Prawet, Thailand, pp. 1644–1649. https://doi.org/10.1109/ICMLAS64557.2025.10968079.
27. Zaimen, K., Brahmia, M. E. A., Moalic, L., Abouaissa, A., & Idoumghar, L. (2022). A Survey of Artificial Intelligence Based WSNs Deployment Techniques and Related Objectives Modeling. IEEE Access, 10, 113294–113329. https://doi.org/10.1109/ACCESS.2022.3217200
28. Srujana, P., Priyanka, J., Gurucharan, K., & Rajanbabu, M. (2024). "Exploring ML Solutions for Challenges in WSN Development: A Comprehensive Survey," *2024 IEEE Wireless Antenna and Microwave Symposium (WAMS)*, Visakhapatnam, India. pp. 1–4. https://doi.org/10.1109/WAMS59642.2024.10527941
29. Gunturu, V. V., Ranga, J., Murthy, C. R., Swapna, B., Balaram, A., & Raja, Ch. (2023). "Artificial Intelligence Integrated with 5G for Future Wireless Networks," *International Congress on Information and Communication Technology*, pp. 1292–1296. https://doi.org/10.1109/ICICT57646.2023.10134364
30. Vergili, I. (2023). Application of AI & ML in 5G Communication. Transactions on Computer Systems and Networks, 149–170. https://doi.org/10.1007/978-981-99-0109-8_9
31. Khan, M. F., Dutta, P., Yadav, A., & Kshirsagar, U. (2023). "Application of AI and ML for Sensor Data Analysis," *7th IET Smart Cities Symposium (SCS 2023)*. Hybrid Conference, Bahrain. pp. 521–526. https://doi.org/10.1049/icp.2024.0981

Chapter 7

Reconfigurable intelligent surfaces

Enabling the future of advanced wireless communications

Sasmita Dash

7.1 INTRODUCTION TO RECONFIGURABLE INTELLIGENT SURFACES

Emerging massive-data applications, such as autonomous vehicles, tactile internet, virtual reality, and augmented reality, have all contributed to the ever-growing communication traffic. MIMO (multiple-input multiple-output), a fundamental technology for 4G and 5G, employs multiple antennas at the transmitter and receiver to achieve spectral efficiency (SE), spatial multiplexing, and link reliability. MIMO with tens to hundreds of antennas (massive MIMO) leverages large antenna arrays for even greater capacity and SE through beamforming and spatial multiplexing, but even these advanced systems are limited in practice by propagation environments and hardware constraints. However, massive MIMO alone cannot overcome the challenges in some propagation scenarios, such as non-line-of-sight (NLoS), high path loss at high frequencies millimeter-wave (mm-wave) and THz bands, and high user mobility. Traditional MIMO systems consider the wireless channel as an uncontrollable medium and only rely on sophisticated signal processing and active antenna design at the transceivers to handle the channel, resulting in increased system complexity, hardware cost, and energy consumption. Moreover, accurate channel estimation for large antenna arrays is challenging and gets worse in dense and dynamic environments, particularly at higher frequencies where propagation is sparse and subject to blockage and fading [1]. Traditional relaying techniques like amplify-and-forward and decode-and-forward relays actively retransmit signals to enhance coverage and throughput; however, the use of relays increases the hardware cost, power consumption, and latency because the relays need full radio frequency (RF) chains and decoding/encoding processes at each relay node. Relaying is not preferred for applications with strict energy or cost constraints because of the computational and power burden [2].

To overcome these limitations, the wireless propagation environment itself can be controlled, rather than simply increasing the number of antennas or adding active network elements. Motivated by this, there has been significant interest in two-dimensional (2D) metamaterials, termed metasurfaces (MSFs), which can manipulate electromagnetic (EM) waves over a wide frequency range from microwaves to THz [3]. MSFs are planar structures made up of sub-wavelength meta-atoms arranged periodically on a surface. Due to their sub-wavelength periodicity, MSF can be described with effective surface parameters, such as impedance or conductivity, which allows unprecedented control over EM wave propagation [4].

By incorporating controllable elements into the unit-cell design and applying corresponding biasing mechanisms, tunable MSF can be realized, and the surface impedance can be adjusted either homogenously or locally to achieve reconfigurable functionalities such as absorption, reflection, transmission, polarization conversion, and anomalous wave steering. MSF can be classified as reflective (one-port) or transmissive (two-port), where reflective

DOI: 10.1201/9781003637721-7

MSF can work as tunable absorbers or reflectors [5], and transmissive MSF allow for controlled transparency or absorption [6]. Since they can be reconfigured, tunable MSF are also known as reconfigurable intelligent surfaces (RIS) or intelligent reflecting surfaces (IRS).

RIS has attracted significant interest for reconfiguring the radio propagation environment intelligently to improve wireless network coverage, capacity, and energy efficiency (EE) and is seen as a fundamental enabler of future 6G communication networks [7]. The main design parameter of RIS-assisted systems is the reflection coefficient of each unit cell, which allows RIS to realize a variety of functions, such as anomalous reflectors, absorbers, beam steerers, and polarizers [8]. An advanced and highly promising realization of RIS is the graphene-based RIS, which can dynamically and continuously tune the phase, amplitude, and polarization of the EM waves by modulating the chemical potential of graphene through electrostatic biasing [9, 57]. Unlike conventional RIS implementations that are based on PIN diodes, varactors, or microelectromechanical system (MEMS) switches, graphene-based RIS uses the electrically tunable surface conductivity of graphene to realize ultra-thin, and fast-reconfigurable RIS, which is highly suitable for the high-frequency, energy-efficient, and compact 6G wireless environments [8, 9, 57]. While conventional MIMO or relay-based solutions actively transmit signals, RIS does not transmit signals but alter the propagation environment around them by controlling the reflection, scattering, or absorption of EM waves. RIS can reduce path loss, convert NLoS scenarios into effective line-of-sight (LoS) paths, and enhance signal quality with low power consumption, without the ability to control the environment [10]. This shift, from controlling transceiver technologies to controlling the wireless medium itself, is a paradigm shift for future wireless systems. In addition to the performance potential in terms of spectral and EE, controlling the propagation environment can also bring additional opportunities in terms of coverage resilience, which is particularly critical in the high-frequency and dense deployment scenarios envisioned for 6G networks. By steering and reshaping radio waves before they reach the receiver, RIS can effectively extend the coverage of the MIMO system and complement MIMO [11].

The ability to dynamically control reflection, transmission, and absorption of EM waves by RIS can enable unprecedented flexibility and adaptability in controlling wireless signals and optimizing wireless communication systems [3]. RIS technology can alter the EM world as we have never been able to before. RIS allows control and programming of the wireless communication channel, making it a powerful tool for wireless communication [12]. Since we can control the illumination of NLoS areas where direct signaling from the anchor node is not possible, RIS is favorable for radio communication and localization, as radio localization typically requires more than one anchor to function, while RIS offers a cost-effective and energy-efficient solution to replace additional anchor nodes and relays [13]. This is because the hardware of RIS can be implemented simply, requiring less power, and making it easy to deploy and maintain [14]. Besides being reconfigurable, RIS has some practical benefits for wireless networks: RIS can be fabricated with low-cost materials such as printed circuit boards, which allows for large-scale and affordable deployment [15], can be used to reduce the required high transmit power while improving overall EE by focusing and redirecting signals toward desired receivers [16], and can reduce interference by generating spatial nulls in unwanted directions, which can enhance signal quality and channel capacity [17].

This chapter presents an overview of RIS as an enabling technology for future and next-generation advanced wireless communication systems. This begins with a description of the operating principles of RIS, including its architecture, wave manipulation mechanisms, operating modes, followed by its localization, then continues to discuss how RIS can enhance the performance of wireless system from perspectives such as coverage enhancement, SE improvement, EE gains, capacity increases, interference mitigation, and physical-layer security. Subsequently, RIS applications such as 5G-Advanced and 6G communications, wireless power transfer (WPT) and energy harvesting (EH), localization and sensing, and integration with complementary technologies, including massive MIMO, internet-of-things (IoT), and intelligent networking are discussed. Finally, the major research challenges and future outlook are presented, positioning RIS as a fundamental technology for future wireless communication systems.

7.2 FUNDAMENTALS AND OPERATING PRINCIPLES OF RIS

MSF is a class of artificial 2D material with an electrically near-zero thickness whose EM properties are designed to be dictated by their geometrical configuration rather than bulk composition [18]. A typical structure of an MSF consists of periodic arrangement of sub-wavelength scattering elements, such as metallic or dielectric inclusions, which can modulate the incident EM waves depending on the spatial distribution and geometry of the scattering element [19]. While under normal conditions, when EM waves strike the boundary between two media, the incident, reflected, and refracted waves are related by Snell's law and the Fresnel equations [19]. An MSF introduces an engineered discontinuity by exploiting the periodic distribution of sub-wavelength scatterers to modify the effective boundary conditions at the interface, resulting in additional phase gradients and resonance shifts, allowing anomalous reflection and diffraction behaviors that violate classical propagation laws and yield controllable phase shifts in the reflected and diffracted waves that are not predicted by conventional EM theory.

RIS is a reconfigurable version of MSF technology that can alter the phase response of each scattering element by applying external stimuli (e.g., electrical biasing or tunable materials) to adjust the physical or EM properties of the unit cells without having to refabricate the structure [20]. The main design challenges of an RIS are to develop an effective control and communication system to address and coordinate a large number of scattering elements in real time, and to achieve precise, wide-range, and continuous phase tunability to accurately control the reflected or diffracted EM waves for dynamic wireless communication.

7.2.1 RIS architecture and components

RIS is a new type of EM surfaces that can actively manipulate the wireless propagation environment to achieve more efficient, flexible, and adaptive wireless communication systems. An RIS comprises a large number of compact, low-cost, and programmable elements, which control EM wave characteristics, such as amplitude, phase, polarization, and, in some cases, frequency, for the signals reflected or transmitted by the surface [21]. Typically, RIS consists of a dense array of nearly passive EM elements in planar or conformal structures, whose collective EM response can be dynamically reconfigured. The basic architecture of an RIS consists of three main components: (i) the reconfigurable surface (meta-atoms), (ii) the control circuitry, and (iii) the communication and power management unit.

The RIS consists of an array of sub-wavelength unit cells, also known as meta-atoms, which are typically fabricated as MSF or metamaterial-inspired structures, allowing independent phase, amplitude, polarization, or frequency response control of reflected or transmitted EM waves. A representation of RIS architecture is shown in Figure 7.1. The reconfigurable surface is the core of the RIS, and it consists of an array of sub-wavelength unit cells, called meta-atoms, that are implemented using MSF or metamaterial-inspired structures, and the tunability is obtained by integrating electronic components, such as PIN diodes, varactor diodes, or MEMS switches, or by employing tunable materials such as liquid crystals and graphene. RIS can achieve a variety of functions, such as beam steering, beam focusing, signal enhancement, anomalous reflection, and interference suppression, by properly configuring the EM response of the individual meta-atoms [8].

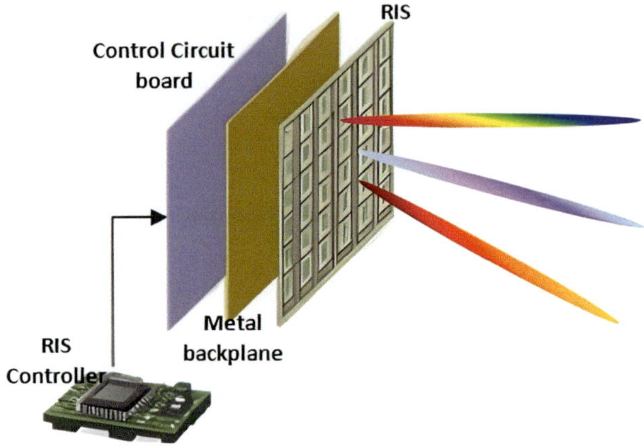

Figure 7.1 RIS architecture.

The control circuitry is the layer of intelligence of the RIS and is responsible for controlling the states of individual unit cells to achieve the desired propagation objectives. The control layer usually uses microcontrollers or field-programmable gate arrays (FPGAs) to run control algorithms and generate biasing signals for the tunable elements, and the control decisions can be computed locally at the RIS or provided by an external network entity such as a base station or edge computing unit based on channel state information (CSI) and system-level optimization strategies [22]. The communication and power management unit coordinates the interaction between the RIS and the wireless network by exchanging control commands, synchronization information, and, in some implementations, limited sensing data via wired or wireless communication links. Because RIS is designed to be very energy efficient, the major source of power consumption is the control and reconfiguration, rather than the signal transmission or amplification. Low-power electronics, battery-assisted operation, or EH (such as solar or RF EH), can be employed to support energy-efficient operation.

RIS reconfiguration is typically realized by jointly controlling the phases of each scattering element, which necessitates the integration of tunable chips within the MSF structure, each chip locally interacting with a scattering element and communicating with a central controller [23]. For example, FPGA-based controllers with PIN diodes are usually used to realize discrete phase states [23], the embedded controller receives reconfiguration requests from external devices, optimizes the phase distribution, and distributes control signals to all tunable elements. Each tunable chip then changes its state to reflect the control signal, thus allowing the corresponding scattering element to reconfigure its EM response. Furthermore, embedded sensors within RIS can sense environmental conditions, so the controller can update configurations autonomously and maintain stable performance in dynamic scenarios. Various types of tunable components have been investigated in the literature. PIN diodes provide binary ON–OFF states that can be used to match or mismatch with free-space impedance, and experimental demonstrations of binary-state MSF unit cells operating at 2.4 GHz using PIN diode–controlled hybrid resonators are presented in [24]. Varactor diodes can provide continuous tunability by changing the capacitance with the applied voltage, allowing a higher phase resolution [21]. Integrated circuits with continuously tunable resistance and capacitance have been proposed to control the phase shifts by tuning the gate voltages, which have resulted in more advanced functionalities such as polarization-insensitive perfect absorption at 5 GHz over wide incidence angles [25].

More advanced architectures combine networks of distributed microcontrollers in the MSF, connected wirelessly to external control devices [26], with each controller translating control commands into local impedance states to achieve the desired EM responses. Others simplify hardware by connecting multiple scattering elements to a single controller chip that assigns additional local resistance and reactance as required. The same principles are used in MSF made of reconfigurable metamaterial strips arranged into grid structures, where tiles of strips are controlled by intra-tile controllers connected through a hierarchical control network [27]. This architectural simplicity, together with low energy consumption and software-defined reconfigurability, makes RIS a key enabling technology for intelligent and programmable wireless environments in 5G-Advanced and future 6G communication systems, which can be flexibly deployed on building facades, indoor walls, ceilings, and urban infrastructure.

7.2.2 Wave manipulation in RIS

The fundamental principle of wave manipulation in RIS is the ability to control the EM response of individual MSF elements with high precision and programmability, where the phase, amplitude, and polarization of the scattered waves can be tailored at the sub-wavelength scale to generate multiple reflected or transmitted wavefronts in different directions simultaneously [28]. The perfect or near-perfect phase control across the MSF enables the spatial and temporal redistribution of reflected power, resulting in highly customized radiation patterns. This exotic power distribution can be further leveraged to encode information in the spatial domain, and thus spatial modulation–based communication paradigms [28].

Early examples of programmable wave manipulation are based on binary-state control of scattering elements with PIN diodes, where the authors optimize a binary coding matrix applied to a large MSF to achieve various EM functionalities, including anomalous reflection, diffusion, beam steering, and beamforming, and demonstrate the dynamic adaptability of RIS by real-time switching among these radiation patterns using an FPGA-based controller. More recently, heuristic optimization techniques have been introduced to further increase the design flexibility, such as using a genetic algorithm to generate arbitrary wave modulations that meet the application-specific radiation requirements [29]. The experimental results at 10 GHz corroborate that the wave manipulation accuracy increases with the MSF aperture size. In addition to spatial control, RIS can also perform simultaneous control of EM waves in both space and frequency domains. Space–time modulated digital coding MSF have been designed to dynamically control reflection angles and power distribution across different frequency components [30], which can provide beam steering, beam shaping, frequency conversion, and scattering-signature control in the 8–10 GHz band [30].

RIS reconfigurability is fundamentally derived from the physical manipulation of the phase response of each scattering element, where the physical parameters of the scattering elements and substrate are tuned by external or ambient stimuli, such as electric, magnetic, optical, and thermal excitations, to control the EM behavior of the MSF globally or locally for advanced wave manipulation. The simplest local tuning approach is to change the physical dimensions of the scattering elements, thereby changing their resonant frequencies and introducing phase shifts. Full phase controllability is realized by unit cell and minimizing undesired reflections through array pattern optimization [8]. Other more practical and scalable methods use electrically controlled tuning mechanisms, such as binary phase switching using PIN diodes and continuous reactance tuning using varactor diodes, where wave manipulation can be achieved by adjusting the biasing voltages of the tunable components, as also done in our work.

Other material-based tuning strategies have also been investigated, such as integrating electronically controllable liquid crystals into MSF unit cells to allow real-time wave manipulation [31] or acoustic and mechanical tuning approaches [32] where the acoustic cell architecture provides enough degrees of freedom to control the refractive MSF, achieving beam redirection angles of more than 60° with efficiencies of 90%. The versatility of RIS-based wave manipulation is further demonstrated by large-scale programmable MSF platforms. Within the Horizon 2020 FET-OPEN program, the VISORSURF project aims to design a hardware platform, called hypersurface, with programmable EM behavior by means of a network of controllable switches [22]. Based on experimental prototypes, programmable microwave MSF are demonstrated to provide multifunctional wave control, such as beam splitting, deflection, and anomalous reflection, through the coordinated control of PIN diodes and varactor-loaded unit cells [33], which can be switched on and off by computer-controlled voltage sources. The RIS-enabled wave manipulation enables simultaneous signal enhancement and direction suppression, which can be used for intelligent control of the wireless environment and for advanced applications in wireless communications, IoT, Industry 4.0, sensing, and 6G systems.

7.2.3 Operation modes of RIS

In contrast to conventional wireless technologies that focus on optimization of the transmitter and receiver, RIS provides a fundamentally new paradigm to allow programmable control of the wireless propagation environment itself. Shaping the EM response of environmental objects distributed throughout the network, RIS can make the wireless channel partially controllable, improving the reliability, coverage, and efficiency of communication. RIS can operate in several distinct modes depending on their physical structure, power consumption, and intended functionality, each with their own hardware requirements and application scenarios.

RIS can be categorized based on structural realization (continuous or discrete MSF), power consumption (passive or active), and most importantly operational behavior, which defines how incident EM waves are manipulated. RIS can be operated in four different modes: reflective, transmissive, hybrid reflecting–sensing, and STAR, each with its own unique advantages and trade-offs in terms of coverage, flexibility, hardware complexity, and EE and each will play a role in the selection of appropriate RIS designs for 5G-Advanced and future 6G wireless networks. Figure 7.2 illustrates different operating modes of RIS.

Reflective RIS: Reflective RIS is the most commonly researched and practically developed operating mode, and in this configuration, the MSF is supported by a conductive ground plane (e.g., copper) to block the wave transmission and all incident signals are reflected. The EM response of each RIS element is considered as a programmable reflector whose surface impedance can be tuned. In a reflective RIS, the reflection coefficient of the nth element can be described as $\alpha^n e^{j\phi_n}$, where $\alpha_n \in [0,1]$ and $\phi_n \in [0,2\pi)$

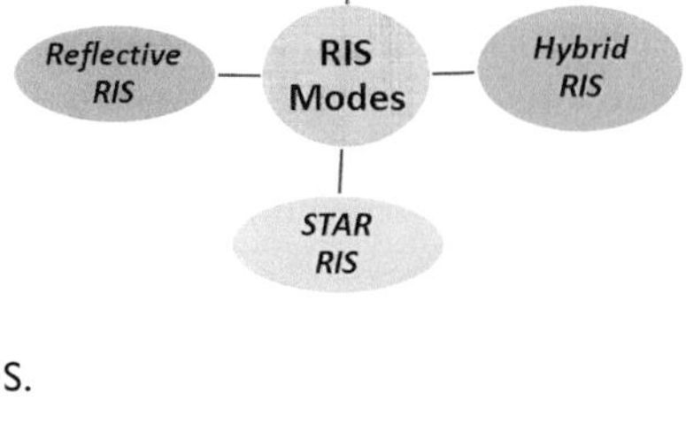

Figure 7.2 Operating modes of RIS.

represent the amplitude and phase response, respectively. With the ability to control these parameters over the surface, RIS can steer beams, increase desired signal paths, decrease interference, and correct extreme path loss. In practice, perfect unit-amplitude reflection ($\alpha = 1$) cannot be achieved, as there is always loss in the circuits; the achievable amplitude–phase resolution depends on the hardware complexity and cost [34]. Reflective RIS implementations in practice typically support one of the following three configurations: (i) constant-amplitude reflection with continuous phase control, (ii) optimized amplitude with continuous phase control, and (iii) constant-amplitude reflection with discrete phase control. While continuous phase tuning exhibits superior performance, discrete phase control is preferred due to reduced hardware complexity and energy consumption. Furthermore, amplitude and phase responses are coupled, and hence phase variation can also lead to amplitude changes.

Transmissive RIS: Transmissive RIS allows incident EM waves to pass through the surface but change their propagation characteristics. Unlike reflective RIS, transmissive RIS does not have a metallic backplane, which means energy can be transmitted across the surface [35]. This operation mode offers more degrees of freedom in wave manipulation, especially in cases where signal coverage is needed on both sides of the surface, and by tuning the impedance of each unit cell, the transmissive RIS can control the phase, amplitude, and polarization of the transmitted wavefront. Transmissive RIS configurations are particularly attractive for indoor deployments, smart windows, and dense urban environments where signal blockage and penetration losses are key challenges.

Hybrid RIS: Traditional RIS designs are based only on passive reflection, which can lead to problems with channel estimation, synchronization, and self-adaptive capability, while hybrid reflecting and sensing RIS can reflect and sense part of the incident signal, so that a portion of the energy is diverted into sensing circuits embedded in the surface, and the other part of the energy is reflected to the receivers, which can estimate the channel parameters, detect the user location and environmental changes, which reduces the dependence of the network entity to update the RIS configuration, which can realize the autonomous and self-adaptive operation mode [36].

Simultaneously transmitting and reflecting (STAR) RIS: STAR-RIS represents an important evolution from the traditional RIS architectures, which can only serve users on the same side of the surface. STAR-RIS can simultaneously transmit and reflect incident signals [37]. STAR-RIS is also supported by novel metamaterials that can operate in both electric polarization and magnetization currents, allowing independent control of reflected and transmitted wave components, which can serve users on both sides of the surface simultaneously, enhancing SE, deployment flexibility, and network densification potential, particularly for future 6G systems, ultra-dense networks, and integrated sensing and communication (ISAC) scenarios.

Active and Passive Operation: From a power-consumption point of view, RIS operation modes can also be divided into passive and active types, where passive RIS only consumes energy for control and reconfiguration and are therefore highly energy-efficient, while active RIS employs amplifying components to compensate for reflection or transmission losses and thus provide improved performance at the expense of higher power consumption and hardware complexity.

7.2.4 RIS-assisted localization

Accurate localization is a critical component of many emerging wireless applications, including autonomous navigation, smart cities, IoT, healthcare monitoring, and emergency services. Conventional localization methods are based on global navigation satellite systems

(GNSS) for outdoor environments and Wi-Fi, Bluetooth, or ultra-wideband (UWB) signals for indoor scenarios, which are limited in accuracy due to multipath propagation, NLoS conditions, signal blockage, and severe attenuation, especially in dense urban and indoor environments. RIS addresses these limitations by actively manipulating the wireless propagation environment through the use of large numbers of programmable passive or semi-passive elements that can control the phase, amplitude, and polarization of incident EM waves to strengthen desired signal paths, reduce interference, and establish beneficial propagation environments between transmitters and receivers.

RIS has emerged as a promising technology for enabling accurate localization in outdoor and indoor wireless environments by intelligently controlling the propagation of EM waves. RIS-assisted localization has been shown to significantly reduce positioning errors by introducing a controllable reflection path and increasing geometric diversity between transmitters, RIS panels, and receivers [38]. The inclusion of RIS within localization systems marks a shift in paradigm toward environment-aware and programmable positioning, although challenges exist in terms of channel state information acquisition, signaling overhead, and real-time optimization, and research suggests that RIS-assisted localization will be a significant component of future 6G networks to provide accurate, energy-efficient, and reliable positioning for next-generation wireless applications [38]. The benefits of RIS-assisted localization manifest differently in outdoor and indoor environments.

In outdoor environments, especially in urban environments where GNSS accuracy is degraded due to signal blocking and multipath, the use of RIS on building walls or roadside structures can enhance the robustness of localization, improving positioning accuracy for vehicular networks, unmanned aerial vehicles (UAVs), and smart city applications by reducing NLoS propagation and strengthening weak signal paths [39, 40]. Additionally, multiple-RIS deployments can facilitate cooperative localization frameworks that provide greater spatial resolution without increasing transmit power or system complexity [40]. RIS also provides more significant assistance for indoor localization because GNSS signals are not available indoors and there is strong multipath fading, and signals reflected by RIS can be steered to user devices to improve the signal-to-noise ratio and the accuracy of angle of arrival (AoA), and Time of Arrival (ToA) estimation [41]. Some recent works have also used machine learning (ML)–based and data-driven methods to optimize the RIS configurations and localization algorithms simultaneously, achieving sub-meter and even centimeter-level accuracy in complex indoor environments such as smart factories and healthcare facilities [41].

7.3 RIS-ENHANCED WIRELESS SYSTEM PERFORMANCE

RIS has become a revolutionary paradigm to increase the performance of wireless systems by providing programmable control over the propagation environment, as compared to conventional wireless systems, which treat the channel as an uncontrollable entity and manipulate EM waves by using tunable meta-atoms to improve signal quality, reliability, efficiency, and security. In this section, the main performance gains introduced by RIS are described, which include signal coverage, spectral and EE, capacity and interference management, and physical layer security (PLS).

7.3.1 Signal coverage enhancement

Signal coverage is one of the major performance metrics in wireless communication systems, and it becomes even more important in mm-wave and THz bands due to severe path loss, blockage, and shadowing that degrade the link reliability, which can be alleviated by the intelligent reflection and redirection of incident signals to intended users by RIS, especially in NLoS scenarios. If the phase shifts of the individual RIS elements are properly configured, the reflected signals can be coherently combined at the receiver with constructive interference, which can lead to an increase in the received signal power by several orders of magnitude [42]. RIS-assisted coverage enhancement is particularly useful for cell-edge regions, severely lossy indoor environments, and high-frequency mm-wave/THz systems that are very susceptible to blockage. Graphene-based RIS is a promising implementation platform for these bands, as it offers tunable surface conductivity, ultra-fast reconfigurability, and strong plasmonic response at THz frequencies [8], whereas conventional metallic or PIN diode–based RIS allows only discrete phase and amplitude control. Graphene RIS is especially suitable for dynamic and adaptive coverage extension in THz communication systems. Additionally, RIS is a low-power and low-complexity solution for coverage enhancement, since it does not need RF chains or power amplification compared to the traditional relays or active repeaters [43].

7.3.2 Spectral and energy efficiency enhancement

The performance metrics SE and EE are important metrics for next-generation wireless networks, especially for 6G and beyond, which aims at massive connectivity and sustainability. RIS can increase SE by enhancing the effective channel gain and signal-to-interference-plus-noise ratio (SINR), and it can achieve beamforming in the passive domain by smartly adjusting the phase, which can spatially focus the signal toward the desired receivers while reducing leakage, thus achieving higher achievable data rates without expanding transmission bandwidth or power [44]. RIS provides an energy-efficient solution because, although RIS elements require a little power to be reconfigured (usually through low-power control circuits), the power consumption of the network can be dramatically reduced by RIS-assisted systems, where the power consumption is orders-of-magnitude lower than active relaying solutions [45]. The main contributors to SE and EE gains are passive beamforming without RF chains, lower transmit power requirements, and better channel conditions to facilitate adaptive modulation and coding. Hence, RIS is considered a green technology that is in line with the energy-efficient design of wireless networks.

7.3.3 Capacity enhancement and interference management

Another major advantage of deploying RIS is that it can enhance the network capacity by dynamically shaping the wireless channel, thus leading to substantial improvements in the sum-rate capacity of multi-user systems via increased desired signal strength through coherent reflection, spatially separating the users via programmable wavefront control, and suppressing inter-user and inter-cell interference. RIS can help the base station to form highly directional beams to support multi-user MIMO (MU-MIMO) systems with more spatial degrees of freedom and to simultaneously serve multiple users [46]. RIS can also be programmed to cancel interference to unintended receivers, enhancing fairness and quality-of-service (QoS). Interference management with RIS is especially useful for ultra-dense networks, cell-free massive MIMO systems, and co-channel spectrum sharing scenarios, where interference-aware channel engineering is difficult to realize by traditional transceiver-based techniques.

7.3.4 Physical layer security enhancement

Due to the vulnerability of wireless environments to eavesdropping, PLS is attracting more attention as a complementary technique to cryptography, and RIS is a promising technology that can enhance PLS by intelligently controlling the propagation of signals to benefit legitimate users and impair eavesdroppers by increasing the received signal power at legitimate receivers, creating destructive interference at potential eavesdroppers, and artificially manipulating channel randomness. This spatial selectivity allows secure communication even if the eavesdropper is near the legitimate receiver [47]. PLS assisted by RIS is particularly attractive since it does not require extra transmit power or artificial noise generation, which renders it an energy-efficient security solution. The results demonstrate that RIS can enhance secrecy capacity, secrecy outage probability, and robustness against passive and active eavesdropping. Therefore, RIS introduces a PLS-by-design dimension that can complement higher-layer security mechanisms.

7.4 RIS APPLICATIONS

RIS is a new enabling technology for beyond-5G and 6G wireless networks, as the combination of ultra-high data rates, extreme SE, sub-millisecond latency, and integrated communication–sensing–energy services requires a redesign of the wireless propagation environment. RIS can turn the radio channel that was once uncontrollable into a controllable and intelligent one by software-defined and nearly passive control of EM waves. In addition, emerging implementations based on graphene and other tunable nanomaterials are also expected to play a key role in mm-wave and THz regimes due to the ultra-fast reconfigurability, strong field confinement, and continuous tunability of graphene, which are particularly desirable for 6G use cases such as dynamic beam steering, ultra-broadband operation, and joint communication and sensing (JCAS). Figure 7.3 presents IRS-assisted wireless communication scenario.

In addition to SE and coverage enhancement in 5G-Advanced and 6G communications, RIS can enable WPT and EH, high-precision localization and sensing, and seamless integration with enabling technologies, including massive MIMO, cell-free architectures, non-terrestrial networks (NTN), artificial intelligence (AI), and THz communications. RIS can be a key enabler of 6G networks in the form of programmable, energy-efficient, and multifunctional wireless ecosystems, particularly with the graphene-enabled and THz-capable forms

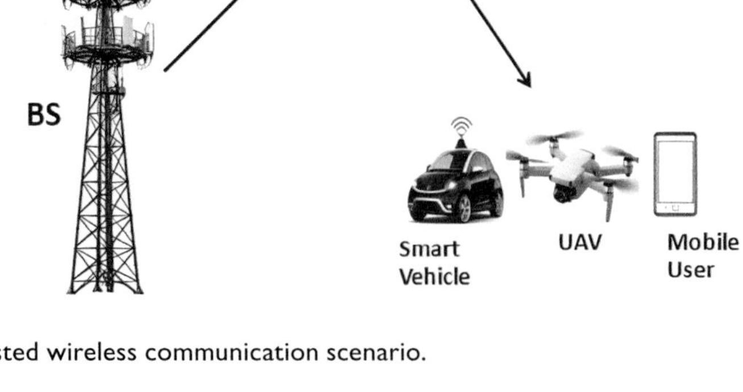

Figure 7.3 IRS-assisted wireless communication scenario.

as research advances toward intelligent, sustainable, and human-centric 6G networks. This section details the principles, benefits, and future research directions of these application domains in the following subsections.

7.4.1 5G-Advanced and 6G communications

RIS is considered a disruptive enabling technology for 5G-Advanced and future 6G communication systems, where the ultra-high data rates, massive connectivity, low latency, and EE requirements challenge the conventional network architecture [48]. Contrary to the conventional wireless systems that only adapt the transmitter and receiver, RIS allows for the programmable control of the radio propagation environment, improving the link reliability and system performance in a cost- and energy-efficient manner [48]. The coverage in 5G-Advanced networks, which will mostly operate in the mm-wave bands, is heavily impacted by signal blockage and severe path loss, especially in dense urban and indoor environments [49]. RIS can mitigate these issues by dynamically reflecting incident EM waves toward intended users, thereby creating favorable NLoS paths and enhancing received signal strength without requiring any additional active radio-frequency chains [49], which is critical for achieving uniform quality of service in heterogeneous network deployments, especially at the cell edge.

RIS can provide significant gain in achievable sum-rate and SINR [48, 50] by dynamically shaping the wireless channel through the phase and amplitude control of the MSF elements, which makes RIS suitable for emerging applications, such as holographic communications, extended reality (XR), and digital twins, and for extreme SE and ultra-massive connectivity, both of which are necessary for 6G. Graphene has emerged as a particularly promising implementation platform for RIS at sub-THz and THz frequencies (widely considered significant spectrum bands for 6G) because of its ability to dynamically and continuously tune the reflection phase and amplitude via its electrically tunable surface conductivity, strong plasmonic response, and ultra-fast reconfigurability (which is difficult to achieve with conventional metallic or diode-based RIS) [8]. The above properties make graphene-based RIS a particularly attractive solution for adaptive beam control, coverage enhancement, and mobility support in THz-enabled 6G networks with low power consumption and compact form factors.

Another main building block of future wireless networks is EE, and RIS is a promising technology because of its nearly passive operation (i.e., RIS consumes much less power than active relays and small cells, yet it can provide significant performance gains [51], and recent studies show that RIS-assisted networks can lower the transmit power requirements to maintain the target QoS levels, which is a major benefit for IoT and battery-constrained devices [51]). Additionally, the combination of AI and ML with RIS control is becoming a key enabler for real-time optimization in highly dynamic 6G environments. Deep learning–based approaches have been demonstrated to optimally configure RIS, even with imperfect channel state information, thereby enhancing data rates and EE [52]. Advanced RIS architectures, such as STAR-RIS, extend deployment flexibility with full-space wireless coverage highly sought after in next-generation networks [52]. RIS is expected to be part of the basis for 5G-Advanced evolution and 6G network design, which will result in intelligent, adaptive, and energy-aware wireless systems. RIS will convert the traditional passive radio environment to an active and controllable entity to improve coverage, capacity, SE, and sustainability, which are key performance targets for future wireless communication ecosystems.

7.4.2 Wireless power transfer and energy harvesting

RIS is a promising technology to enable WPT and EH in next-generation wireless networks, including 5G-Advanced and 6G ecosystems [53]. Conventional WPT methods are based on inductive or radiative coupling mechanisms, which suffer from limited transfer range, conversion efficiency, and scalability. RIS provides a new paradigm to actively control the propagation environment to concentrate EM energy toward the targeted energy harvesters or device receivers, which improves both power delivery and EH efficiency without adding power-hungry active components at every surface location. RIS can be integrated into WPT systems using the concept of simultaneous wireless information and power transfer (SWIPT) frameworks in which RF signals are used for both information transfer and energy transfer [52]. In RIS-assisted SWIPT, by adjusting the RIS phase shifts and reflection coefficients, the radio environment is designed to maximize the received power at the energy receiver, which can overcome the severe path loss effects caused by far-field power transfer [54].

More advanced RIS designs, such as STAR-RIS and beyond diagonal RIS (BD-RIS), enhance the EH performance. RIS elements perform both EH and information reflection concurrently to achieve better trade-offs between the harvested power and communication rate and to manipulate the cascade channel for RF WPT more flexibly to improve the harvested power in both LoS and NLoS propagation environments, respectively [55]. Aside from improving WPT to terminal devices, RIS technology has also been investigated for self-powered RIS operation. An RIS that harvests energy from incident RF signals or dedicated energy sources can potentially operate independently of the external grid power, which can be critical for large-scale RIS deployments in dense urban or remote settings where wired power supply is not feasible. The RIS can harvest sufficient energy to sustain its control and reconfigurability functions under certain deployment and power splitting protocols, making it a viable option for large-scale RIS deployments in dense urban or remote settings. Both experimental and analytical studies have shown that the harvested power can be greatly improved at the energy receivers, and the system reliability and sustainability can be improved by using RIS to realize the pervasive deployment of self-charging IoT devices and wireless sensors in smart environments [56]. RIS-assisted WPT and EH are a promising technology integration in future networks that can achieve the improvement of both communication and energy sustainability.

7.4.3 Localization and sensing

Localization and sensing are core capabilities for modern and future wireless networks, with applications in indoor positioning, autonomous systems, augmented reality, smart manufacturing, and environmental monitoring. Traditional localization and sensing techniques are based on direct LoS signal paths and dense infrastructure (i.e., multiple base stations or dedicated radar systems), so their performance is significantly degraded in NLoS environments with severe multipath propagation, blockage, and coverage holes. RIS enable improved localization robustness in complex environments by intelligently shaping the radio propagation environment to create virtual LoS links and increase the spatial resolvability of wireless channels [57]. RIS enhances localization accuracy as an additional passive and controllable environmental anchor, as its reflecting elements can be dynamically adjusted to control key localization parameters such as signal propagation delay, AoA and angle of departure (AoD), increasing the Fisher information available for position estimation [57]. Therefore, RIS can steer EM energy toward users or targets and wavefront shaping to make small variations in time and angle measurements more discernible at the receiver, further enhancing position and orientation estimation under adverse propagation conditions [57, 58].

In addition to its role in enhancing environmental perception, RIS can also improve the detectability of objects and scatterers that are shadowed by other objects or suffer from multipath fading in the context of ISAC systems, allowing for joint optimization of sensing and communication functionalities that are critical to emerging perception-driven 6G services such as autonomous navigation and intelligent transportation systems [40]. Recent studies show that RIS can be effective in localization and sensing applications, such as RIS-assisted localization schemes inspired by radar and time-difference-of-arrival (TDOA) principles that have been shown to detect targets more effectively and with higher accuracy in obstructed and cluttered environments compared to conventional passive sensing [59], as well as hybrid RIS architectures that integrate sensing capabilities directly into the surface elements to measure incident wave characteristics, which enables direct AoA estimation and real-time environmental awareness. Another promising direction is near-field RIS-assisted localization, which utilizes RIS in the near field of users or objects and can exploit the high spatial variation of the EM field to achieve high-precision localization. It has been shown that practical issues such as mutual coupling and hardware impairments can significantly degrade localization performance, and joint calibration and estimation frameworks have been proposed to ensure robust and accurate position inference [60].

Additionally, ML and deep learning approaches have been introduced to RIS-enhanced localization systems to improve performance in complex indoor environments by learning from RIS-controlled channel measurements to obtain more accurate LoS/NLoS classification and localization inference. Besides enhancing accuracy, RIS-enabled localization and sensing can also facilitate opportunistic and collaborative positioning strategies that do not require as much dense infrastructure or GPS, resulting in lower cost and energy for deployment, which shows the potential of RIS to transform the wireless environment from a passive medium to an active part of localization and sensing [61].

7.4.4 Integration with other technologies

In addition to the standalone benefits of RIS, they have synergistic benefits when integrated with other advanced wireless technologies, such as massive MIMO, NTN, mm-wave/THz communications, SWIPT, cell-free architectures, and AI-driven network optimization, which all serve to enhance performance, coverage, and efficiency and are a central enabler of the performance requirements and functional flexibility expected of 5G-Advanced and 6G systems.

Integration with Massive MIMO: Massive MIMO is a significant component of current wireless networks that can offer high SE and robust spatial multiplexing, but it is challenged by hardware complexity, pilot contamination, and propagation blockages at higher frequencies. When RIS is integrated with massive MIMO, the propagation environment is reconfigurable and the system can shape channel conditions to increase spatial degrees of freedom and channel rank for performance enhancement in MU-MIMO [62]. RIS can also provide passive beamforming gains to reduce the complexity of the high-complexity active antenna arrays at the base station.

Integration with NTN: To achieve ubiquitous connectivity, NTN, such as satellite, high-altitude platform stations (HAPS), and UAVs, are crucial for reliable communication links in remote areas or disaster scenarios, as the integration of RIS with NTNs can alleviate link outages resulting from signal blockage and atmospheric attenuation by dynamically steering incident signals from NTNs toward ground users [50].

Integration with mm-wave and THz Communications: Although the mm-wave/THz bands provide an enormous amount of bandwidth, the propagation characteristics of these high-frequency regimes are characterized by high path loss and sensitivity to blockage [63, 64]. As a result, RIS integration is an effective solution to enhance beam steering and establish virtual LoS [4, 5], thus enabling practical deployment of mm-wave/THz communications to provide ultra-high-speed data services, backhaul/fronthaul links, and extended coverage.

Integration with SWIPT: In the age of massive connectivity, wireless devices require both communication and energy solutions. RIS can be combined with SWIPT frameworks to simultaneously optimize wireless information transfer and EH, with the ability to steer power toward EH circuits while maintaining high information throughput and reducing energy consumption at the transmitter, thus paving the way for sustainable IoT applications [65]. Such a synergy is especially beneficial for low-power IoT nodes and sensors that need to be recharged periodically without wired infrastructure.

Integration with Cell-Free and Distributed Networks: Cell-free and distributed network architectures do not adhere to the concept of cells and are based on cooperative transmissions from geographically distributed access points. Integration of RIS in cell-free networks improves the spatial diversity and minimizes the interference between the users by shaping the effective multi-access channel and can be considered as a virtual access point that cooperatively interacts with distributed antennas to enhance the service uniformity and QoS for all users [66]. This integration also supports scalability and performance in dense urban and massive IoT deployments.

Integration with AI and Network Optimization: Future networks will have dynamic requirements, and thus, advanced network optimization and real-time adaptation are important to achieving the requirements. RIS surfaces generate a high-dimensional control space that can be managed by AI, ML, and deep reinforcement learning, and adaptive configuration of RIS elements based on evolving network states, user mobility, and interference conditions to optimize signal quality, EE, and throughput, which can realize autonomous network management, self-optimizing coverage, and contextual awareness [67].

7.5 RESEARCH CHALLENGES AND OUTLOOK

This section focuses on limitations, open research problems, and practical challenges of RIS-assisted localization that need to be addressed for the reliable, scalable, and cost-effective deployment of RIS-based localization systems in future 6G networks. Although RIS can achieve substantial performance enhancement in terms of localization accuracy through the programmable control of the wireless propagation environment, it also has several inherent limitations.

The main challenge is the cost and complexity of deployment, including the installation, calibration, and maintenance of multiple RIS panels in a large-scale localization system, which requires careful infrastructure planning and may introduce high financial and operational overhead. Most existing RIS-assisted localization frameworks assume LoS or quasi-LoS communication between the RIS, transmitters, and receivers, while signal paths in realistic environments can be blocked by walls, furniture, and moving objects, degrading localization accuracy and robustness. While RIS can enhance signal propagation, highly reflective or obstructed scenarios may still lead to NLoS conditions, causing residual localization errors, which constitutes an open research challenge to effectively counteract the NLoS effect. A third critical limitation is the power supply and connectivity requirements of RIS devices, since although RIS elements are typically passive or semi-passive, the associated control circuitry, sensing modules, and communication interfaces all need constant power and reliable backhaul connectivity, which pose serious logistical challenges for large-scale deployments or in infrastructure-constrained environments.

RIS-assisted localization also faces the challenge of scalability: with a larger number of users, devices, and RIS panels, the coordination, optimization, and real-time configuration of RIS elements will become more complicated and will likely require re-optimization for dynamic environmental changes, user mobility, and evolving service requirements, which incurs system overhead and maintenance effort. In dense deployments with multiple RIS panels, interference and coexistence issues can occur, especially when RIS operates in shared spectrum with other wireless systems [68]. Uncoordinated RIS reflections can also cause interference to neighboring RIS panels or conventional communication links, and therefore effective interference management, spectrum sharing strategies, and cooperative RIS operation are required to achieve reliable performance. However, RIS remains a very promising technology for localization and integrated sensing–communication in 6G networks, and overcoming the aforementioned limitations through improved channel modeling, NLoS-aware localization algorithms, scalable control architectures, energy-aware designs, and intelligent optimization techniques will be critical to the practical realization of RIS-assisted localization, which also provides rich avenues for future research, making RIS a key technology for environment-aware and intelligent wireless systems.

Channel Estimation and Acquisition: Since RIS relies on phase adjustments based on propagation characteristics, accurate CSI is critical to the operation of RIS. However, since RIS does not have direct RF chains, conventional estimation methods are not applicable, and estimating cascaded channels (i.e., transmitter-RIS and RIS-receiver links) would incur significant overhead and complexity, especially in multi-user environments. Therefore, techniques that reduce the training overhead while ensuring estimation accuracy are pivotal to achieving scalable RIS, such as advanced pilot design, compressive sensing approaches, and low-complexity iterative algorithms [69].

Hardware Impairments and Realistic Modeling: Realistic modeling should also include power consumption, switching speed, and reliability in diverse wireless environments. Practical RIS hardware deviates from ideal assumptions; for instance, phase shifts may only be quantized to a finite number of discrete values, and mutual coupling among RIS elements can lead to unintended distortions. If not accounted for in design and optimization routines, such hardware impairments can degrade performance, calling for accurate and practical models that include phase quantization, element coupling, and surface imperfections [70].

Passive Beamforming Optimization: Since the unit-modulus phase shift constraints of RIS elements, the coupling between RIS and active beamformers, and the multi-user interference lead to non-convex problems, finding globally optimal solutions is computationally infeasible for large surfaces, especially for the case of real-time adaptability, which necessitates research on low-complexity algorithms, distributed beamforming design, and ML–assisted optimization [67].

Integration with Emerging Technologies: As RIS is integrated with other complementary technologies such as massive MIMO, NTN, mm-wave and THz bands, and AI-enabled network management, cross-domain challenges emerge. For instance, the synchronization between RIS and high-mobility nodes (e.g., UAVs or vehicles) is difficult to handle due to the rapid channel variations. Joint optimization strategies need to be devised to guarantee the robust performance of these integrated systems with manageable overhead and energy consumption [62].

Deployment Strategies and Network Architecture: The achievable gains in coverage and performance with RIS depend on the deployment strategies (e.g., where to place RIS panels and how many to deploy) which are still an open problem, particularly in heterogeneous scenarios involving indoor/outdoor environments, urban street canyons, and non-uniform user distributions. Network architectures with co-design of RIS, relays, conventional base stations, and cell-free nodes should be designed to balance performance enhancement with cost, latency, and complexity.

Security, Privacy, and Robustness: Although RIS can enhance PLS by strengthening legitimate communication links while weakening potential eavesdropper links, practical adversarial scenarios make this challenging, and attackers may tamper with RIS configurations or exploit vulnerabilities in control signaling to compromise communication privacy. Therefore, robust and secure RIS control protocols, authentication mechanisms, and cyber-physical defense strategies are required to maintain system integrity.

Standardization and Practical Implementation: It is critical that standardization of RIS is achieved to move from research prototypes to commercial systems, including defining control interfaces, signaling protocols, and performance evaluation benchmarks aligned with evolving 5G-Advanced and 6G specifications, real hardware testbeds and field trials, and standard metrics for evaluating RIS performance (e.g., phase resolution impact, effective reflectivity, and channel coherence). The prospect for RIS research is promising and expands into multi-disciplinary areas, such as RIS-assisted JCAS, intelligent control via AI for dynamic environments, RIS-aided integrated localization and mapping, and energy-harvesting RIS for self-sustainable surfaces. With the continued advancement of ML and optimization algorithms, self-optimizing RIS configuration and adaptation to environmental context can be envisioned, and RIS in conjunction with massive MIMO, non-orthogonal multiple access (NOMA), and cell-free network paradigms are anticipated to be an integral component of 6G ecosystem design.

7.6 CONCLUSION

As wireless networks evolve into 6G and beyond, RIS is expected to be a pivotal enabler for intelligent, adaptive, secure, and energy-efficient communications. Continued interdisciplinary research and coordinated standardization efforts will be critical to translating RIS from a theoretical concept to a technology widely deployed to shape the future of wireless communications. This chapter presented an in-depth and systematic overview of RIS as an enabler for next-generation wireless communication systems. Starting with the basic principles of RIS operation, the chapter discussed the underlying architecture, constituent components, wave manipulation mechanisms, and different operating modes that enable RIS to dynamically control EM propagation, as well as RIS-assisted outdoor and indoor localization, which demonstrates how RIS can change the environment into an active part in signal propagation and spatial awareness. The impact of RIS on wireless systems performance was subsequently investigated [26], showing that properly controlling the radio environment with RIS can boost coverage area especially in NLoS or limited-coverage situations, as well as SE, energy-efficiency, system capacity while suppressing interference and physical-layer security. In addition to SE and coverage enhancement in 5G-Advanced and 6G communications, RIS can enable WPT and EH, high-precision localization and sensing, and seamless integration with enabling technologies, including massive MIMO, cell-free architectures, NTN, AI, and THz communications. The chapter also outlined several open research challenges that need to be addressed for practical and large-scale RIS deployment. The research outlook indicates that further progress in hardware design, EM modeling, learning-assisted optimization, and cross-layer system integration will be important to fully harnessing the potential of RIS in real-world networks. RIS has great potential to be used for 6G networks in the form of programmable, energy-efficient, and multifunctional wireless ecosystems, particularly with the graphene-enabled and THz-capable forms as research advances toward intelligent, sustainable, and human-centric 6G networks.

REFERENCES

1. Pavia, J.P., Velez, V., Souto, N., Silva, M.M. da and Correia, A. (2024) System-level assessment of massive multiple-input–multiple-output and reconfigurable intelligent surfaces in centralized radio access network and IoT scenarios in sub-6 GHz, mm-wave, and THz bands. *Applied Sciences*, 14(3), 1098.
2. Wang, M., Zhao, B., Li, W., Jin, M. and Jin, S.-N. (2025) Full-duplex relaying systems with massive MIMO: Equal gain approach. *Symmetry*, 17(5), 770.
3. De Lima, C., et al. (2021) Convergent communication, sensing and localization in 6G systems: An overview of technologies, opportunities and challenges. *IEEE Access*, 9, 26902–26925.
4. Pendry, J.B., Schurig, D. and Smith, D.R. (2006) Controlling electromagnetic fields. *Science*, 312(5781), 1780–1782.
5. Qu, C. (2015) Tailor the functionalities of metasurfaces based on a complete phase diagram. *Physical Review Letters*, 115(23), 235503.
6. Li, Y., Lin, J., Guo, H., Sun, W., Xiao, S. and Zhou, L. (2020) A tunable metasurface with switchable functionalities: From perfect transparency to perfect absorption. *Advanced Optical Materials*, 8(6), 1901548.
7. Akyildiz, I.F., Kak, A. and Nie, S. (2020) 6G and beyond: The future of wireless communications systems. *IEEE Access*, 8, 133995–134030.
8. Dash, S., Psomas, C., Krikidis, I., Akyildiz, I.F. and Pitsillides, A. (2022) Active control of THz waves in wireless environments using graphene-based RIS. *IEEE Transactions on Antennas and Propagation*, 70(10), 8785–8797.
9. Dash, S., Liaskos, C., Akyildiz, I.F. and Pitsillides, A. (2020) Graphene hypersurface for manipulation of THz waves. *Materials Science Forum*, 1009, 63–68.
10. Gudla, V.V., Kumaravelu, V.B., Imoize, A.L., Castillo Soria, F.R., Sujatha, A.B., Kennedy, H.S.J., Jadhav, H.K., Murugadass, A. and Sur, S.N. (2025) Performance analysis of RIS-assisted millimeter-wave massive MIMO system under 3GPP 5G channels. *Information*, 16(5), 396.
11. Zhou, X. (2025) Research on channel modeling and performance optimization of RIS-assisted massive MIMO systems. *Applied and Computational Engineering*, 141, 89–97.
12. Chepuri, S.P., Shlezinger, N., Liu, F., Alexandropoulos, G.C., Buzzi, S. and Eldar, Y.C. (2023) Integrated sensing and communications with reconfigurable intelligent surfaces: From signal modeling to processing. *IEEE Signal Processing Magazine*, 40(6), 41–62.
13. Jian, M., et al. (2022) Reconfigurable intelligent surfaces for wireless communications: Overview of hardware designs, channel models and estimation techniques. *Intelligent and Converged Networks*, 3(1), 1–32.
14. Albanese, A., Sciancalepore, V. and Costa-Pérez, X. (2021) First responders got wings: UAVs to the rescue of localization operations in beyond-5G systems. *IEEE Communications Magazine*, 59(11), 28–34.
15. He, J., Wymeersch, H., Sanguanpuak, T., Silven, O. and Juntti, M. (2020) Adaptive beamforming design for mmWave RIS-aided joint localization and communication. In: *Proceedings of the IEEE WCNC Workshops*, pp. 1–6.
16. Hu, S., Rusek, F. and Edfors, O. (2017) Cramér–Rao lower bounds for positioning with large intelligent surfaces. In: *Proceedings of the IEEE VTC-Fall*, pp. 1–6.
17. Zeng, S., Zhang, H., Di, B., Han, Z. and Song, L. (2021) Reconfigurable intelligent surface assisted wireless coverage extension: RIS orientation and location optimization. *IEEE Communications Letters*, 25(1), 269–273.
18. Bonod, N. (2015) Large-scale dielectric metasurfaces. *Nature Materials*, 14, 664–665.
19. Chen, H.-T., Taylor, A.J. and Yu, N. (2016) A review of metasurfaces: Physics and applications. *Reports on Progress in Physics*, 79(7), 076401.
20. Liaskos, C., Nie, S., Tsioliaridou, A., Pitsillides, A., Ioannidis, S. and Akyildiz, I.F. (2018) A new wireless communication paradigm through software-controlled metasurfaces. *IEEE Communications Magazine*, 56(9), 162–169.
21. Kim, H.K., Lee, D. and Lim, S. (2016) Frequency-tunable metamaterial absorber using a varactor-loaded fishnet-like resonator. *Applied Optics*, 55(15), 4113–4118.
22. Liaskos, C., et al. (2015) Design and development of software-defined metamaterials for nanonetworks. *IEEE Circuits and Systems Magazine*, 15(4), 12–25.
23. Yang, H., et al. (2016) A programmable metasurface with dynamic polarization, scattering and focusing control. *Scientific Reports*, 6, 35692.
24. Kaina, N., Dupré, M., Fink, M. and Lerosey, G. (2014) Hybridized resonances to design tunable binary phase metasurface unit cells. *Optics Express*, 22(16), 18881–18888.

25. Liu, F., et al. (2019) Intelligent metasurfaces with continuously tunable local surface impedance for multiple reconfigurable functions. *Physical Review Applied*, 11(4), 044024.
26. Abadal, S., et al. (2017) Computing and communications for the software-defined metamaterial paradigm. *IEEE Access*, 5, 6225–6235.
27. Kouvaros, P., Kouzapas, D., Philippou, A., Georgiou, J., Petrou, L. and Pitsillides, A. (2018) Formal verification of a programmable hypersurface. In: *Proceedings of the International Workshop on Formal Methods for Industrial Critical Systems*, pp. 83–97.
28. Basar, E. (2016) Index modulation techniques for 5G wireless networks. *IEEE Communications Magazine*, 54(7), 168–175.
29. Zhou, Y., et al. (2018) Design of phase gradient coding metasurfaces for broadband wave modulating. *Scientific Reports*, 8, 8672.
30. Lei, Z., et al. (2018) Space–time-coding digital metasurfaces. *Nature Communications*, 9, 4334.
31. Foo, S. (2017) Liquid-crystal reconfigurable metasurface reflectors. In: *Proceedings of IEEE APS/URSI*, pp. 2069–2070.
32. Li, J., Shen, C., Rubio, A.D., Tretyakov, S. and Cummer, S. (2018) Systematic design and experimental demonstration of bianisotropic metasurfaces for scattering-free manipulation of acoustic wavefronts. *Nature Communications*, 9, 1342.
33. Huang, C., Zhang, C., Yang, J., Sun, B., Zhao, B. and Luo, X. (2017) Reconfigurable metasurface for multifunctional control of electromagnetic waves. *Advanced Optical Materials*, 5(22), 1700485.
34. Ozturk, C., Keskin, M.F., Wymeersch, H. and Gezici, S. (2023) RIS-aided near-field localization under phase-dependent amplitude variations. *IEEE Transactions on Wireless Communications*, 22(8), 5550–5566.
35. Zeng, S., et al. (2021) Reconfigurable intelligent surfaces in 6G: Reflective, transmissive, or both? *IEEE Communications Letters*, 25(6), 2063–2067.
36. Alexandropoulos, G.C., et al. (2024) Hybrid reconfigurable intelligent metasurfaces: Enabling simultaneous tunable reflections and sensing for 6G wireless communications. *IEEE Vehicular Technology Magazine*, 19(1), 75–84.
37. Mu, X., Liu, Y., Guo, L., Lin, J. and Schober, R. (2023) Simultaneously transmitting and reflecting RIS-aided wireless communications. *IEEE Transactions on Wireless Communications*, 22(11), 7819–7835.
38. Hassouna, S., et al. (2024) RIS-assisted near-field localization using a practical phase-shift model. *Scientific Reports*, 14, 18342.
39. Li, J., et al. (2025) RIS-assisted anti-spatial-aliasing direct localization in NLOS scenarios. *Scientific Reports*, 15, 5123.
40. Wang, M., et al. (2024) Localization performance analysis and algorithm design of RIS-assisted D2D systems. *Sensors*, 24(11), 3694.
41. Abdullah, O., et al. (2025) Transformer-based multi-modal indoor localization in RIS-assisted wireless networks. *IEEE Open Journal of the Communications Society*, 6, 1–15.
42. Wu, Q. and Zhang, R. (2019) Intelligent reflecting surface enhanced wireless network. *IEEE Transactions on Wireless Communications*, 18(11), 5394–5409.
43. Di Renzo, M., et al. (2020) Smart radio environments empowered by reconfigurable intelligent surfaces: How it works, state of research, and the road ahead. *IEEE Journal on Selected Areas in Communications*, 38(11), 2450–2525.
44. Basar, E., et al. (2019) Wireless communications through reconfigurable intelligent surfaces. *IEEE Access*, 7, 116753–116773.
45. Björnson, E., Özdogan, Ö. and Larsson, E.G. (2020) Intelligent reflecting surface versus decode-and-forward: How large surfaces are needed to beat relaying? *IEEE Wireless Communications Letters*, 9(2), 244–248.
46. Huang, C., et al. (2019) Reconfigurable intelligent surfaces for energy efficiency in wireless communication. *IEEE Transactions on Wireless Communications*, 18(8), 4157–4170.
47. Cui, M., Zhang, G. and Zhang, R. (2020) Secure wireless communication via intelligent reflecting surface. *IEEE Wireless Communications Letters*, 8(5), 1410–1414.
48. Basar, E., Alexandropoulos, G.C., Liu, Y., Wu, Q., Jin, S., Yuen, C., Dobre, O.A. and Schober, R. (2023) Reconfigurable intelligent surfaces for 6G: Emerging hardware architectures, applications, and open challenges. *IEEE Communications Magazine*, 61(6), 84–90.
49. Iqbal, M., Ashraf, T., Zubair, M., et al. (2025) A comprehensive survey on reconfigurable intelligent surfaces (RIS) and STAR-RIS for next-generation wireless networks. *Discover Applied Sciences*, 7, 1253.
50. Worka, C.E., Khan, F.A., Ahmed, Q.Z., et al. (2024) RIS-assisted non-terrestrial network-based 6G communications: A contemporary survey. *Sensors*, 24(21), 6958.

51. Wu, Q. and Zhang, R. (2020) Towards smart and reconfigurable environments: Intelligent reflecting surface aided wireless network. *IEEE Communications Magazine*, 58(1), 106–112.
52. Megahed, A., Abd El-Haleem, A.M., Elmesalawy, M.M., et al. (2025) Deep learning optimization of STAR-RIS for enhanced data rate and energy efficiency in 6G wireless networks. *Scientific Reports*, 15, 26311.
53. Peng, H. and Wang, L.-C. (2023) Energy harvesting reconfigurable intelligent surface for UAV based on robust deep reinforcement learning. *IEEE Transactions on Wireless Communications*, 22(10), 6826–6838.
54. de Figueiredo, F.A.P. (2023) Unlocking the power of reconfigurable intelligent surfaces: From wireless communication to energy efficiency and beyond. *Applied Sciences*, 13(21), 11750.
55. Shi, W., Wu, Q., Xiao, F., et al. (2024) Weighted sum power maximization for STAR-RIS-aided SWIPT systems with nonlinear energy harvesting. *Science China Information Sciences*, 67, 162303.
56. IEEE Vehicular Technology Magazine (2025) Perpetual reconfigurable intelligent surfaces through in-band energy harvesting: Architectures, protocols, and challenges. *IEEE Vehicular Technology Magazine*, 20(1), 24–33.
57. Dash, S., Liaskos, C., Akyildiz, I.F. and Pitsillides, A. (2019) Wideband perfect absorption polarization-insensitive reconfigurable graphene metasurface for THz wireless environment. In: *Proceedings of the IEEE Microwave Theory and Techniques in Wireless Communications (MTTW)*, Riga, Latvia, 1–2 October.
58. Emenonye, D., Dhillon, H.S. and Buehrer, R.M. (2024) Fundamentals of RIS-aided localization in the far-field. *IEEE Transactions on Wireless Communications*, 23(4), 3408–3424.
59. Zhao, Y., et al. (2024) 5G reconfigurable intelligent surface TDOA localization algorithm. *Electronics*, 13(12), 2409.
60. Chockalingam, A., Divya, M. and Moh, S. (2024) Joint channel estimation and localization for RIS-assisted mmWave systems under hardware impairments. *IEEE Transactions on Wireless Communications*, 23(12), 11984–11998.
61. Li, Y. and Yu, W. (2024) Localization in multipath environments via active sensing with reconfigurable intelligent surfaces. *IEEE Communications Letters*, 28(9), 2061–2065.
62. Zeng, M., Zhang, R. and Niyato, D. (2024) Reconfigurable intelligent surfaces assisted multi-user massive MIMO: Channel estimation, beamforming, and performance analysis. *IEEE Transactions on Wireless Communications*, 23(6), 4461–4475.
63. Cui, Y., Wei, W. and Zhang, R. (2024) RIS-aided millimeter-wave communications: Channel estimation, beam training, and beamforming design. *IEEE Wireless Communications*, 31(4), 78–85.
64. Han, C., Jornet, J.M. and Akyildiz, I.F. (2024) Terahertz communications: A complete overview of signal models, standardization, and integration with reconfigurable intelligent surfaces. *IEEE Communications Surveys & Tutorials*, 26(3), 1850–1887.
65. Zhao, N., Chen, P., Chen, Y. and Elkashlan, M. (2025) Simultaneous wireless information and power transfer with reconfigurable intelligent surfaces: A structured framework. *IEEE Internet of Things Journal*, 12(8), 7564–7579.
66. Zhang, J., Xu, X., Liu, Y. and Haenggi, M. (2025) Cell-free massive MIMO with reinforced reconfigurable intelligent surfaces for distributed wireless networks. *IEEE Transactions on Communications*, 73(1), 312–326.
67. Duan, H., Chen, J., Shi, Y. and Yang, L. (2025) AI-empowered reconfigurable intelligent surface optimization for 6G wireless networks. *IEEE Communications Magazine*, 63(1), 92–98.
68. Nguyen, T.V., Nguyen, D.N., Di Renzo, M. and Zhang, R. (2023) Leveraging secondary reflections and mitigating interference in multi-RIS aided wireless networks. *IEEE Transactions on Wireless Communications*, 22(1), 502–517.
69. Yu, X., Xu, D. and Schober, R. (2022) Reconfigurable intelligent surface assisted communications with limited phase shifts: Reflection optimization. *IEEE Transactions on Communications*, 70(8), 4936–4951.
70. Zappone, A., Di Renzo, M. and Debbah, M. (2023) Overhead and scaling laws for RIS-aided wireless networks. *IEEE Transactions on Wireless Communications*, 22(4), 2110–2126.

Chapter 8

5G and beyond wireless networks

Bijayalaxmi Panda, Manaswinee Madhumita Panda, Laxminarayan Dash, Chhabi Rani Panigrahi, and Bibudhendu Pati

8.1 INTRODUCTION TO 5G

A fifth generation (5G) mobile network is represented as 5G. It signifies the modern global wireless technology following first generation (1G) to fourth generation (4G) networks. 5G brings forth a novel kind of network that can link almost all people and things together—this includes devices, objects, and machinery. This wireless technology aims to deliver multi-gigabit peak data speeds, ultra-low latency, greater reliability, expanded network capacity, enhanced availability, and a more uniform user experience for an increased number of individuals. Improved performance and efficiency allowed for the development of new applications and industries in the 5G network [1]. It provides faster upload and download speeds, a more stable connection, and enhanced performance. It offers more upload and download speeds, a smoother connection, and improved capacity than previous networks. 5G is considerably faster and more reliable than currently trendy 4G networks and has the potential to revolutionize the manner in which we make use of the internet to gain applications, social networks, and information [2]. For example, autonomous vehicles, advanced gaming software, and streaming media that provide real-time streaming all require extremely reliable, high-speed connections to data, and all of them are going to benefit heavily from 5G access.

The growing demand for internet connectivity, along with advancements in technologies, such as artificial intelligence (AI), the Internet of Things (IoT), and automation, is fuelling a surge in data generation. The amount of data being generated has increased exponentially and is projected to reach hundred of zettabytes within the next decade. The existing mobile infrastructure was not originally built to handle such vast amounts of data and requires investment. In the meantime, thanks to low latency, wide bandwidth, and high speed, 5G technology, for example, can be used for applications such as cloud-based traffic management, drone delivery services, video calls, and games that are as fast as those that are played on a computer gaming system [3]. 5G has a plethora of uses, ranging from easy e-commerce transactions to instant emergency services, educational processes, and a more mobile workforce. The technology can have a significant impact on both the global economy and the lives of regular people.

5G technology is the latest development in the technology of mobile communication networks. It aims at ultra-data rates, hyper-efficiency, low-latency, and massive network connections. It is more efficient than the older generations of networks such as second generation (2G), third generation (3G), and 4G. It also provides unprecedented benefits to the industry, the IoT, and automation in smart cities [4]. Figure 8.1 shows the cellular system standards and generation timeline.

Massive Multiple-Input Multiple-Output (MIMO) is a tool that enriches the gesture strength due to the ability of a larger number of antenna elements to pass information to the single user. In addition, beam forming is another technique that can be used to direct the signal strength to the user while avoiding interruption. Another potential improvement lies in system slicing,

 DOI: 10.1201/9781003637721-8

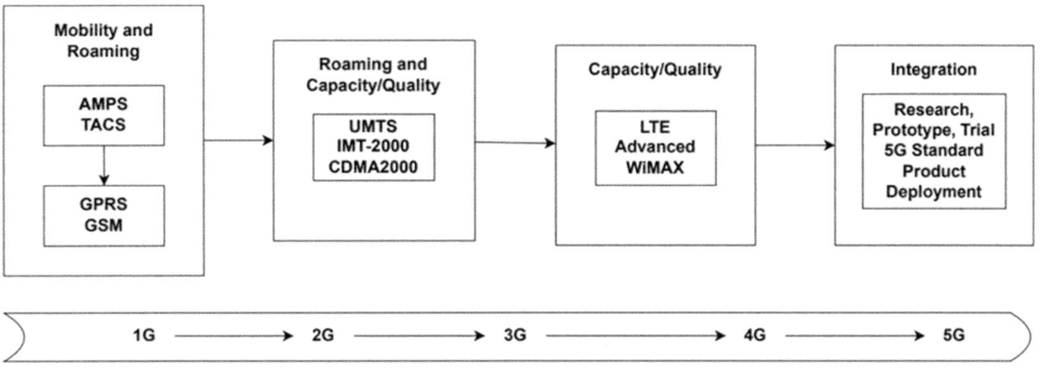

Figure 8.1 Cellular system standards and generation timeline.

enabling the creation of multiple virtual internets within a single system for different users [5]. Furthermore, another important area is the users. Another important area is edge computing which is a new form of computing that brings computing to the user and so enhances the response time and reduces latency. Small cell systems, large cell towers, and high-speed internet fibre-optic cables are what make up 5G networking. In large cities, the signal strength is increased and bypasses the challenges of high-frequency waves' range in small cells [6]. They facilitate the fast movement of data. In terms of efficiency and flexibility, the new systems have also significantly impacted them. This is made possible through cloud systems and software-defined networks. Table 8.1 displays the fifth-generation evolution.

Table 8.1 Evolution of 5G

Type	*Deployment*	*Technologies and Standards*	*Features*
First generation	In the 1980s, analog telecommunication introduced	• AMPS (Advanced Mobile Phone Systems)—NMT (Nordic Mobile Telephone)	• Voice calls—Seamless NMT for unified data and messaging
Second generation	In the 1990s, digital cellular organized	• CDMA (Code-Division Multiple Access), GSM (Global System for Mobile Communications)/EDGE (Enhanced Data rates for GSM Evolution), TDMA (Time-Division Multiple Access)	• Voice- SMS text messages- Low-rate data
Third generation	In 2000, first broadband deployed	• CDMA2000 1X/Evolution-Data Optimized (EVDO)- Universal Mobile Telecommunications Service (UMTS)/ High-Speed Packet Access (HSPA)- Worldwide Interoperability for Microwave Access (WiMAX)	• 144 Kbps to 2 Mbps - Speeds for indoors- Rich content Enabling
Fourth generation	Installed in 2010	• Long-Term Evolution	• 1 Gbps starts from 100s of Mbps - Streaming abilities and videos
Fifth generation	First installed in 2018	• ITU, IMT, 3GPP known as International Telecommunication Union, International Mobile Telecommunications, - 3rd Generation Partnership Project -2020 clear technical objects are developing 5G specifications	• 3× developed spectral effectiveness than 4G-Highest downlink quantities up to 20 GBps

As compared to the preceding generation, the 5G comes with ten Gbps data transfer, less energy consumption, less than 1 millisecond of time consumption, and the highest capacity of 1 million devices per square kilometre. By utilizing these capabilities, it can support the autonomous driving, remote surgeries, 4K/8K video streaming, augmented reality, and virtual reality. In addition to that, the future improvements might include terahertz (THz) communication for fast data transmission, coverage by the satellite, mobile networks optimized by AI, quantum encryption, and rural coverage [7]. The 5G mobile technology is a significant improvement in telecom evolution and innovation in all aspects. In foresight of this invention, it will enable contemporary industries as well as changing the whole world.

Wireless communication started during the time of wireless signal using radio waves. Expansion leads to radio broadcasting, radar technology, satellite, etc. The evolution of mobile broadband came in the modern era. The advanced features are highlighted below.

8.1.1 Key features of 4G network

4G is an improved cellular network that is faster than the 3G, and it enables smartphone users to engage in high-quality video calls, play games, watch videos in high definition, and download data faster without interruptions. Figure 8.2 displays the 4G network architecture.

The 4G architecture involves the following components.

eNodeB (Evolved Node B): The eNodeB serves as the 4G site where a user equipment (UE) establishes its connection. It delivers the radio interface for the UE.

MME (Mobility Management Entity): It refers to the Mobility Management Entity. The MME verifies the identity of the UE. Additionally, it keeps track of the UE's location, meaning it knows the UE's whereabouts. Additionally, it selects the suitable source gateway (SGW) and PGW (Packet Data Network Gateway) to be assigned to this UE.

SGW: The SGW serves as a data anchor for the UE during transitions between eNodeBs and facilitates data transfer between the eNodeB and PGW.

PGW: An abbreviation for "PDN Gateway". The PGW serves as the connection point between the LTE network and the PDN.

HSS (High-Speed Steel): The HSS is a critical database in 4G networks responsible for managing user authentication.

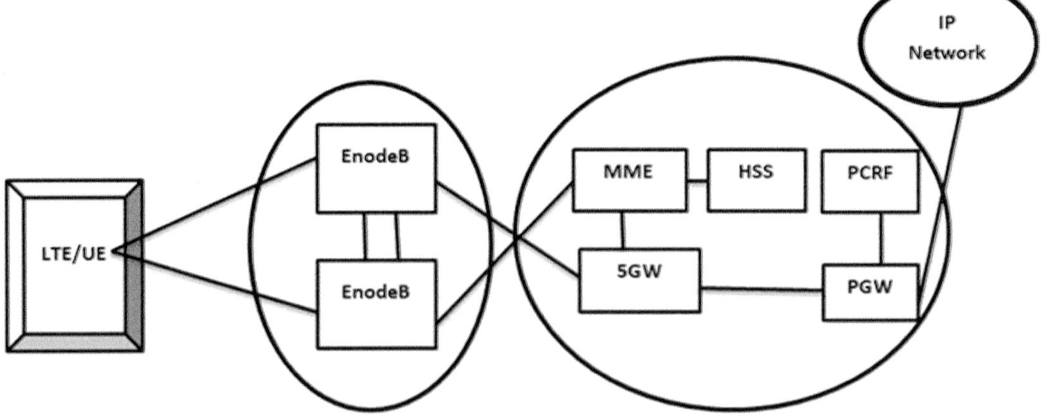

Figure 8.2 Architecture of 4G network.

V

W

X

Z

For Product Safety Concerns and Information please contact our EU representative GPSR@taylorandfrancis.com Taylor & Francis Verlag GmbH, Kaufingerstraße 24, 80331 München, Germany

Implementations: The implementations of 4G include high-speed internet for fast browsing and downloads; HD and 4K video steaming without buffering; real-time multiplayer gaming with low-latency; high-quality VoIP and video calls on apps like Zoom and WhatsApp; and IoT connectivity for smart devices for automation and public Wi-Fi expansion in cafes, malls, or transport hubs.

8.1.2 Key features of 3G network

3G stands for third-generation technology in wireless mobile telecommunications. 3G is an improved form of the 2G network and provides higher transmission rates, a higher data capacity, and a variety of new services that were previously unavailable. 3G was envisaged to provide a stronger data and voice communication experience. 3G came into being in the late 1990s, with the first commercial 3G networks appearing in the early 2000s [8]. It has higher voice than 2G networks, a higher data speed, and better security for data transfers. Figure 8.3 displays the 3G network architecture.

Components of 3G networks include base stations (NodeBs), radio network controller (RNC), and UE. The transfer speed for reaching the data is 2 Mbps which improves bandwidth that helps to send and receive huge email messages.

The various 3G network applications include the following:

Mobile internet access: Mobile internet access refers to connecting to the internet over a mobile network rather than a conventional connection. It enables users to access the internet, use apps, stream films, and engage in other online activities via cellular technologies.

Voice and video calls: 3G networks use circuit-switched technology to provide both voice and video conversations, with video calls made possible by faster data transmission rates than 2G networks.

Video streaming: 3G mobile communication technology brought faster data rates than 2G, allowing real-time multimedia applications such as video streaming, video calls, and mobile TV. Data speeds can range from 384 kbps to 2 Mbps, depending on network circumstances. Video streaming over 3G entails the continuous transmission of compressed video data from a server to a mobile device. To minimize interruptions, it uses buffering and adaptive approaches.

Smart-home applications: Wireless connection allows smart-home products to be remotely monitored and controlled. Prior to the widespread adoption of Wi-Fi, 4G, and IoT standards, 3G technology was crucial in enabling remote smart-home communication since it offered it.

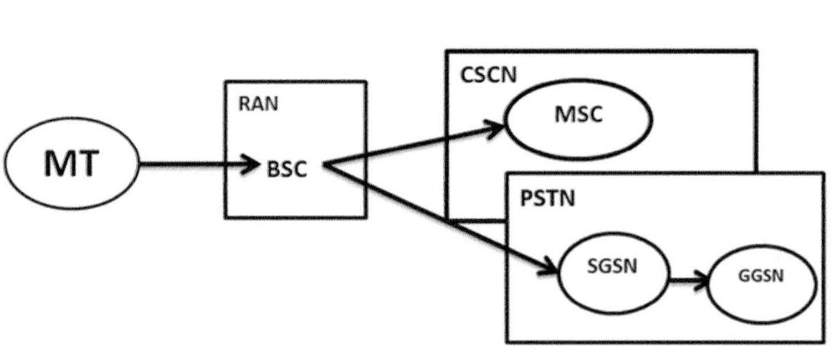

Figure 8.3 Block diagram of 3G network architecture.

Mobile data access (up to 2 Mbps): It always on connectivity, reliable coverage in urban and rural areas' 3G-enabled early smart-home systems even without broadband or Wi-Fi.

Wireless sensor networks (WSNs): These are made up of low-power sensor nodes that monitor physical conditions. Integrating WSN with a 3G cellular network enables sensor data to be transferred over long distances via wide-area mobile access.

8.1.3 Key features of 2G network

2G was a replacement for first-generation cellular technology, facilitating the transmission of communications and data. It was the first digital network. The 2G network was established in the early 1990s and came to the forefront of people's communication habits, offering high-quality communications. The 2G network was also able to transmit information.

The changing from analog to digital started here. The technologies like Global System for Mobile Communications (GSM) and Code-Division Multiple Access (CDMA) are used. The implementation focuses on the following approaches.

Improved call quality and security: The encrypted voice transmissions improved the calls' clarity and eliminated the chances of eavesdropping.

Lower power consumption: It provides a more battery-efficient operation, which will allow you to use it longer.

Utilization of battery spectrum: Compared to first-generation networks, it enabled more users to share the same bandwidth on a cell tower.

Support for roaming: Standardized protocols were introduced to allow for international roaming.

Applications of 2G network involve voice calling with digital clarity, text messaging (SMS), basic mobile banking and alerts via SMS, early mobile-based authentication systems (like OTPs), rural and backup communication systems in developing regions.

8.2 WORKING PROCEDURE OF 5G

In this section, 5G is explained based on familiarity with the following basic components of the technology: 5G New Radio (NR) is the air interface connecting customers, machines, and devices for data sharing. 5G NR uses low- and mid-frequency bands (sub-7 GHz, FR1) and high-frequency bands (above 24 GHz, FR2/mmWave). 5G incorporates all previous cellular spectrum plus additional sub-7 GHz and higher frequencies. This expanded spectrum addresses the throughput and bandwidth limitations of earlier technologies. The throughput and bandwidth constraints of previous technologies are addressed by this enlarged spectrum [9].

The "brain" of the 5G network is the 5G Next Generation Core (5G NGC), which controls and oversees the network, its services, and the connected devices. The cloud-aligned, service-based architecture (SBA) that pervades all 5G operations and interactions, including authentication, security, session management, and end-device traffic aggregation, is leveraged by the new 5G core as defined by the 3rd Generation Partnership Project (3GPP). With virtualized software functionalities that can be deployed in an agile and dispersed fashion, the 5G core also highlights Network Function Virtualization (NFV) as a fundamental architectural feature [10].

Multi-access edge computing (MEC) is among the significant pillars for solving the tough KPIs of 5G, especially in the case of low latency and bandwidth effectiveness. MEC is also forecasted to play a key role in the development of mobile networks into versatile service platforms for industry and other specific customer segments.

IP Multimedia Subsystem (IMS) offers IP-based multimedia services, which enables the mobile operators to offer a next-generation rich multimedia service. IMS offers access to IP-based services irrespective of the access network.

8.2.1 Society helping 5G

Although 5G networks are expected to create millions of jobs and trillions of dollars, their positive social effects go much beyond their financial effects. It includes the following:

Shrewd cities: IoT devices are used in intelligent urban environments to gather real-time data about people, infrastructure, and traffic. By using this data, urban planners may make more educated decisions, reduce emissions, enhance public services, ease traffic, and improve air quality. The development of 5G technology may be the impetus behind the real interconnection of the world's largest cities.

Healthcare: There will be significant benefits for healthcare technologies from the deployment of 5G technology. For example, its reduced latency will allow high-definition video to be sent in real time, perhaps paving the way for remote surgery to become a standard practice. Ingestibles and wearables will become more common, giving medical providers useful feedback data. Additionally, real-time monitoring will help doctors detect disease symptoms early and allow for more individualized treatment for patients.

Surroundings: It is possible that the deployment of 5G technology may help lower global emissions. Compared to earlier networks, it has the advantages of reduced power usage and increased transmission efficiency. 5G will also make it easier to monitor pollutants, water and air quality, and other environmental indicators in real time. Additionally, in order to encourage effective resource use and lower pollution, it will assist the development of smart buildings, smart grid projects, electric vehicles, and remote work initiatives.

8.2.2 Operation of 5G

5G technology uses cell sites that send data via radio waves, just as earlier cellular networks. These cell locations are connected to networks by cable or wireless technology. Modifying the information encoding technique improves 5G technology's functionality by significantly increasing the range of airwaves available to carriers.

OFDM (orthogonal frequency division multiplexing): 5G technology relies heavily on OFDM. With less latency and more flexibility than LTE networks, this modulation technology encodes high-frequency airwaves incompatible with 4G.

Smaller towers: Unlike 4G and previous cellular technologies, which used isolated mobile towers, 5G technology uses tiny transmitters that are strategically placed on buildings and other structures. Multiple devices will be able to connect at faster speeds thanks to the network management capabilities from these tiny cell sites.

Slicing of network: Several separate virtual networks can be established on common infrastructure by mobile network carriers using 5G technology. Every network slice can be customized for different services and business models, such as streaming or enterprise operations. The varied needs of different industries can be met by putting in place a set of 5G network functionalities tailored to each unique use case or business model. Device efficiency is increased and user experience stability is improved by this division of services.

8.3 CHALLENGES IN DEPLOYING 5G

There are numerous technological, financial, and legal obstacles to overcome in the deployment of 5G networks. Figure 8.4 shows the main challenges in organizing 5G. A summary of the main challenges is presented as follows.

8.3.1 Infrastructure requirement

To meet the speed, latency, and capacity requirements of the technology, 5G infrastructure needs a mix of new and improved components. This comprises a strong core network, a high-capacity backhaul, an appropriate transport network, and radio access networks (RAN), comprising microcells and small cells [11]. Moreover, to facilitate smooth 5G communication, fibre-optic networks and improvements to current cellular infrastructure are required. Figure 8.5 displays the key problems associated with ultra-dense networks in 5G.

Dense network for small cells, fibre backhaul: For coverage of tiny geographic regions (tens to a few hundred metres), small cells are low-power, short-range wireless transmission systems. Microcells, picocells, and femtocells are among them. 5G has a poor penetration rate through obstructions and a limited range, particularly in the millimetre-wave (mmWave) bands [12]. Numerous tiny cells must be placed in close proximity to one another in order to sustain high-speed, low-latency connections, particularly in metropolitan settings. calls for a significant infrastructure build-out, including buildings, poles, and lamps, which calls for coordination with local authorities and regulatory permits. The link between the small cell (or base station) and the main network is referred to as backhaul. Fibre-optic cables are the recommended backhaul media in 5G because

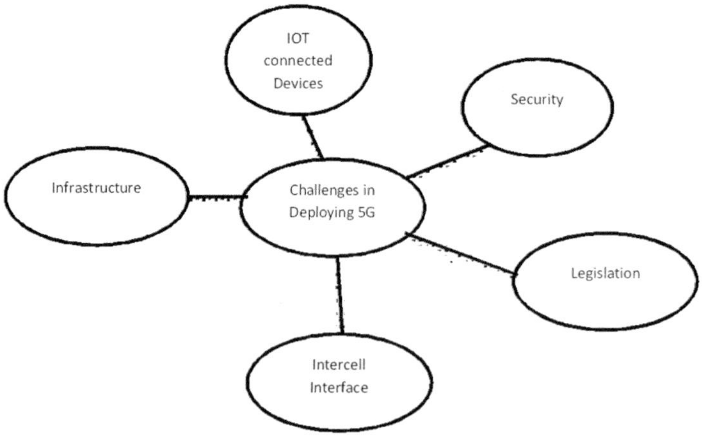

Figure 8.4 Challenges in deploying 5G.

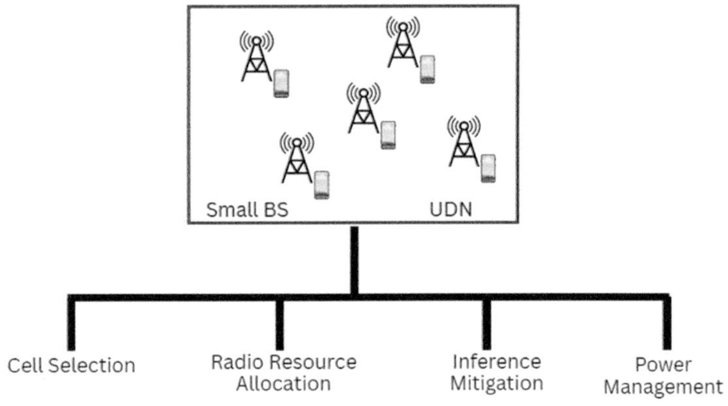

Figure 8.5 Key problems with ultra-dense networks in 5G.

of their low latency and high bandwidth. Fibre is required to effectively move the massive volumes of data generated and sent by 5G networks from cells to data centres and the internet [13]. Fibre installation is costly and time-consuming, particularly in places with a high population density or difficult geography. Additionally, it might entail excavating roads and other infrastructure, necessitating cooperation with local governments.

Costs of 5G deployment: Installation of small cell due to manpower, site leasing, and hardware costs, deploying thousands of tiny cells throughout cities, is expensive. Backhaul of fibre-optic takes a lot of civil work to install or upgrade fibre networks to link all those cells to the central network [14]. The supply of power to every cell needs a reliable power source, which raises expenses, particularly in busy or distant locations.

Equipment and hardware: 5G base stations compared to 4G, new base stations are costlier and sophisticated. Edge computing units achieve ultra-low latency; data processing must be positioned closer to the user (edge computing), which calls for more funding [15]. For core network upgrades to accommodate 5G features, operators must switch to a more software-based and virtualized core.

Operational costs: Site acquisition and permits can be costly and time-consuming to negotiate site leases and obtain municipal approval. Maintenance and monitoring costs of continuous maintenance, security, and performance management increase with the amount of infrastructure [16].

Spectrum licencing fees: High auction prices of telecom corporations must place competing bids, frequently spending billions, when governments auction off spectrum bands, particularly mmWave and mid-band [17]. Recurring charges of annual licence costs for spectrum use are also levied in certain nations.

Fragmented spectrum: Non-contiguous pieces of spectrum may be purchased by operators, which is less effective and would necessitate a larger investment in aggregating equipment [18].

8.3.2 5G spectrum management and availability

The range of electromagnetic frequencies used for wireless signal transmission is known as the radio spectrum. Three primary spectrum bands are used by 5G such as low-band (sub-1 GHz): Slower speeds and wide coverage. Speed and range are balanced in the mid-band

(1–6 GHz) (e.g., 3.5 GHz). High-band (beyond 24 GHz/mmWave): Extremely fast, but with limited penetration and short range [19].

The main issues with availability of 5G network are as follows:

Restricted frequency bands: Mobile communication can only use a limited number of frequencies.

Pre-allocated bands: The military, television broadcasting, satellite, and other industries currently employ a large number of spectrum bands [20].

Uneven global allocation: Different nations assign distinct frequency ranges, which has an impact on roaming and device compatibility worldwide.

8.4 COMPETITION FOR SPECTRUM RESOURCES WORLDWIDE

The increasing international conflict between nations, businesses, and organizations for access to and control of parts of the electromagnetic spectrum, especially the radio frequency (RF) bands used for wireless communication, is known as the "global competition for spectrum resources" [21]. As a result of the growing demand for spectrum from a variety of technologies, including 5G and sixth generation (6G) networks for mobile phones, communication by satellite, broadband and Wi-Fi services, the IoT, defence and military systems, weather forecasting, and remote sensing.

Necessity of spectrum: Spectrum is a highly desired asset because it is a finite resource that cannot be extended [22]. Innovation and GDP are greatly boosted by wireless technologies. Spectrum is necessary for secure communications, military, and monitoring in nations [23]. Innovation in fields like driverless cars, smart cities, and AI-powered networks is fuelled by spectrum access.

Players and stakeholders worldwide: Governments oversee the distribution of national spectrum and facilitate international coordination through organizations like International Telecommunication Union (ITU). Tech businesses, satellite providers, and telecom corporations for commercial access [24] for reasons of national security, specific spectrum bands are needed.

5G network speed: The 5G network speed is the rate at which information may be sent and received via a 5G wireless connection. Theoretically, upload speeds can reach 10 Gbps and download speeds can reach 20 Gbps on a 5G network. However, depending on the type of 5G spectrums, the number of devices connected, the distance from the cell tower, and the interference from barriers, users may experience different speeds from the actual 5G mobile network. Figure 8.6 shows the speed of different generations [25].

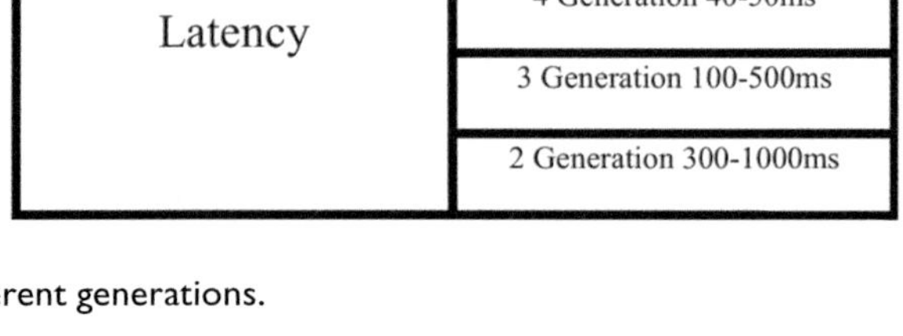

Figure 8.6 Speed of different generations.

8.5 SECURITY AND PRIVACY

Strong encryption, better authentication, and a resilient network architecture are the main ways that 5G networks provide better security and privacy features than earlier generations. However, new security threats are also brought about by the complexity of 5G, which includes technologies like network slicing and virtualization [26].

Encryption method: Strong encryption, better authentication, and a resilient network architecture are the main ways that 5G networks provide better security and privacy features than earlier generations [27]. However, 5G's added complexity—which includes technologies like network slicing and virtualization—also brings with it new security risks.

Enhanced authentication: To confirm user identification and stop spoofing or illegal access, 5G integrates improved authentication techniques such as SIM-based authentication and authorization.

8.5.1 Protecting data and ensuring secure connections in 5G networks

The growing number of connected devices, faster data speeds, and more use cases in 5G networks make data protection and secure communications a top need [28]. An outline of the main tactics and tools employed to deal with these issues is provided below:

Improved encryption systems: Encrypting the user and control planes: Stronger encryption technologies, including 256-bit AES, are used in 5G to safeguard user and control data. Ensuring that control messages have not been altered during transmission is known as integrity protection [29].

AI and Machine Learning (ML) to identify threats: Real-time detection of anomalous activity or traffic patterns is made possible by AI/ML models, which enables the prompt detection of attacks such as DDoS, spoofing, unauthorized entry [30].

Security of network slicing: Security settings may differ for each network slice (specialized virtual network). Enables customization according to the sensitivity of the application (e.g., IoT sensors versus autonomous vehicles).

Architecture of zero trust: Assumes that by default, no area of the network is secure. Demands on-going user and device permission and authentication [31].

Protocols for secure authentication: Authentication and key agreement or 5G-AKA: A strong mechanism for device and user authentication.

Cloud and edge security: Edge computing creates new endpoints, which need to be secured. The endpoints are Secure APIs, containerization security, and edge trust anchors.

8.6 REGULATORY AND POLICY ISSUES IN 5G NETWORKS

5G network development and operation are governed by a complicated web of national and international regulations and policies. These problems have a big impact on how quickly, widely, and successfully 5G is adopted around the world [32].

International coordination spectrum harmonization: The ITU and other international organizations are essential to the standardization of 5G frequency bands. In addition to facilitating international roaming and economies of scale for equipment makers, this attempts to guarantee compatibility between nations [33]. By providing member nations with recommendations and rules to standardize spectrum allocation, regional organizations like the Asia-Pacific Telecommunity (APT) also support spectrum management at the regional level.

Standards technologies: International technical standards must be created in order for 5G to function flawlessly across national borders. Organizations like the 3GPP have played a major role in defining these standards, which ensure that different networks and equipment can work together. Additionally, the ITU is leading the development of IMT-2020, a collection of 5G standards that are applicable everywhere [34].

International deployment: Since 5G offers cross-border applications like remote healthcare and driverless autos, international cooperation is needed to standardize laws and guarantee seamless service delivery. Contracts relating to data flows, security protocols, and responsibility are becoming increasingly important.

Global issues resolving: International cooperation is also necessary to address challenges like cyber security assaults that could easily cross international borders in interconnected 5G networks [35]. Collaboration is necessary to establish security standards and incident response protocols.

Local regulatory requirements: Spectrum licencing and allocation: National regulatory agencies must licence and assign spectrum to mobile carriers in order to construct 5G. Spectrum assignment methods like auctions or "beauty contests" (which grant licences based on criteria like rollout speed and network performance) can have a significant impact on the pace and quality of 5G implementation [36]. Market dynamics and regional conditions often dictate the specific frequency bands and licencing terms.

Social equity and universal access: Some regulatory frameworks have provisions ensuring that all residents, especially those residing in rural or impoverished areas, can benefit from 5G's advantages [37]. To promote deployment in regions with weaker economic viability, this may mean requiring operators to provide coverage or providing financial incentives [38]. Regulators may also need to ensure that all patients have access to 5G-enabled services, such as remote healthcare, regardless of their location or socioeconomic status.

8.7 ENERGY EFFICIENCY AND ENVIRONMENTAL IMPACT

Regarding the impact on the environment and energy efficiency, the deployment of 5G networks presents both potential and concerns. Although 5G technology itself offers the potential for greater energy efficiency per bit of data transmitted compared to previous generations, the densification of networks and the increased number of connected devices could lead to a significant increase in overall energy consumption if not managed sustainably [39].

Resolving the energy consumption of denser networks increased infrastructure: 5G networks require a denser deployment of small cells and base stations in order to achieve the promised high speeds and low latency, especially when employing higher frequency bands like millimetre wave (mmWave). Naturally, this increase in the number of active network nodes may lead to higher energy demand.

Massive MIMO antennas: While MIMO antennas are necessary for 5G's expanded coverage and capacity, their use may end up consuming more power per base station sector.

Legacy network coexistence: In the early stages of deployment, 5G networks usually coexist with 4G and even older networks. Keeping these parallel infrastructures in place until older networks are removed could lead to greater overall [40].

In order to lessen the environmental impact of 5G networks over their entire lifespan, sustainable deployment techniques are crucial.

Optimized network planning: By carefully planning to reduce the number of new cell sites and maximize the use of existing infrastructure through tactics like co-location, the environmental impact of construction and resource consumption can be reduced [41].

Smart grid integration: 5G may be crucial for the creation of smart grids, which enhance energy distribution and more effectively integrate renewable energy sources, leading to broader environmental benefits.

Regulatory standards and incentives: Governments and regulatory bodies may play a significant role in establishing environmental standards, offering rewards for energy-efficient behaviour, and promoting sustainable 5G network deployment strategies [42].

8.8 NEXT GENERATION OF 5G

6G comes after 5G. With promises of higher speeds, reduced latency, and more dependability, 6G offers notable improvements over 5G and opens the door for new use cases and applications. By facilitating smooth and instantaneous communication, it is anticipated to facilitate technologies such as smart cities, intelligent transportation, and connected industries.

8.8.1 Vision for 6G

Ideal for important uses such as self-driving cars, remote operations, and real-time applications through the tactile internet combined ML, and ubiquitous AI results in networks that self-optimize and self-repair [43]. Using a predictive approach for allocating resources, adapting quality of service based on user circumstances, reducing dependency on central cloud servers through the use of edge AI, remote procedures, self-driving automobiles, real-time optics via the tactile internet.

ML and pervasive AI combine for networks that optimize and repair themselves, allocating resources in a predictive manner. Quality-of-service adaptation according to user context and utilizing edge AI to lessen reliance on central cloud servers.

Key technologies for 6G: Using advanced AI and ML techniques to autonomously manage, optimize, and adapt communication networks—especially mobile and wireless networks like 4G, 5G, and beyond—is known as "Advanced AI/ML for Self-Optimizing Networks (SONs)". SONs are networks with the ability to self-configuration and enhance performance through self-optimization. Find and correct errors (self-healing), with little assistance from humans. Because of their size and complexity, they are essential in contemporary mobile networks.

Use cases of AI/ML in SON: Forecasting traffic and balancing loads to prevent congestion, ML models predict traffic demands. Finding anomalies: AI detects unusual behaviour network faults automatically. In order to conserve energy, ML assists in dynamically shutting off underutilized base stations. Reduced lost calls and seamless handovers are guaranteed by predictive models. Automatically modifying network settings for best performance, such as antenna tilt and power levels.

Quantum computing and communication: Superposition, entanglement, and quantum tunnelling—three concepts from quantum mechanics—are used in quantum computing to carry out calculations in radically novel ways [44].

Quantum bits or qubits are capable of simultaneously representing 0 and 1: Allows for exponential speedups and parallel processing for specific challenges (e.g., cryptography, optimization, and AI).

Quantum transmission for 6G protection: Unbreakable encryption is offered by Quantum Key Distribution (QKD) for data transferred over 6G networks. It helps in the protection of vital financial systems, healthcare data, infrastructure, etc. and operates in free-space/terrestrial wireless situations as well as fibre-based ones.

Challenges of quantum-6G integration: The development of quantum hardware is still in its infancy; it requires low temperatures and is noise-sensitive. Entanglement distribution and quantum repeaters are not yet widely used. There is a demand for hybrid classical-quantum infrastructure.

Digital twin: A digital twin is a real-time digital replica of a physical entity (e.g., person, machine, system, and environment). It continuously receives data from its physical counterpart via sensors and IoT and enables simulation, prediction, monitoring, and optimization in real time. Digital twins are real-time digital copies of physical objects, such as people, machines, systems, and environments. It constantly gets data from its physical counterpart through sensors and the IoT. Real-time monitoring, prediction, modelling, and optimization are made possible.

8.9 APPLICATIONS OF 6G

6G applications include real-time models of buildings, energy grids, and traffic systems in smart cities.

Healthcare: Patient-specific digital twins for remote diagnosis and prognostic treatment.
Industry 4.0: Digital factory models for automation and predictive maintenance.
Autonomous cars: Improve decision-making by simulating real-world situations [45].

Real-time models of buildings, electricity networks, and traffic systems are examples of smart cities.

Telepresence: Training, education, and fully immersive holographic meetings.
Healthcare: Holographic visualization for remote surgical aid. Sports, gaming, and live holographic concerts are examples of entertainment.
Retail: Virtual try-ons and 3D product previews.

8.9.1 6G holographic communications

Real-time 3D presence and engagement are made possible by holographic communication, which provides realistic volumetric video in addition to flat screens and virtual reality. Imagine holograms à la Star Wars for meetings, video chats, teaching, etc. It demands accurate representation of motion, perspective, and depth [46].

8.9.2 Brain computer interface

Using next-generation wireless communication (6G) to enable brain-computer interfaces (BCIs) with previously unheard-of speed, latency, and dependability is known as BCI in 6G. Applications such as telepresence, brain data streaming, the metaverse, and cyber-physical systems with BCI integration are anticipated to be powered by 6G.

Extremely low latency: Allows brain impulses to be used to control gadgets in real time. Vital for medical applications such as remote surgery and neuro-prosthetics.

Elevated data rates: Massive amounts of data are produced by brain signals. Cloud-based processing and real-time transmission will be made possible by 6G.

Extremely high connectivity: Enables multi-user BCI systems or brain-to-brain communication. Supports smart city decentralized neural networks.

Intelligence on the edge: AI and edge computing will be included in 6G, which will speed up and improve BCI. Lessens the requirement for intensive local processing.

Advanced healthcare solutions: It is anticipated that the healthcare industry will undergo a significant transformation with the introduction of the 6G networks, particularly through applications such as real-time remote procedures. The possible areas in which how 6G makes next-generation healthcare possible are as follows:

- *Role of 6G function in healthcare:* 6G will serve as the connecting core for healthcare in the future. For time-sensitive interventions and remote robotic surgery, ultra-low latency (~1 Ms or less) is essential.
- *High data throughput (>1 Tbps):* Required for sending holograms or high-resolution 3D medical imaging [47].
- *Massive connectivity:* Enables wearables, patient monitoring systems, and IoT medical equipment to be used by large populations.
- *Integration of edge AI:* Enables real-time bio signal and image processing without cloud latency.
- *Worldwide connectivity without regard to location*: One of 6G's most ambitious claims is to provide truly global, borderless connectivity, bridging the digital divide and enabling seamless communication wherever you are on Earth (and even beyond).
- *Integration of satellites and terrestrial:* LEO (Low Earth Orbit) satellites and terrestrial networks will be combined in 6G. Imagine SpaceX's Starlink, but with direct integration with your phone or IoT gadget.
- *Intelligent surfaces that can be reconfigured (RIS)*: Intelligent materials that focus, bounce, and guide signals to difficult-to-reach locations.
- *Spectrum of high frequencies (THz bands):* Offers a vast amount of bandwidth for quick and clear communication.
- *AI-assisted network administration*: Even in networks that are thick or changing, intelligent routing and load balancing are possible [48].
- *Autonomous ecosystems and virtual space:* A world where people, machines, and systems live together in self-governing, intelligent ecosystems is what 6G networks promise to bring about. In this world, virtual presence will become so real-time and immersive that physical presence may no longer be necessary. In the context of 6G, an autonomous ecosystem is a self-governing network of linked gadgets, automobiles, robots, drones, sensors, and AI systems that work together without involving humans and make choices based on data that is current. Dynamically adapt and heal yourself using ultra-fast, low-latency 6G networks to communicate.

8.9.3 Challenges for 6G

Although 6G has enormous potential to revolutionize technology, society, and communication, it also presents a number of difficult problems, including ethical, sociological, technological, and financial ones. This is a concise and organized breakdown:

THz communication signal loss: Because of air absorption, THz signals are severely attenuated.

Short range: Propagation distance is constrained at higher frequencies.

Limitations in hardware: Requires new components and equipment (e.g., THz antennas, modulators, and amplifiers).

8.9.4 Rights, privacy, and regulatory concerns in hyper-connected societies

Privacy issues: Deep exposure of personal data: Wearables and BCIs gather sensitive biometric, emotional, and neurological data. A "transparent human" issue arises when location, routines, and even emotions are tracked in real time.
Overload in surveillance: Always-on connectivity could be abused by businesses or governments for widespread spying. Without deep exposure of personal data: Wearables and BCIs collect private neurological, emotional, and biometric information.

8.9.5 Ethical issues

The ethical issues involved are as follows:

Automation versus autonomy: Human agency may be weakened if AI and self-governing systems make more judgements.
Manipulation and mental privacy: It is possible to decipher or alter thoughts and emotions with BCIs. Brings up issues of mental autonomy, free will, and moral advertising.
"Virtual You" and digital identity: How can we prevent misuse of your AI-powered speech clone, hologram, or avatar? Identity theft and deepfakes will advance to a completely new level. Without protections, "smart" cities could turn into "spy" cities [49].
Ownership of data and consent: Uncertain terms of service frequently cause users to unintentionally code rights.

8.9.6 Security issues

Hacking into implants or neural interfaces could result in actual physical or mental harm. Connected homes, hospitals, or automobiles could be exploited via IoT vulnerabilities.

Legal and regulatory difficulties: In the absence of international standards, diverse nations have varying privacy and technology regulations, which lead to gaps and data tourism, and cross-border data governance mechanisms are necessary.
Regulation of spectrum and infrastructure: Satellite-terrestrial communication and the regulation of the shared THz band are still in their infancy [50].
Equity and the digital divide: Rural, low-income, or underserved communities might not be included in hyper-connected benefits. In order to be deployed ethically, accessibility and inclusivity must be included.

8.10 CONCLUSION AND FUTURE WORK

5G, the fifth generation of mobile networks, promises ultra-low latency, huge device connection, increased reliability, and speeds up to 100× faster than 4G. These capabilities mark a fundamental shift in how businesses function, how services are provided, and how people engage with technology. In this chapter, important areas are highlighted that 5G has changed and evaluated the societal ramifications like industry and manufacturing 4.0, telemedicine and healthcare, transportation and smart cities, entertainment and media, agriculture and environmental monitoring, retail and logistics, education and collaboration, and cross-sector considerations. It will help society in various areas like by providing smart factories and automation, integrating edge computing in industry, and manufacturing. It also provides haptic feedback, remote surgery, wearable technology in the healthcare sector, autonomous vehicles, IoT system enabled in traffic management in the transportation sector, precision farming in agriculture, smart retail and supply chain in retail management, virtual classrooms for education, and data privacy security with digital equity for consideration across sector.

Although 5G networks are still being deployed worldwide, the groundwork for sixth-generation (6G) technology anticipated around 2030 is already in motion. Preliminary research and prototype trials forecast that 6G could deliver terahertz-frequency operation, nano scale latency, converged sensing and data links, AI-first network design, ultra-dense, and energy-smart device support. Rolling out 5G and laying the foundation for 6G depend on global partnerships and agreed-upon frameworks which is known as international collaboration and unified standards coordinated spectrum policies, consistent technical specifications, joint research initiatives and public–private alliances, and equitable access goals. 5G transcends mere speed boosts; it underpins next-generation innovation in transportation, healthcare, agriculture, retail, education, and beyond. Its promise lies in ultra-responsive connections, massive device support, and edge-based intelligence. Yet to fully harness 5G's benefits, stakeholders must address digital inclusion, security, regulatory coherence, and environmental responsibility, ensuring technology uplifts all communities in the years ahead.

REFERENCES

1. Metre, P.B., Kalnoor, G., Mahesh, G. and Gowrishankar, S. (2025) A comparative study of network slicing techniques for effective utilization of channel for 5G and beyond 5G networks. *IEEE Access*, 13.
2. Villa, D., et al. (2025) X5G: An open, programmable, multi-vendor, end-to-end, private 5G O-RAN testbed with NVIDIA ARC and OpenAirInterface. *IEEE Transactions on Mobile Computing*, 24.
3. Tarafder, P., Chun, C., Ullah, A., Kim, Y. and Choi, W. (2025) Channel estimation in 5G-and-beyond wireless communication: A comprehensive survey. *Electronics*, 14(4), p. 750. https://doi.org/10.3390/electronics14040750
4. Yang, Y., Vaseem, M., Wang, R., Makki, B. and Shamim, A. (2025) A fully screen-printed vanadium-dioxide switch-based wideband reconfigurable intelligent surface for 5G and beyond. *IEEE Transactions on Microwave Theory and Techniques*. https://doi.org/10.1109/TMTT.2025.3546554
5. Batista, A.P., Ayub, M.S., Adasme, P., Begazo, D.C., Shad, M.R., Saadi, M., Rosa, R.L. and Rodríguez, D.Z. (2025) A methodology for estimating radiofrequency signal attenuation from rainfall and atmospheric gases in 5G and beyond networks. *IET Networks*, 14(1), p. e70000. https://doi.org/10.1049/ntw2.70000
6. Suram, S., Miryala, R.S.B. and Miryala, R.K. (2025) 5G communication technologies in business: Applications and real-world cases. In: *5G/6G Advancements in Communication Technologies for Agile Management*. Hershey, PA: IGI Global Scientific Publishing, pp. 1–20.
7. Batewela, S., Liyanage, M., Zeydan, E., Ylianttila, M. and Ranaweera, P. (2025) Security orchestration in 5G and beyond smart network technologies. *IEEE Open Journal of the Computer Society*, 6.
8. Stoltidis, A., Choumas, K. and Korakis, T. (2025) Active queue management in 5G and beyond cellular networks using machine learning. *Computer Communications*, 236, p. 108108.
9. Khan, I., Ghoshal, M., Angjo, J., Dimce, S., Hussain, M., Parastar, P. and Widmer, J. (2025) How mature is 5G deployment? A cross-sectional, year-long study of 5G uplink performance. *Computer Communications*, 237, p. 108153.
10. Ashraf, N., et al. (2025) Performance analysis of SWIPT-assisted cooperative NOMA network with non-linear EH, interference and imperfect SIC. *IEEE Access*, 13.

11. Tran-Dang, H. and Kim, D.S. (2025) Digital twin-empowered intelligent computation offloading for edge computing in the era of 5G and beyond: A state-of-the-art survey. *ICT Express*, 11.
12. Enwereonye, U.P., Shahraki, A.S., Alavizadeh, H. and Kayes, A.S.M. (2025) Physical layer security techniques for grant-free massive machine-type communications in 5G and beyond: A survey, challenges, and future directions. *Computer Networks*, 264, p. 111268.
13. Riaz, H., Öztürk, S., Aldirmaz-Colak, S. and Çalhan, A. (2025) A handover decision optimization method based on data-driven MLP in 5G ultra-dense small cell HetNets. *Journal of Network and Systems Management*, 33(2), pp. 1–27.
14. Farzaneh, B., Shahriar, N., Al Muktadir, A.H., Towhid, M.S. and Khosravani, M.S. (2024) DTL-5G: Deep transfer learning-based DDoS attack detection in 5G and beyond networks. *Computer Communications*, 228, p. 107927.
15. Chu, T.T., Labiod, M.A., Augustin, B., Mathialahan, K. and Mellouk, A. (2025) Empirical evaluation of QUIC-based software-defined service migration in multi-access edge computing over 5G networks. *Journal of Network and Systems Management*, 33(2), pp. 1–25.
16. Jacobson, T., Ding, J. and Zou, H. (2025) Predicting 5G throughput with BAMMO: A boosted additive model for data with missing observations. *Journal of the Royal Statistical Society: Series C (Applied Statistics)*, 74(1), pp. 249–273.
17. Alam, M.J., Chugh, R., Azad, S. and Hossain, M.R. (2024) Optimizing cell association in 5G and beyond networks: A modified load-aware biased technique. *Telecommunication Systems*, 87(3), pp. 731–742.
18. Szabó, G., Pető, J. and Vidács, A. (2025) Deployment options of AI components for network resource management in 5G-enabled agile industrial production cell. *International Journal of Communication Systems*, 38(3), p. e5983.
19. Shirsath, V.A. and Chandane, M.M. (2025) Beyond the basics: An in-depth analysis and multidimensional survey of programmable switch in software-defined networking. *International Journal of Networked and Distributed Computing*, 13(1), p. 8.
20. Charpentier, V., Slamnik-Kriještorac, N., Landi, G., Caenepeel, M., Vasseur, O. and Marquez-Barja, J.M. (2024) Paving the way towards safer and more efficient maritime industry with 5G and beyond edge computing systems. *Computer Networks*, 250, p. 110499.
21. Verma, G. and Ali, S.T. (2025) Non-orthogonal multiple access (NOMA) for 5G wireless networks: A performance evaluation. *International Journal of Sciences and Innovation Engineering*, 2(1), pp. 7–14.
22. Yadav, S. and Rishi, R. (2024) Joint mode selection and resource allocation for cellular V2X communication using distributed deep reinforcement learning under 5G and beyond networks. *Computer Communications*, 221, pp. 54–65.
23. Ravi, B. and Verma, U. (2025) Spectrum allocation in 5G and beyond intelligent ubiquitous networks. *International Journal of Network Management*, 35(1), p. e2315.
24. Masaracchia, A., Van Huynh, D., Duong, T.Q., Dobre, O.A., Nallanathan, A. and Canberk, B. (2025) The role of digital twin in 6G-based URLLCs: Current contributions, research challenges, and next directions. *IEEE Open Journal of the Communications Society*, 6.
25. Rajesh, K. and Vetrivelan, P. (2025) Comprehensive analysis on 5G and 6G wireless network security and privacy. *Telecommunication Systems*, 88(2), p. 52.
26. Berkani, M.R.A., et al. (2025) Advances in federated learning: Applications and challenges in smart building environments and beyond. *Computers*, 14(4), p. 124.
27. Haq, A.U., Sefati, S.S., Nawaz, S.J., Mihovska, A. and Beliatis, M.J. (2025) Need of UAVs and physical layer security in next-generation non-terrestrial wireless networks: Potential challenges and open issues. *IEEE Open Journal of Vehicular Technology*, 6.
28. Rachakonda, L.P., Siddula, M. and Sathya, V. (2024) A comprehensive study on IoT privacy and security challenges with focus on spectrum sharing in next-generation networks (5G/6G/beyond). *High-Confidence Computing*, 4(2), p. 100220.
29. Nadaf, S.R., Yarlagadda, S.B., Rath, H.K. and Mishra, G. (2025) A cognitive resource-aware orchestration of network services for 5G and beyond. In: *Proceedings of the 26th International Conference on Distributed Computing and Networking*, pp. 1–10.
30. Yang, J., Arya, S. and Wang, Y. (2024) Formal-guided fuzz testing: Targeting security assurance from specification to implementation for 5G and beyond. *IEEE Access*, 12, pp. 29175–29193.
31. Hamroun, C., Fladenmuller, A., Pariente, M. and Pujolle, G. (2025) Intrusion detection in 5G and Wi-Fi networks: A survey of current methods, challenges and perspectives. *IEEE Access*, 13.
32. Manikandan, A., Babu, R.T.S., Ganesh, S.J. and Sanjay, T. (2025) Generative adversarial networks-based adaptive modulation and coding for next-generation 5G communication systems. *Discover Applied Sciences*, 7(2), p. 91.

33. Lohan, P., Kantarci, B., Ferrag, M.A., Tihanyi, N. and Shi, Y. (2024) From 5G to 6G networks: A survey on AI-based jamming and interference detection and mitigation. *IEEE Open Journal of the Communications Society*, 5, pp. 3920–3974.
34. Alam, M., Ahmed, N., Matam, R. and Barbhuiya, F.A. (2024) Analyzing the suitability of IEEE 802.11ah for next-generation Internet of Things: A comparative study. *Ad Hoc Networks*, 156, p. 103437.
35. Bhandari, A., Sharma, R. and Singh, A. (2024) Latency-optimized C-RAN in optical backhaul and RF fronthaul architecture for beyond 5G networks: A comprehensive survey. *Computer Networks*, 247, p. 110459.
36. Teng, F., Cui, T., Zhou, L., Gao, Q., Zhou, Q. and Li, W. (2024) Programmable synthetic receptors: The next generation of cell and gene therapies. *Signal Transduction and Targeted Therapy*, 9(1), p. 7.
37. Lilhore, U.K., Dalal, S. and Simaiya, S. (2024) A cognitive security framework for detecting intrusions in IoT and 5G utilizing deep learning. *Computers & Security*, 136, p. 103560.
38. Tsourdinis, T., Chatzistefanidis, I., Makris, N., Korakis, T., Nikaein, N. and Fdida, S. (2024) Service-aware real-time slicing for virtualized beyond 5G networks. *Computer Networks*, 247, p. 110445.
39. Marinova, S. and Leon-Garcia, A. (2024) Intelligent O-RAN beyond 5G: Architecture, use cases, challenges, and opportunities. *IEEE Access*, 12, pp. 27088–27114.
40. Ahmed, S.F., Alam, M.S.B., Afrin, S., Rafa, S.J., Taher, S.B., Kabir, M. and Gandomi, A.H. (2024) Toward a secure 5G-enabled Internet of Things: A survey on requirements, privacy, security, challenges, and opportunities. *IEEE Access*, 12, pp. 13125–13145.
41. Rafique, W., Barai, J., Fapojuwo, A.O. and Krishnamurthy, D. (2024) A survey on beyond 5G network slicing for smart cities applications. *IEEE Communications Surveys & Tutorials*, 27.
42. Kousias, K., Rajiullah, M., Caso, G., Alay, O., Brunstrom, A., Ali, U. and Di Benedetto, M.G. (2024) Empirical performance analysis and ML-based modeling of 5G non-standalone networks. *Computer Networks*, 241, p. 110207.
43. Duong, T.V.T. (2024) An improved method of AODV routing protocol using reinforcement learning for ensuring QoS in 5G-based mobile ad hoc networks. *ICT Express*, 10(1), pp. 97–103.
44. Vallappil, A.K., Rahim, M.K.A., Khawaja, B.A. and Iqbal, M.N. (2024) A miniaturized metamaterial-loaded switched-beam antenna array system with enhanced bandwidth for 5G applications. *IEEE Access*, 12, pp. 6684–6697.
45. Kaur, R., Bansal, B., Majhi, S., Jain, S., Huang, C. and Yuen, C. (2024) A survey on reconfigurable intelligent surface for physical layer security of next-generation wireless communications. *IEEE Open Journal of Vehicular Technology*, 5, pp. 172–199.
46. Behera, B., Mishra, S., Alsharif, M.H., Uthansakul, P. and Uthansakul, M. (2024) A metasurface-inspired printed monopole antenna for 5G and RF energy harvesting application. *Engineering Science and Technology, an International Journal*, 51, p. 101638.
47. Patel, U., Upadhyaya, T., Sorathiya, V., Pandya, K., Alwabli, A., Dave, K. and El-Shafai, W. (2024) Split ring resonator geometry-inspired crossed flower-shaped fractal antenna for satellite and 5G communication applications. *Results in Engineering*, 22, p. 102110.
48. Islam, S., Atallah, Z.A., Budati, A.K., Hasan, M.K., Kolandaisamy, R. and Nurhizam, S. (2024) Mobile networks toward 5G/6G: Network architecture, opportunities and challenges in smart cities. *IEEE Open Journal of the Communications Society*, 6.
49. Rahman, H., Shah, U.M., Riaz, S.M., Kifayat, K., Moqurrab, S.A. and Yoo, J. (2024) Digital twin framework for smart greenhouse management using next-generation mobile networks and machine learning. *Future Generation Computer Systems*, 156, pp. 285–300.
50. Caleb, S., Padmapriya, G., Nandhini, T.J. and Latha, R. (2025) Revolutionizing fault detection in self-healing networks via multi-serial cascaded and adaptive networks. *Knowledge-Based Systems*, 309, p. 112732.

Chapter 9

IoT in wireless communication

Neha Singh, Supriya Aggarwal, and Sasmita Dash

9.1 INTRODUCTION

Internet of Things (IoT) relies on a simple yet intelligent idea: small devices used in daily life can acquire, share, and compute information wirelessly. The idea or protocols for different machine connections already existed for centuries. However, IoT provides far more than just connections. The large-scale networking, autonomy, and intelligence are the key factors of IoT-based connections and communications. Several devices can communicate which each other at a very large scale without human intervention, which is the backbone of IoT development, providing continuous, enormous data flow between these small devices and the digital world. Wireless communication for IoT is an important enabling technology for deploying several devices at a very large scale to operate in different environments. The important characteristics of IoT, "connect anything, anytime, anywhere and anyone" make it the "nervous system" of a smart digital world in the current era. The wireless links are the nerves of this nervous system, which facilitates signal transmission through different devices deployed in home automation systems, industrial IoT (IIoT)-based systems, smart autonomous vehicles, and other domain-specific IoTs. In initial experiments, several limitations, such as short range, low speed, and high cost, were faced by wireless sensors radio-frequency identifications (RFIDs). Gradual technological advancements in enabling technologies in IoT and supporting wireless protocols addressed these challenges, leading to the interconnection of billions of devices together [1].

IoT is in the center where everyday devices are connected through different communication technologies like Bluetooth, LPWAN (low-power wide area network), cellular, and wireless fidelity (Wi-Fi) for making an intelligent ecosystem for homes, industries, transportation, healthcare, agriculture, and many more different sectors. The continuous information exchange between devices in real time is also depicted using flowing links. The progression in communication protocols for IoT is the culmination of decades of advancements in popular technologies like cellular infrastructure, short-range communication such as Wi-Fi and Bluetooth, and other energy-efficient specialized protocols designed to handle low-power and low-data efficiently.

The historical evolution of different wireless IoT technologies has been shown in Figure 9.1. Starting from early low-power wireless sensor networks in the 1990s, the emergence of popular short-range standards such as Bluetooth and Wi-Fi in the 2000s provides energy-efficient communication for consumer and industrial devices. Zigbee and LPWAN technologies were introduced in the 2010s. Technologies such as LoRaWAN and Sigfox are specifically designed to sustain for longer durations and cover wider ranges [2]. Since 2015, enhancements in long-term evolution (LTE) and 5G deployments for IoT applications have been continuously optimizing for reliable, high-speed, and ultra-low latency communication in diverse sectors like intelligent transportation systems and

 DOI: 10.1201/9781003637721-9

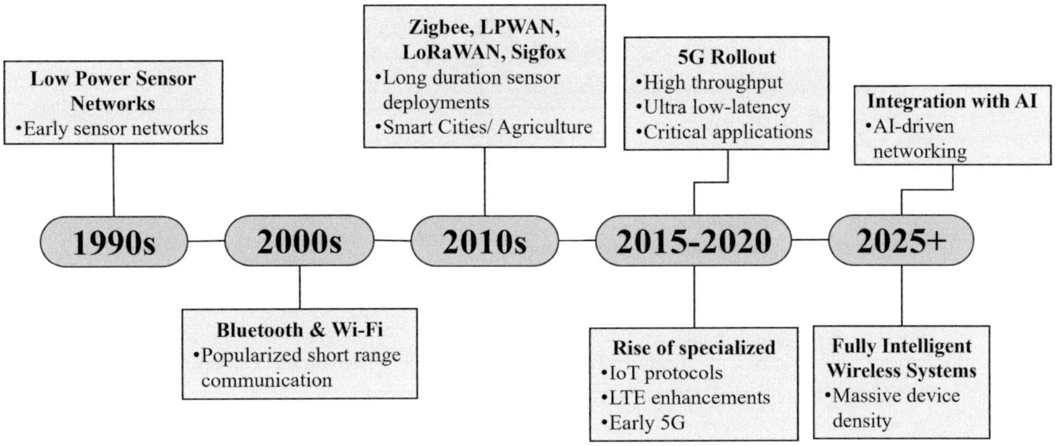

Figure 9.1 Technological evolution of communication protocols for IoT from the early era to the future 6G communication.

remote medical services. The next-generation wireless networks 6G and beyond promise an artificial intelligence (AI)-driven, fully intelligent ecosystem for massive device densities [3, 4]. The large-scale interconnections, ability to scale up to a massive density, and adaptability in diverse environments are the three foundational characteristics of IoT, and each characteristic requires different communication protocols and network architecture. Table 9.1 shows a brief comparative analysis of these characteristics for different IoT wireless technologies.

Connectivity: Instead of peak bandwidth, the focus is on optimization for reliable and ultra-low-power communication links for small and frequent packet transmission. Short-range protocols such as Bluetooth low energy (BLE) and Zigbee facilitate IoT devices with multi-year battery life with sufficient data transmission speed up to 1–2 Mbps. For long-range communication, other technologies such as LPWAN and cellular IoT are used, which provide long-distance coverage in kilometers with very low-power and high-speed data rate in kbps.

Scalability: IoT network architectures also support the growth of tens of connected nodes to scale up to millions. Mesh networks, like Zigbee, efficiently support the interconnection of hundreds of nodes locally, and cellular enhancements like narrowband-IoT (NB-IoT) and LTE for machines (LTE-M) enable scalability up to 1 million devices per

Table 9.1 Comparison of essential characteristics of IoT and related technology

Characteristic	*Focus*	*Key technologies*	*Performance metrics*	*Example use case*
Connectivity	Low-power, reliable small-packet links	BLE, Zigbee, LPWAN, NB-IoT	BLE: 1–2 Mbps @ ~12 mA LoRa: <50 kbps @ ~45 mA	Wearable health monitors reporting vital signs
Scalability	Massive device density support	Zigbee mesh, NB-IoT, LTE-M, 5G mMTC	NB-IoT: up to 1 M devices/km^2 Zigbee: 100s nodes	Smart city air-quality sensor networks
Adaptability	Dynamic operation across environments	LoRaWAN ADR, NB-IoT band switching	LoRaWAN ADR: 0.3–50 kbps, adjusting to link quality	Agricultural soil-moisture sensors in remote fields

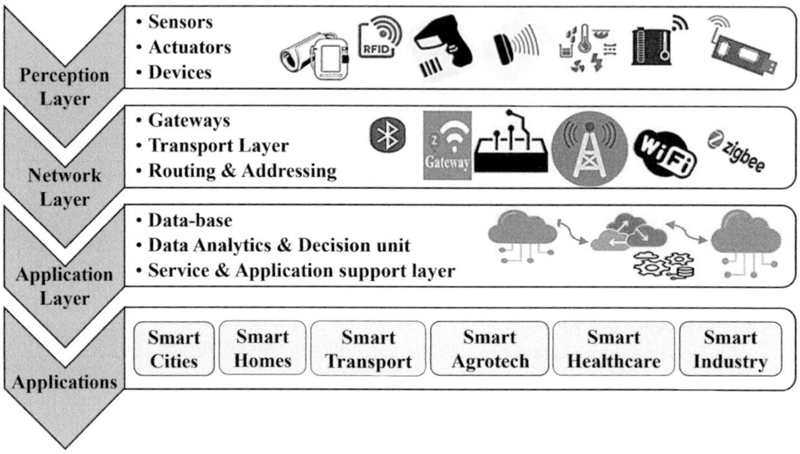

Figure 9.2 Three-layer architecture for IoT wireless communication.

square kilometer. These techniques also ensure efficient spectrum use while preventing congestion.

Adaptability: These characteristics of IoT ensure the reliable operation of devices across diverse environments like urban, rural, mobile, and indoor-outdoor setups by utilizing different features like adaptive data rates (ADRs) in LoRaWAN and deep indoor coverage in NB-IoT. Figure 9.2 shows the three-layer IoT wireless communication core architecture.

Perception Layer: Using low-power wireless standards, BLE, RFID, etc., sensors and actuators interact with the physical environment. Sensors collect data, such as temperature, motion, and location, and transmit these raw data efficiently [5].

Network Layer: Using short-range and long-range wireless communication protocols, this layer facilitates the communication infrastructure between the perception layer and the application layer devices. For short-range communication, local area networks (LANs) like Wi-Fi (IEEE 802.11), Zigbee, and BLE, and for large-range communications, wide area networks such as LPWAN (LoRaWAN, Sigfox) and cellular (NB-IoT, LTE-M, 5G) [6] are used.

Application Layer: This is the final layer of IoT architecture that provides processing of the data using cloud analytics, edge analytics, and a user interface [1].

Energy efficiency is the primary requirement of IoT wireless standard design. Several battery-operated devices function autonomously for months or years in remote and rural areas. Recent IoT wireless standard protocols are designed to support duty cycling, adaptive transmission power, and energy-harvesting capabilities. The optimized modulation techniques and a good network management capability facilitate LoRaWAN to support battery life beyond ten years while covering a range of up to 10 km [2].

Table 9.2 summarizes the different IoT wireless technologies based on range coverage and application requirements. The primary requirement for short-range communications is low latency, high availability, and ultra-low-power usage. BLE, Zigbee, and Wi-Fi provide range coverage up to 300 m, mainly used for confined areas like smart homes and wearables. Long-range communication for IoT requires long battery life with wide area coverage. NB-IoT and LTE-M are such standards. The future advancements in 6G standards for IoT applications will enable AI-based self-optimization of transmission parameters. It leads

Table 9.2 Comparative analysis of IoT wireless standard protocols for range, data rate, and power requirements

Technology	*Freq (MHz)*	*Range (m)*	*Rate (kbps)*	*Power*	*Battery life*	*Applications*
Bluetooth low energy (BLE)	2400	10–100	1000–2000	Very low (12.5 mA)	1–3 years	Wearables, smart home devices
Zigbee (IEEE 802.15.4)	2400/868/915	10–100	20–250	Low (40 mA)	2–5 years	Home automation, industrial sensors
Wi-Fi (IEEE 802.11)	2400/5000	50–300	1000–600,000	High (50+ mA)	Hours–days	Smart appliances, high-data IoT
LoRaWAN	433/868/915	2000–15,000	0.3–50	Very low (<50 mA)	2–10 years	Smart cities, agriculture
Sigfox	868/902	3000–50,000	0.1–0.6	Ultra low	8–20 years	Asset tracking, remote monitoring
NB-IoT	700–2100	1000–35,000	20–250	Low	5–10 years	Smart meters, utilities
LTE-M (Cat-M1)	700–2100	1000–35,000	375–1000	Medium	3–5 years	Connected vehicles, wearables

to simultaneously maximizing the three foundational IoT network characteristics, namely, connectivity, scalability, and adaptability. It is also expected to be highly scalable, supporting more than 1 million per square kilometer device densities and ultra-low latency. Applications such as autonomous swarms, real-time digital twins, and pervasive intelligence in Industry 5.0 will become feasible [4].

While the wireless communication laid down the foundation of IoT, its true impact extends the capabilities far beyond traditional wired approaches. It enables a completely new era of different applications, IoT devices deployment frameworks, and efficient and economically supported models, which would be difficult or very expensive with wired infrastructure [7]. Table 9.3 compares the fundamental distinctions between wired and wireless IoT approaches.

While wired technologies need high infrastructure planning, trenching, and continuous maintenance, the profound importance of wireless communication in IoT lies in the ease of deployment. Wireless IoT facilitates quick and flexible deployments across diverse environments [3, 8] like smart agrotech. The deployment of thousands of soil sensors in big farmlands can be completed in a few days rather than months. In disaster management, temporary sensors can be deployed immediately without infrastructure dependencies. Therefore, IoT deployments for diverse environments, from smart cities to remote

Table 9.3 Fundamental differences between wired and wireless IoT approaches

Deployment factor	*Wired IoT*	*Wireless IoT*	*Significance*
Initial deployment time	Weeks to months	Hours to days	Faster deployment
Infrastructure cost	High (cables, labor)	Low (devices only)	Cost reduction
Geographic coverage	Limited by cable reach	Unlimited (cellular/LPWAN)	Global capability
Mobility support	None (fixed locations)	Full (vehicles, wearables)	Enables mobile applications
Scalability	Constrained by ports/ cables	Massive (millions/km^2)	More devices
Maintenance	Physical access required	Over-the-air updates	Remote management
Disaster resilience	Vulnerable to damage	Self-healing networks	Improved reliability

agricultural fields, rely heavily on wireless technologies, which would be highly complex with wired infrastructure. Remote healthcare, autonomous transportation systems, and tracking of environmental parameters in real time would not be possible without wireless IoT standards [9, 10]. In wired infrastructure, complexity and cost are increased with additional connections, whereas wireless systems add more intelligence and cost effectiveness with more IoT device deployment. More connected vehicles lead to more accuracy in traffic management systems. Maintenance systems based on predictive technology also enhance reliability by adding more data from more machines. Such growth is possible because of scalable wireless protocols like NB-IoT and 5G's mMTC (massive machine-type communication) [3]. Future 5G and 6G IoT standards will provide more reliable, ultra-low latency, massive device scalability, and AI-based optimization techniques [4]. With these features, such high-quality wireless protocols will further enhance reliability in critical applications like autonomous systems, digital twins, and remote surgeries [7]. Therefore, it can be understood that wireless communication protocols are the backbone of modern IoT, which transforms it from an individual sensor as a data collecting unit into an intelligent, interconnected infrastructure [8].

9.2 WIRELESS COMMUNICATION STANDARDS AND TECHNOLOGIES IN IoT

Wireless communication technologies lay the foundation for seamless data transmission between devices without using physical wires. Electromagnetic waves like radio waves are used to exchange information among various connected devices, which may or may not require internet connectivity. This enables mobility of the devices as the need for physical wires is eliminated. Issues like connecting plug compatibility and wire lengths are permanently eradicated. Furthermore, inherent complexities in installation and wiring processes are tremendously reduced. Overall, wireless technologies offer greater flexibility to connect devices on the fly and reduce deployment costs. These simplify and expedite data collection, processing, and exchange among a vast number of devices. Depending on the IoT requirements like bandwidth, range, energy consumption, reliability, and security, various technologies are available. At a glance, these technologies can be classified as shown in Figure 9.3. Typically, this division is based on the range of communication, power consumption, and data rates. Short-range communication is mostly preferred for local connectivity, while long-range communication is used for wide coverage and cellular networks for global and reliable connections. The selection of appropriate wireless communication technology plays a key role in determining the system performance, scalability, and energy efficiency of any IoT system. This section discusses different wireless communication technologies used in IoT networks, their operational characteristics, and deployment considerations.

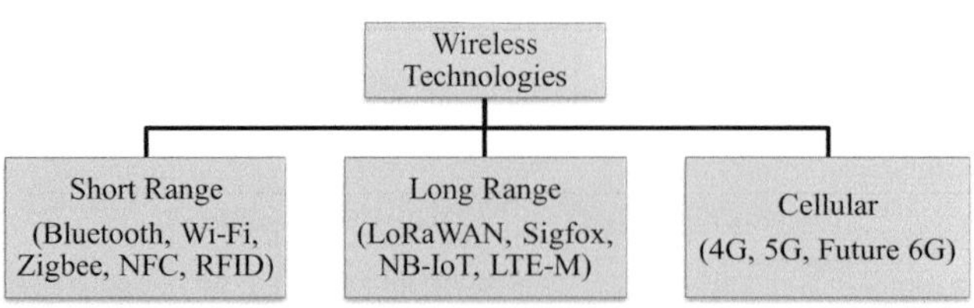

Figure 9.3 Different wireless communication technologies in IoT.

9.2.1 Short-range communication

Short-range communication is suitable for distances typically less than 100 m. It is a quick and reliable technique, even though the signal quality is affected by obstacles and interference sources. It provides a convenient platform for data transmission and information exchange in home networks, smart homes, and industrial automation. The technologies used in short-range wireless communication are Bluetooth, BLE, Wi-Fi, ZigBee, NFC (near-field communication), and RFID.

9.2.1.1 Bluetooth

Bluetooth is considered a power-efficient wireless technology for small-scale customer IoT applications; a very effective technology for device-to-device file transfer, wireless speakers, earphones, and headsets. Also, it is one of the best alternatives to connect to keyboards, mice, fax machines, and printers in offices and homes. Among IoT devices, it is widely used in fitness trackers, smart watches, and home automation devices. In the current state-of-the-art scenario, it is omnipresent in smartphones and consumer electronics. Bluetooth is based on a master–slave architecture where a single master can communicate with up to seven slave devices in a piconet. Essentially a one-to-one wireless connection, it uses 2.4-GHz radio frequency bands. A very short-range communication technique, typically in the range of 10 m, with a data rate of 1 Mbps. Bluetooth is standardized by IEEE as IEEE 802.15.1. It uses frequency-hopping spread spectrum communication technique. Bluetooth is a packet-based protocol, which divides the data to be transmitted into packets where each packet is transmitted over one of 79 designated Bluetooth channels. Each Bluetooth channel is 1 MHz wide. Bluetooth is prone to severe interference in the vicinity of Wi-Fi, other Bluetooth devices, or microwaves. Rapid channel switching up to 1600 hops/s tends to mitigate the interference effects. Bluetooth devices are often profiled, meaning they follow predefined communication protocols and features to ensure devices manufactured by different vendors work together. For Bluetooth devices to communicate with each other, it is necessary that they belong to the same profile. Some of the profiles in Bluetooth devices are A2DP (advanced audio distribution profile), HFP (hands-free profile), and HID (human interface device).

Though Bluetooth implements security features like authentication and key-generation using SAFER+block cipher, Bluetooth devices to be paired should have the same PIN, which is generated using the E22 algorithm. In spite of these security features, Bluetooth is still vulnerable to security attacks like bluesnarfing (stealing data) and bluejacking (sending unwanted messages). Classic Bluetooth consumes more power and is suitable for high-speed data transfers like headphones. The high-energy consumption led researchers to develop a lower energy variant commonly known as BLE. Due to its ultra-low-power consumption, it is widely used in small battery IoT devices like medical wearables, smart watches, and sensors. It is marketed by the Bluetooth Special Interest Group (Bluetooth SIG), which designed BLE to operate on the same frequency bands as classic Bluetooth but with a simpler modulation scheme. BLE uses Gaussian Frequency Shift Keying (GFSK), which encodes data by shifting the carrier frequency up or down. BLE offers a significant reduction in power consumption, which is around tens of milliwatts as opposed to 1 W of Classic Bluetooth. Recent versions of BLE have evolved significantly to upgrade mesh capabilities and reduce energy consumption. The connectivity offered by BLE 5.0 and 5.1 in open areas extends up to 240 m and offers a data rate of 2 Mbps. Additionally, power consumption is significantly reduced to around 10–15 mW during data transmission [11]. Recent advances in BLE have mesh networks that can connect up to 32,000 nodes on a single network, making it suitable for large-scale indoor IoT applications [12].

9.2.1.2 *Wi-Fi*

Wi-Fi stands for wireless fidelity. In the English language, fidelity refers to faithfulness. Therefore, as the name suggests, it provides a reliable wireless communication between devices connected to a LAN or the internet. It is a very effective technology commonly used in homes, offices, public spaces, and enterprises. Devices connected over the network communicate with a wireless access point (WAP) known as a router, which acts as a bridge to the wired network. Wi-Fi offers wide availability with extensive infrastructure support and is suitable for devices requiring constant and high-bandwidth connections like video streaming. As Wi-Fi standards have evolved, they have provided users with faster data transmission rates, longer ranges, and more reliable and secure connectivity. Typically, Wi-Fi follows the IEEE 802.11 standard, operating on radio bands in the range 2.4, 5, and 6 GHz. The foundational standard offered a speed of 2 Mbps, which has now increased to over 11 Mbps for Wi-Fi 6. However, the theoretical speeds are over 10 Gbps for Wi-Fi 6, which is much higher than what is observed by the user. Factors like network congestion, interference, range, and distance from the access point almost make it impossible to achieve theoretical speeds at the user end. It generally uses an orthogonal frequency division multiplexing (OFDM) modulation scheme. Generally, a Wi-Fi network is known to support a range of 50–100 m.

Wi-Fi 5 is an ideal choice for high-bandwidth applications like smart cameras and multimedia-enabled sensors in IoT. While Wi-Fi 6 is built for a large number of connected devices, it is suitable for IoT applications in buildings, factories, and offices. The latest Wi-Fi standard deployed in 2024; Wi-Fi 7 is designed for an extremely high data rate of up to 40 Gbps theoretically. Designed for speed and low latency, it mitigates the network congestion issues in dense Wi-Fi networks. Wi-Fi HaLow is an IEEE 802.11ah standard, particularly for extended-range and low-power applications to connect a large number of devices. It operates on sub-1-GHz bands like 868 MHz in Europe and 902 MHz in North America and uses OFDM as a core modulation scheme but supports a range of modulation and coding schemes. It uses a lower-order modulation scheme like binary phase shift keying (BPSK) to achieve long-range communication up to 1 km. Given the constraint of low-power consumption, it is a Wi-Fi technique targeted for IoT applications. Protocols like WPA2 and WPA3 provide network security for devices connected using Wi-Fi. The use of Wi-Fi is so pervasive that its network performance is hampered in crowded areas like stadiums, concert halls, public transportation hubs, and even homes when more and more devices are connected to the same router simultaneously. Wi-Fi 6 is the sixth-generation Wi-Fi technology, and 6E is an extension, which adds 6-GHz spectrum to the already existing frequency bands. Due to less interference in this band, it is able to provide low latency (below 10 ms), which makes Wi-Fi 6E an ideal choice for high-bandwidth IoT applications [13]. It is able to provide more resilient networks to support mission-critical applications. Overall, it significantly enhances the services supported by the existing Wi-Fi networks [14].

9.2.1.3 *Zigbee*

Zigbee gets its name from the zig-zag dance pattern (known as waggle dance) bees use to communicate with other hive members for food. It is a low-power and low-data-rate mesh network designed for IoT applications. The mesh connection enables devices to communicate over extended distances by relaying signals through intermediate nodes. It thereby enhances network reliability and coverage due to self-healing routes, which prevent single-point failures. However, other topologies like tree and star are also supported. An ideal technology when small data packets are required to be transmitted intermittently with minimal energy use. It follows the IEEE 802.15.4 specification. Globally, it operates on 2.4-GHz frequency bands but in Europe it uses 868 MHz, while 915 MHz is used in America. Low data rates limited to

250 kbps make it unsuitable for high-bandwidth applications. It employs 128-bit AES encryption for network authentication and confidentiality. Smart homes, healthcare devices, and sensors are a few IoT applications that could use Zigbee. It normally supports connectivity up to 100 m. In 2.4-GHz band, it uses offset quadrature phase shift keying (OQPSK) with direct sequence spread spectrum (DSSS), while in MHz frequency bands, it uses BPSK.

Zigbee devices are characterized as connectors, routers, and end-devices. Connectors facilitate data transfer among routers. Routers are used for data transfer between routers and end-devices. It is widely deployed for monitoring and controlling operations in IoT, like smart meters, smart grids, and medical wearables. As the requirements for IoT applications related to latency and power become more and more stringent, Zigbee continues its dominance in both industrial and home automation sectors due to its self-healing mesh topology. The latest Zigbee PRO 2023 enhances security and provides support for bands beyond 2.4 GHz. Zigbee limits power consumption, where during active transmission mode, the current remains stable at 15–25 mA, while during the sleep mode, the current remains below 3 μA [15].

9.2.1.4 NFC

NFC can connect electronic devices in a close neighborhood only. It supports a very short range of a few centimeters (<5 cm). It uses a 13.56-MHz frequency in the ISM band for communication with data rates in the range 105–424 kbps. It is based on the concept of inductive coupling between electromagnetic coils for transmitting data among devices. NFC devices can be classified as either active or passive. Passive devices are capable of only transmitting data to other devices without any processing capabilities. Hence, these might not require any power source for their operation. Active devices, however, are capable of both sending and receiving data and even processing it into actionable information. NFC initiator generates a radio-frequency field, which powers the passive devices using coupling to enable unidirectional data transmission. The various ways of communication using NFC are:

Card emulation: Tap the cards for automatic payments, identification, and access.
Reader/writer: Read the information stored on a card or a tag.
Peer-to-peer: To transfer information among devices, the devices have a capability to switch from active to passive and vice versa, depending on the requirement of the data transfer operation.
Charging mode: Wherein the devices in the radio-frequency fields can be charged using coupling, like earphones and watches.

NFC devices communicate using amplitude shift keying (ASK) modulation technique with varying modulation depth depending on their nature (passive or active). While the passive devices operate on a low-modulation depth of 10%, the active devices operate at 100% modulation depth to guarantee effective communication. NFC is widely used in applications like card payments, smart advertising and marketing, digital keys, and public transportation. Given the very short distance of communication between NFC devices, these are not supported using any security protocols. But this does not mean that communication is not prone to security breaches; for example, contactless payments are subject to ghost-tapping and cloning.

9.2.1.5 RFID

RFID is a superset of NFC, offering longer ranges up to a few 100 m. As the name suggests, radio frequency is used for the identification of objects. It is widely used to search, identify, and track resources and hence finds wide application in areas like supply-chain management, logistic operations, livestock tracking, and cashless transactions. Moreover, in the medical field, nowadays hospitals use RFID-enabled wristbands for patient tracking during the treatment process. The complete system requires an RFID tag (known as a transponder), a reader (transceiver), and an antenna. RFID works at various frequency bands with ultra-high frequency (UHF-RFID) at 868 MHz or 912 MHz, high frequency (HF-RFID) at 13.56 MHz, and low frequency (LF-RFID) operating in the range 125–134 kHz. The modulation techniques used are ASK for the reader to tag and tag to reader using backscatter modulation, where the tag modulates its impedance according to the signal to be transmitted.

9.2.2 Long-range communication

Long-range communication technology is to enable wireless connections between devices within a distance of tens of kilometers. These technologies are targeted to be scalable for wider networks connecting billions of devices. With an enormous number of devices connected to the same network, it is necessary to prevent data loss due to interference. Furthermore, the data rates are relatively lower than those of short-range technologies to achieve a higher range. Mostly, long-range techniques operate in sub-GHz frequency bands to increase the range of connections. LoRaWAN (LPWANs), Sigfox, NB-IoT (narrow-band IoTs), LTE-M, also known as LTE Cat-M1, are long-range wireless technologies commonly employed in IoT applications.

9.2.2.1 LoRaWAN

LPWAN known as LoRaWAN is a long-range communication technology similar to Bluetooth. A LPWAN is designed for long-range communication at a low bit rate between IoT devices. It is capable of transferring small data packets over longer ranges with low energy consumption. LoRaWAN uses chirp spread spectrum (CSS) technology, where the data is encoded on radio waves using chirp pulses. It is a technique similar to the way dolphins and bats communicate. It operates under sub-GHz frequency bands, 868 MHz in Europe, and 915 MHz in North America. It can achieve data rates up to 30 kbps. These low data rates make it unsuitable for high bandwidth applications like phone calls, audio–video streaming, and real-time data monitoring. Mostly suited for time-insensitive data requiring low-bandwidth data communications, and are quite effective for data-tracking applications and facility management/monitoring, for example, smart agriculture, smart cities, asset tracking, and industrial automation.

LoRaWAN uses a star of stars topology where the end devices communicate with gateways using LoRa (long range), enabling long range in urban environments. Extremely low power consumption makes it suitable for devices with small batteries. It is also capable of supporting thousands of devices, thereby offering wider scalability. LoRaWAN architecture includes end devices, gateways, network sensors, and application servers to provide a scalable and flexible IoT network. Using this technique, devices in the range of 10 km can easily be connected to each other. It ensures secure communication using 128-bit AES encryption standard. By 2024, LoRaWAN will have seen widespread adoption in over 170 countries. Recent advancements in LoRaWAN 1.1 have upgraded security features and roaming capabilities [16]. ADR algorithms can now achieve up to 15% better energy efficiency compared to their counterparts based on fixed configurations [17].

9.2.2.2 Sigfox

Sigfox is named after a French company that developed and operates the technology. It offers low-power, low-cost, and long-range communication for simple IoT applications. Its network architecture is based on a star topology where small data packets are directly sent to Sigfox base stations, which are further forwarded to a centralized cloud platform. The ultra-narrowband modulation techniques employed in Sigfox are differential BPSK (DBPSK) and GFSK, as it operates in the sub-GHz band with 868 MHz in Europe and 902 MHz in America. It can easily be used to cover large areas and facilitates underground signal communication as well. However, specific Sigfox hardware and subscription are required for deployment. Additionally, the extremely low data rate of 100 bps renders it suitable only for basic messaging. Nevertheless, IoT networks that require sending small amounts of data intermittently over long distances with low energy consumption use Sigfox. Metering devices, environment monitoring, and asset tracking are some IoT applications where long-range communication to cloud data storage is achieved using Sigfox. But an in-built firewall prevents devices from communicating directly via the internet protocol. Sigfox also supports device authentication using the 128-bit AES encryption standard.

9.2.3 NB-IoT

NB-IoT is a wireless technology for wide and reliable coverage supported by low-power consumption. It operates within a narrow bandwidth of 200 kHz, enabling seamless coexistence with LTE and GSM networks. A higher uplink data rate up to 250 kbps is supported using single carrier frequency division multiple access (SC-FDMA), while the downlink data rate is lower, about 20–60 kbps, and utilizes the OFDM modulation technique. NB-IoT is equally beneficial for indoor and outdoor applications as it supports excellent coverage, dense networks, and longer battery lives, being a low-power technology. Given its harmonious conjunction with GSM and LTE networks, it leverages already established cellular security protocols to ensure privacy, confidentiality, authentication, integrity, etc. However, lower data rates limit its applicability. Moreover, additional cellular subscription costs make it expensive for deployment. Nonetheless, as it capitalizes on the infrastructure laid for mobile networks, NB-IoT proves relatively cost-effective in terms of resource utilization. The various applications of NB-IoT include:

a. smart meters and their remote monitoring: for oil, gas, electricity, etc.
b. smart agriculture: to optimize resources, irrigation, and farming practices
c. smart cities for waste management, parking areas
d. wildlife management and conservation: track movements, habitats of animals
e. healthcare and patient monitoring through wearables within a periphery.

Even with energy efficiency advantages offered by NB-IoT, the low data rate makes it unsuitable for applications like video or audio streaming, which require high data rates for seamless transmission. Furthermore, for devices with high mobility, establishing a connection will consume more power, defeating the inherent benefit of NB-IoT and also rendering it unsuitable for devices with high mobility. Typically, it supports connectivity up to 10 km. NB-IoT deployment has expanded after integration with 5G, offering enhanced coverage and reduced latency. The latest release supports up to 1 million devices per square kilometer, with power consumption as low as 15 mA for data transmission [18]. Deep indoor penetration improves up to 20 dB over previous versions, only to further enhance the suitability of NB-IoT for basement and underground applications [19].

9.2.3.1 LTE-M (Cat-M1)

It stands for LTE-Ms, also known as LTE Cat-M1 (enhanced machine type communication). It is a cellular technology that provides a higher data rate and mobility as compared to NB-IoT. To coexist with LTE networks, it uses a bandwidth of approximately 1.4 MHz and operates in regular LTE networks. High data rate of 1 Mbps enables it to support data-intensive IoT applications in the range of 10 km. Lower power consumption with higher coverage and speed makes it advantageous for use in many mobile IoT devices. But it is less suited for deep indoor or underground coverage. It still requires a cellular network subscription, which adds to the deployment cost. Basically built around the 4G cellular network, it encompasses the benefits like low-power consumption, lower deployment cost, with a reasonable long-range connectivity, which makes it suitable for use in IoT applications. As LTE-M inhabits the cellular network environment, the various security protocols used in cellular networks for authentication, data integrity, privacy, etc., are available to ensure safe and reliable data transmission. The energy savings approaches employed in LTE-M are PSM (power saving mode) and extended discontinuous reception (eDRX), which allows connected devices to remain in the sleep mode unless required to wake up for data transmission. The single antenna approach in LTE-M simplifies design and prioritizes efficiency and cost-effectiveness for IoT devices. LTE-M uses adaptive modulation techniques from quadrature phase shift keying (QPSK) to 64-QAM (quadrature amplitude modulation), depending on the signal quality. QPSK is preferred for weaker signal connections, while QAM is preferred for stronger signals.

9.2.4 Cellular networks

Given the consumer mobile market, cellular networks are popularly used as they offer reliable communication and very high bandwidth to support both audio and video streaming/calling. With the introduction of ultra-low latency 5G and 6G, wireless can effectively support time-sensitive industrial automation, real-time medical support, public safety, and surveillance. Cellular networks are able to provide IoT services without additional infrastructure as these use mobile networks for connection and are capable of delivering coverage, flexibility, security, and performance for modern applications. Operating IoT devices by leveraging the existing mobile network resources serves as a source of widespread use with limited extended costs. The cellular networks of 4G and 5G are currently in operation, while 6G and AI-powered IoT are viewed as future upcoming trends in cellular networks. With the demand for technology, cellular networks facilitate connecting multiple devices like cars, audio speakers, televisions, monitoring devices, and sensors for reliable and secure communication. In today's technologically advanced world, almost every person on earth has access to a cellular network, enabling them to use IoT devices connected to cellular networks conveniently. IoT applications that require network coverage over multiple regions have to use cellular networks. As cellular technology advanced from 2G, 3G, 4G, and further to the currently employed 5G networks, it has immensely enhanced the capabilities of data transmission. From text messaging, then mobile internet, followed by high-definition audio and video streaming, the services provided by cellular networks have exhibited tremendous advancement. The leap in data speed is accompanied by enhanced security features.

9.2.4.1 4G LTE

The 4G marks a substantial transformation in cellular technology with migration to IP-based networks, where LTE is the core underlying technology. The data transmission technology changed from circuit-switched network in 2G/3G to a packet-switched network in 4G. Enhanced data speeds ranging from 1 Mbps to 10 Gbps fulfill the requirements for modern multimedia transmissions. Such high data speeds inherently warrant high power consumption; therefore, they are mostly used to support real-time applications like surveillance, infotainment, and telemedicine. The networks of 4G employ OFDM modulation techniques, which divide the carrier into multiple sub-carriers where each carrier can be modulated using QAM or QPSK. It supports adaptive modulation and coding (AMC), a technique that enables dynamic switching between different modulation techniques depending on the signal quality, channel conditions, and use locations to guarantee a reliable connection. A completely different architecture of 4G networks makes it non-backward compatible with 2G/3G networks.

9.2.4.2 5G

The latest cellular network technology is capable of offering ultra-fast data rates, low latency, and massive connectivity. It operates on the same frequencies as its predecessors, 3G and 4G, and provides much stronger and more reliable communication. The evolution of 5G technology has had a significant influence on technologies that empower self-driving cars, gaming systems, and video streaming. Furthermore, it has showcased its power to change the face of communication, how people interact and use the internet, social media, etc. With an average speed of 400 Mbps, the data rates offered by 5G are 10× higher than those offered by 4G. Use of advanced communication technologies enables it to have wide coverage and supports dense network connections. Low latency makes it an ideal choice for time-critical IoT applications. The physical footprint, i.e., the size of transmitters, is much smaller than preceding technologies, which aids hassle-free installation. It is able to operate on much higher bandwidths; the sub-3-GHz band is extended to 100-GHz radio spectrum. The frequency band is divided into three categories: low-band (up to 1 GHz), mid-band (1–6 GHz), and high-band (above 24 GHz). The network of 5G combines OFDM with AMC schemes, which helps to achieve low bit error rates. Being the latest among all wireless technologies, the deployment costs are the highest, which makes it expensive. Moreover, the high power consumption renders it unsuitable for IoT devices with small batteries. The huge amount of data and, with a large cohort of connected devices, 5G is transforming the world. Fully autonomous cars, smart buildings and infrastructure, industrial automation, augmented reality, and virtual reality are a few application areas that have flourished because of the data transfer capabilities offered by 5G. Even though 5G is considered state-of-the-art, it currently faces issues, which need addressed. Security breaches, potential cyberattacks on networks, and malware are certain aspects that require attention. Currently, end-to-end encryption is lacking. Network slicing and creating a virtual network expose 5G to cyberattacks, hacking, and other security breaches. Though pilot projects have shown considerable success in adopting 5G, expansion for global coverage is expensive and time-consuming, which is an even bigger problem when targeting remote areas. The high-frequency waves used for data transmission have limited penetration depth, which means they are prone to blockage and higher attenuation, restricting their coverage by limiting the cell radius. The new radio (NR) of 5G for IoT applications enables ultra-reliable low-latency communication (URLLC) with latencies below 1 ms and 99.999% reliability [20]. mMTC supports device densities exceeding 1 million per km^2 [21].

9.2.4.3 Future 6G/AI-powered IoT

Every ten years has seen a change in the communication technology. As people plan on deploying 5G, a platform for developing 6G has already commenced. The network of 6G will not only be a radio wave technology for communication, but a smart wireless technology leveraging on AI, green technology, and integrated sensing. As the number of wireless devices has already outnumbered the population, the need for higher data rates, reliable, and secure connectivity is continuously rising. The network needs to be scalable with intelligent computing and sensing, along with efficient bandwidth usage, to fulfill the application requirements. Like 5G was designed to coexist with 4G, the NR access technology for 6G should coexist with 5G. The multi-rat spectrum sharing (MRSS) should be backward compatible with the dynamic spectrum sharing (DSS) of 5G to ensure a smooth transition from 5G to 6G with minimal infrastructure changes. With the incorporation of AI in network design, 6G targets more intelligent and efficient radio access networks (RANs). Building official standards of 6G would warrant careful analysis of waveforms, channel coding, AI integration, architecture design, and migration issues. In order to deploy 6G by 2030 (as expected), it is necessary that all standards and specifications concerning 6G networks be ratified by 2028. Commercial deployment of 6G can only proceed after successful trials and interoperability testing. The network of 6G aims to provide data rates which are 10× higher than 5G, amounting to 1000 Gbps, and a latency of 100 µs which is 1/10th the 5G latency. Its target is to explore the sub-THz band (up to 300 GHz) for wireless communication among billions of devices at high speed. The frequency bands will be categorized as low-band (below 1 GHz), mid-band (1–24 GHz), high-band (24–92 GHz), and sub-THz frequency band (92–300 GHz). The new applications that would focus on 6G will be immersive extended reality (XR), mobile holograms, and ultra-low energy IoT services. As the frequency increases, the wavelength decreases; effective communication will require specialized antennas for beamforming. The communication of 6G is still in its initial development stage, with some very fundamental and technical challenges to overcome, like high propagation path loss and water vapor absorption. It would be interesting to see how the standards for 6G evolve to enable lightning-fast data transmission. At a glance, a brief comparison of different wireless technologies discussed above: short-range protocols (10–100 m), long-range low-power protocols (1–50 km), and emerging protocols is demonstrated in Table 9.4.

The choice of wireless protocol for IoT applications depends on the following multiple factors:

Range Requirements: BLE/Zigbee suitable for local areas, LoRaWAN/Sigfox for wide-area coverage, and NB-IoT/LTE-M for mobility.
Data Volume: Wi-Fi supports high-throughput applications, while LPWAN serves better for periodic small payloads.
Power Budget: Sigfox for ultra-low power consumption, BLE for a balanced trade-off between energy consumption and data rates.
Infrastructure: Cellular networks for extensive coverage, LoRaWAN for custom deployments, particularly for low-power applications.

Table 9.4 Salient features of different wireless communication technologies

Protocol	*Range*	*Data rate*	*Power (TX)*	*Battery life*	*Deployment cost*	*Security level*	*Recent updates*
BLE 5.1	10–240 m	2 Mbps	12 mA	2–5 years	Low	AES-128	Mesh, positioning
Zigbee 3.0	10–100 m	250 kbps	20 mA	3–7 years	Medium	AES-128	OTA updates, unified
Wi-Fi 6E	50–300 m	9.6 Gbps	300–1000 mA	Days-weeks	Low	WPA3	6 GHz, low latency
LoRaWAN 1.1	2–15 km	50 kbps	45 mA	5–15 years	Medium	AES-128	Enhanced security
NB-IoT R17	1–35 km	250 kbps	15 mA	10+ years	High	256-bit	5G integration
LTE-M R16	1–35 km	4 Mbps	80 mA	5–10 years	High	256-bit	Enhanced coverage
Sigfox 0G	3–50 km	600 bps	25 mA	15+ years	Low	Proprietary	Global roaming

Security Needs: Cellular networks provide secure data transmission for critical applications, though proprietary protocols like SigFox can be used for low-power security.

Recent studies have shown that hybrid deployments combining multiple protocols achieve up to 30% better coverage and 25% improved energy efficiency in comparison to single-protocol network implementations [22].

9.3 WIRELESS NETWORKS IN IoT ARCHITECTURE

In IoT systems, wireless networks transfer data smoothly from physical sensors to analysis tools. The focus of this section is the IoT architecture layers and necessary details about the integration of different radios, systems, protocols, and gateways for a reliable, scalable, and energy-efficient IoT infrastructure deployment.

9.3.1 Perception layer

This layer makes the base where actuators and sensors directly interact with the physical world. Technologies for this layer are based on reliability, low power [5], and low latency requirements and include radios suitable for their operating conditions [23]. The devices operating in this layer and commonly used technologies, along with their use cases and key features, are summarized in Table 9.5.

9.3.2 Network layer

In this layer, the data collected by sensors is transmitted to gateways or central processing units over extended ranges. Various wireless standards are combined to achieve the requirements of connectivity, throughput, and energy usage in current IoT deployments [3]. Wi-Fi standards like IEEE 802.11 a/b/g/n/ah provide high bandwidth over short distances, which makes them suitable for indoor applications such as camera feeds, firmware updates, and rich sensor data [2]. However, it is not well suited for sensors in the physical layer due to its high energy demand. Table 9.6 summarizes commonly used technologies in the network layer.

Table 9.5 Technologies used in perception layer

Category	*Technology/device*	*Primary function*	*Key design features*	*Typical applications*
Sensors	Temperature, humidity, pressure, light, accelerometer, ECG, gas, soil moisture sensors	Measurement of environmental or physiological parameters	High sensitivity, low power, miniaturized, digital output support	Smart agriculture, healthcare, environment monitoring
Actuators	Motors, valves, relays, servos, LEDs	Perform actions based on control commands	Fast response, precision, low energy requirement	Industrial control, robotics, smart homes
Identification devices	RFID tags, NFC modules	Object recognition and tagging	Passive/active operation, short-range communication	Inventory tracking, logistics, access systems
Embedded platforms / Edge nodes	Arduino, Raspberry Pi, ESP32, STM32, BeagleBone	Local processing, device control, pre-processing	Programmable, multi-sensor integration, edge analytics	Edge computing, smart node control
Short-range communication interfaces	BLE, Zigbee, RFID modules	Device-to-gateway data exchange	Low energy, short-range (10–100 m), compact transceivers	Wearables, smart building nodes

Table 9.6 Widely used technologies in the network layer

Technology/standard	*Category*	*Typical range*	*Data rate*	*Energy profile*	*Design focus/key features*	*Typical applications*
Wi-Fi (IEEE 802.11 a/b/g/n/ah)	Short/medium range (WLAN)	Up to 100 m (indoors)	10 Mbps–1 Gbps	High	High-throughput, IP-based, supports large data (video, firmware updates)	Smart homes, industrial monitoring, indoor cameras
LoRaWAN	LPWAN (sub-GHz)	2–15 km	0.3–50 kbps	Very low	Chirp spread spectrum modulation, long-range, ultra-low power	Agriculture (soil moisture), environmental monitoring, smart cities
Sigfox	LPWAN (ultra-narrowband)	3–40 km	100 bps (uplink only)	Ultra-low	Long-range, minimal bandwidth, low data payload	Remote metering, asset tracking, environmental sensors
NB-IoT (narrowband IoT)	Cellular LPWAN	1–10 km	26–250 kbps	Low	Operates in licensed LTE bands, deep indoor coverage, 200-kHz bandwidth	Smart metering, healthcare sensors, static IoT nodes
NR RedCap (5G reduced capability)	5G (3GPP release 17)	<5 km	Up to 150 Mbps	Moderate	Targets mid-tier IoT devices, reduced complexity vs. full 5G	Industrial sensors, smart wearables
5G NR (full)	Cellular broadband	100 m–several km	Up to 10 Gbps	Moderate–high	Ultra-reliable, low latency (~1 ms), massive connectivity	Autonomous vehicles, telemedicine, smart manufacturing
5G NR (full)	Cellular broadband	100 m–several km	Up to 10 Gbps	Moderate–high	Ultra-reliable, low latency (~1 ms), massive connectivity	Autonomous vehicles, telemedicine, smart manufacturing
6G (emerging)	Next-gen cellular	Global	Multi-Gbps	Adaptive	AI-driven network optimization, sub-ms latency, integrated sensing	Real-time robotics, holographic communication, intelligent automation

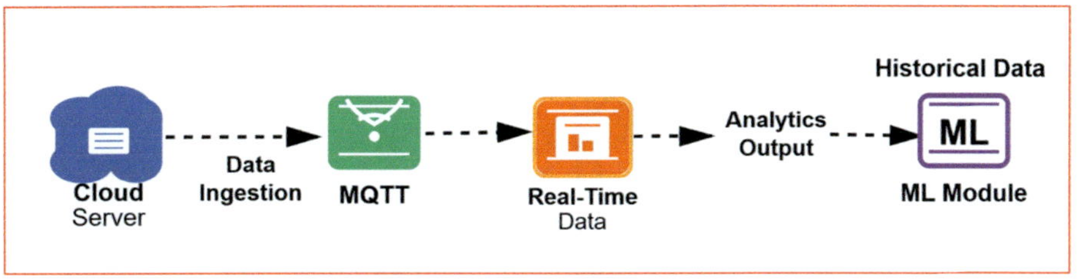

Figure 9.4 Application layer (data flow from sensors through gateways and analytics to user dashboards).

9.3.3 Application layer

In this layer, raw sensor data is processed and transformed into actionable intelligence. This layer particularly uses cloud platforms, edge computing, and AI-driven intelligence. It enables forecasting and predictions of events like equipment breakdowns, automating related decisions such as opening valves, and sending notifications like health alerts. Figure 9.4 highlights the flow of data from sensors to gateways and then to the processing unit. This layer depends on lightweight, efficient protocols tailored for constrained environments, enabling billions of devices to operate seamlessly. Table 9.7 summarizes the commonly used technologies in the application layer.

9.4 APPLICATION OF IoT IN WIRELESS COMMUNICATION

Wireless IoT empowers tangible solutions that touch every aspect of modern life. From city streets to farm fields, diverse wireless technologies come together to solve real problems. In this section, we explore six key domains—smart cities, homes, healthcare, industry, transportation, and agriculture—showing which radios are used, how systems are built, and the benefits they deliver. Figure 9.5 shows the general architectural flow of IoT-based systems.

Table 9.7 Widely used technologies in the application layer

Category	*Technology/platform*	*Primary function*	*Key features/design focus*	*Typical applications*
Cloud IoT platforms	AWS IoT core, Microsoft Azure IoT Hub, Google Cloud IoT, IBM Watson IoT	Centralized data storage, analytics, and device management	Scalable, supports MQTT/HTTP, security, and data governance	Smart cities, connected infrastructure
Edge computing frameworks	EdgeX Foundry, AWS Greengrass, Azure IoT Edge	Localized data processing and control	Low latency, offline operation, reduced bandwidth use	Industrial IoT, real-time automation
Data visualization and orchestration tools	Grafana, Power BI, Node-RED	Dashboard visualization and system control	Interactive dashboards, workflow automation	Remote monitoring, smart grids
AI/ML frameworks	TensorFlow, PyTorch, Scikit-learn, Keras	Predictive modeling, anomaly detection, pattern learning	Deep learning support, integration with cloud/edge	Healthcare analytics, fault detection
Application protocols	MQTT, CoAP, RESTful APIs	Data transfer and device communication	Lightweight, asynchronous, interoperable	Communication between IoT nodes and servers
Security and authentication	OAuth 2.0, SSL/TLS, X.509 certificates	Data confidentiality and secure access	Encryption, identity management, trust verification	Device security, cloud access control

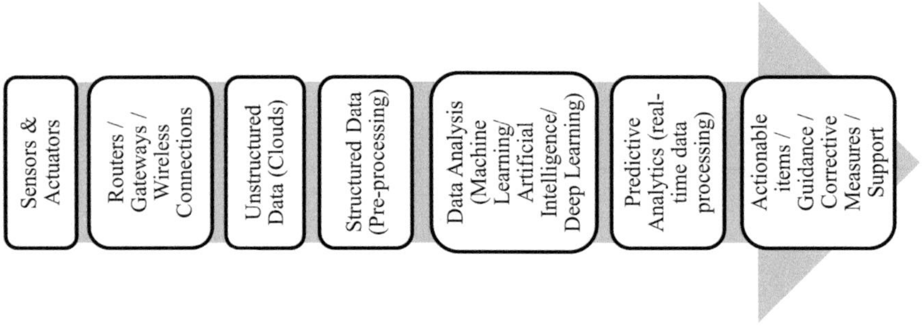

Figure 9.5 Basic flow of IoT architecture–sensors to actionable items.

9.4.1 Smart cities and infrastructure

Smart cities exploit wireless IoT services to make urban life safer, cleaner, and more efficient; for example, imagine a streetlight that brightens only when it senses a pedestrian or a moving vehicle, parking space assistance through phone apps. These conveniences rely on a network of sensors, gateways, connectivity options, and analytic platforms working seamlessly. At the base level, NB-IoT and LoRaWAN sensors monitor air quality, noise levels, and traffic flow. Air-quality sensors are scattered across the neighborhood, reporting readings over LoRaWAN periodically. Thereby, enabling real-time pollution maps that help city planners identify green zones and mitigate health risks [9]. NB-IoT parking sensors detect vehicle presence and instantly report availability, cutting cruising time by approximately 30%, and lowering emissions from idling cars [24].

Streetlights are equipped with BLE sensors and motion detectors that appropriately dim or brighten them based on real-time activities like pedestrian movement and speeding vehicles. These adaptive lighting systems, with the help of wireless communication technologies, are capable of reducing the energy consumption by up to 50%. Further guarantee pedestrian safety at night [25] and avoid accidents by judiciously lighting the areas to detect vehicular traffic. Video cameras equipped with edge-AI are able to detect accidents or crowding and transmit required data using cellular networks. The alert signals are then generated for immediate action by control centers. As a result, the response times are drastically reduced, enabling rapid emergency response [25]. IoT services in smart cities are based on different backhaul networks, which stitch multiple layers together, like fiber-to-gateway connects LoRaWAN concentrators; cellular towers handle NB-IoT and 5G; and citywide Wi-Fi covers public spaces. The data streams converge on cloud platforms that run digital-twin models and AI analytics, guiding timely actions. Predictive algorithms forecast traffic jams to adjust signal timings dynamically. This also helps in pre-emptive maintenance of road surfaces to mitigate risks arising from the formation [26]. A summarized study of different subsectors in smart city IoT applications is tabulated in Table 9.8.

Table 9.8 Smart city IoT applications

Application	*Wireless tech*	*Function*	*Impact*
Air-quality monitoring	LoRaWAN	Real-time pollution mapping	Reduction in unhealthy zones
Smart parking	NB-IoT	Instant space availability	Faster parking times
Adaptive street lighting	BLE + LTE/5G	Motion-based dimming	Energy savings
Public safety surveillance	5G URLLC	Low-latency video analytics	<10 ms alert latency
Traffic management	Cellular + Wi-Fi	Dynamic signal control	Reduction in congestion

9.4.2 Smart homes and buildings

Homes and commercial buildings are conducive places for personal and professional lives. Smart homes and buildings incorporate small wireless sensors and smart devices that talk to central hubs, learn routines, and adjust environments automatically. Imagine a home that knows when you leave and automatically locks the doors and lowers the thermostat. An office senses a meeting-room occupancy to adjust the lighting, AC temperature, and ventilation for comfort and energy savings.

In most setups, BLE and Zigbee are used to connect smart door locks, motion detectors, and smart plugs in a mesh or star network. A BLE-enabled door lock will recognize your smartphone as you approach and unlock without a key. Zigbee light bulbs and occupancy sensors communicate to adjust lighting in empty rooms. This reduces energy consumption by cutting lighting costs by approximately 20% [27]. Wi-Fi thermostats linked to cloud platforms learn your schedule to pre-heat or cool rooms whenever needed, ultimately trimming the HVAC bills [28] and reducing financial burden. Houses and buildings equipped with IoT devices seamlessly blend security and convenience. CCTV cameras ensure continuous monitoring by using WPA3-protected Wi-Fi to stream videos to mobile apps [29]. Voice assistants like Alexa, Siri, and Google Assistants use AI-based natural language processing techniques along with local devices to communicate over secure MQTT connections to facilitate hands-free tasks like blinds adjustment and playing music [29]. Multi-story buildings are equipped with water-leak sensors that send data over LoRaWAN gateways to mitigate flooding risks and enable timely action before damage occurs [30, 31]. Figure 9.6 shows different aspects of smart-home IoT applications: lighting controls, security systems, climate management, appliance automation, etc., while their details are summarized in Table 9.9. A typical smart-home network has BLE locks and sensors, Zigbee lights, Wi-Fi cameras, and MQTT bridges all feeding a cloud dashboard where rules and analytics run.

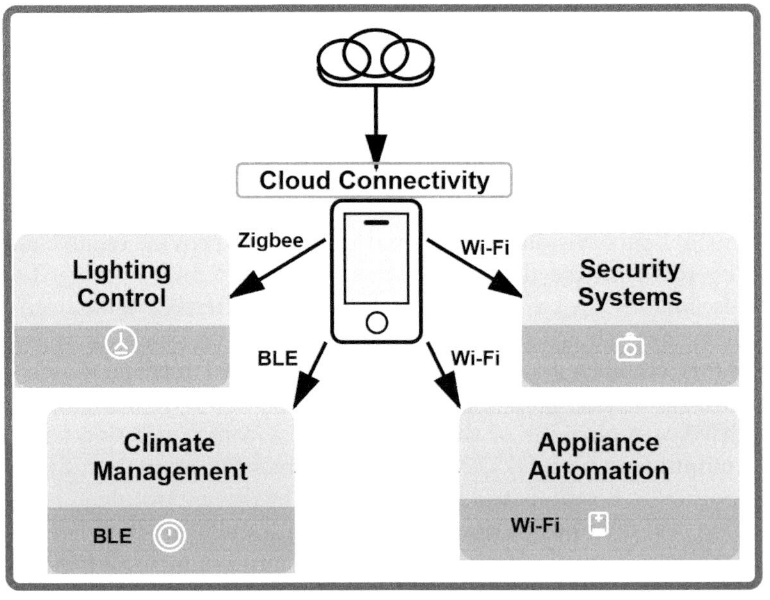

Figure 9.6 Smart home and building IoT application.

Table 9.9 Smart home and building IoT applications

Component	*Wireless tech*	*Function*	*Benefit*
Door locks and entry	BLE	Keyless access	Hands-free entry
Lighting and switches	Zigbee	Occupancy-based control	Lighting savings
Thermostat and HVAC	Wi-Fi	Schedule learning	HVAC cost reduction
Security cameras	Wi-Fi (WPA3)	Encrypted video streaming	24/7 remote monitoring
Voice control	MQTT over TLS	Cloud-based commands	Seamless hands-free operation
Leak and environment	LoRaWAN	Long-range sensor alerts	Early flood detection

9.4.3 Healthcare and wearables

Wireless-enabled IoT services are transforming the healthcare sector by using wearable devices as monitoring systems to promote healthy living [32]. Wearables like a smartwatch are capable of detecting heart irregularities and a CGM (continuous glucose monitor) to prevent diabetic emergencies. Smart IoT health devices and sensors combined with efficient wireless technologies have created a continuous care environment between patients and healthcare providers [33], but unlike other IoT domains, healthcare demands an ultra-reliable communication service, which ensures strict privacy and immediate response under critical conditions [34]. A perfect balance of power efficiency and real-time connectivity is the reason why BLE dominates the personal healthcare wearables [35]. BLE-enabled fitness trackers like Fitbit, Garmin, and Oura ring 4 can monitor heart rate, sleep patterns, and activity levels for weeks on a single charge and also seamlessly sync the data with smartphone health apps [36]. Smart glucose monitors instantly transmit blood sugar levels to smartphones. This enables instant insulin adjustment and prevents any dangerous episodes [37]. Blood pressure cuffs periodically send measurements to cloud platforms where AI algorithms detect hypertension patterns, alerting doctors and patients whenever and wherever human intervention becomes necessary. Wearable ECG patches are the pinnacle of BLE health monitoring as these mitigate life risks due to heart attacks or cardiac arrests [37]. These lightweight devices stick to the chest to continuously monitor the heart rhythm for several days or weeks. The acquired data can be wirelessly stored in clouds or transmitted to cardiologists who can spot arrhythmias in real time. Studies have found that BLE-enabled ECG monitoring detects early warning signs and reduces cardiac event response time to a few minutes, which is crucial for saving lives [38].

Remote Patient Monitoring Systems: For patients suffering from chronic conditions and require constant supervision, NB-IoT and LoRaWAN provide reliable and long-range connectivity to fulfill the demands of healthcare system [39] NB-IoT health monitors can also operate in basements, rural areas, or wherever cellular coverage exists, making them ideal choice for geriatric care and people living alone [40, 41]. Smart pill dispensers connected via NB-IoT alert for missed medical doses, track adherence patterns, and even contact emergency services upon detecting unusual behavior [42]. LoRaWAN is also one of the major wireless communication technologies used in health monitoring networks [37]. Temperature sensors, motion detectors, and air quality monitors in hospitals and homes create a health environment map. These maps are critical in assessing health risks and serve as a foundation in policy-making related to health hazards, identifying green zones, and planning medical facilities. For example, LoRaWAN sensors can detect pollutant spikes and trigger ventilation systems, which can benefit patients with COPD experiencing air quality issues; under extreme circumstances, these can also alert nearby medical teams for timely assistance [25].

Table 9.10 Healthcare and wearable IoT applications

Device/application	*Wireless tech*	*Function*	*Impact*
Vibration and temp sensors	NB-IoT	Predictive maintenance	Less downtime [43]
Asset/location tracking	LoRaWAN	Real-time warehouse mapping	Inventory loss reduction [44]
Collaborative robots	Private 5G	URLLC for motion control	Throughput gain [45]
Vision inspection systems	Wi-Fi 6E	HD video streaming to edge	Defect detection accuracy [46]
Tool and part beacons	BLE	Automated inventory checks	Productivity increase [47]

Critical Care and Emergency Response: The technology of 5G and related wireless technologies has transformed healthcare by enabling real-time telemedicine with ultra-low latency [38]. Moreover, these have not only made remote surgeries feasible but have also played a crucial role in their successful execution. IoT-enabled networks have facilitated remote medical and emergency consultations. Ambulances are now equipped with 5G to live stream HD video of patients, allowing doctors to begin treatment even before patient arrival. This has played a vital role in capitalizing on the golden hour to save patients' lives under critical situations. Remote ICU monitoring via wireless technologies enables specialist physicians to oversee critical patients in rural hospitals and provide expert care globally [39]. Wearable emergency devices use instant, fast, and reliable connectivity to send alerts in case of life-threatening situations [41]. For example, fall detectors worn by elderly patients can summon help, cardiac monitors can transmit emergency alerts to first responders with precise location data etc. During clinical trials, it has been shown that 5G-enabled emergency response systems drastically reduce average response time, which plays a vital role in improving survival rates [42]. Different applications of IoT in healthcare are summarized and illustrated in Table 9.10, and the architectural flow of different healthcare scenarios using IoT devices/sensors and wireless technologies are depicted in Figure 9.7.

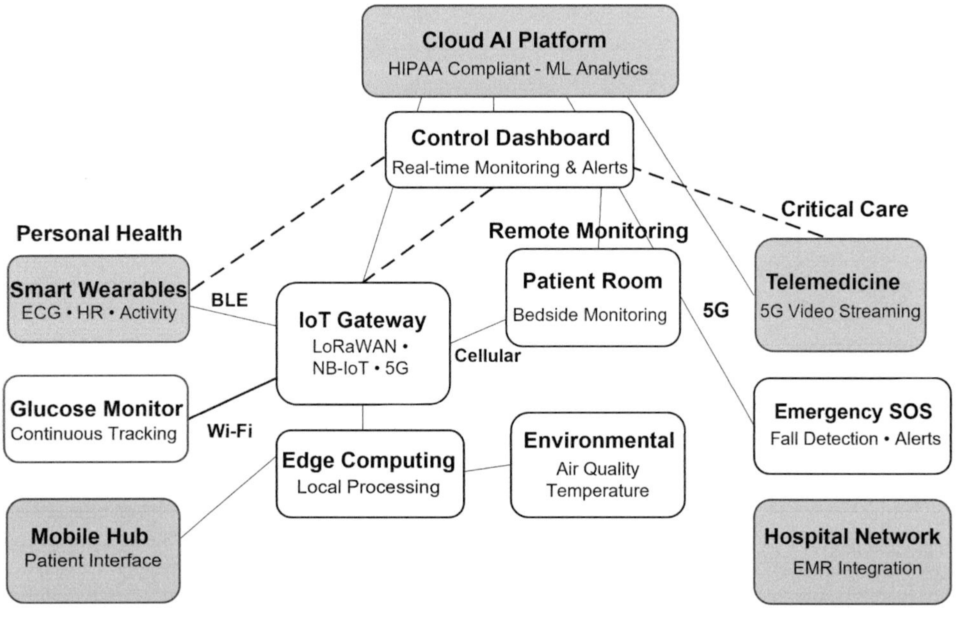

Figure 9.7 Healthcare and wearable IoT deployment architecture.

9.4.4 Industrial IoT

IIoT leverages wireless communication technologies to enhance operational efficiency, safety, and predictive capabilities of manufacturing plants, warehouses, and supply chains [48]. By connecting sensors, actuators, and controllers via low-latency, resilient radio links, IIoT facilitates real-time monitoring of equipment health, asset tracking, and automation of complex processes [49]. Unlike consumer IoT, IIoT demands deterministic and robust performance, security, and seamless integration of industrial systems [50].

Wireless Technologies and Use Cases: Within smart factories, NB-IoT serves as an enabling technology for vibration and temperature sensors to report machine health metrics over cellular networks. This facilitates timely maintenance to reduce unplanned downtime and prevent equipment failure [43]. On the other hand, LoRaWAN networks cover large warehouses to track pallets and forklifts and provide periodic location updates. These are ultra-low-power sensors designed to operate efficiently with multi-year battery life [44]. Private networks deliver URLLC for collaborative robots to enable coordinated motion control with extremely low cycle times [45]. Wi-Fi 6E hotspots connect high-bandwidth vision systems used in quality inspection. These devices stream HD videos to edge servers for enabling automatic defect detection [46]. BLE-based sensors and devices are used to support automated inventory management, which shortens search times [47].

Operational Benefits: By integrating various wireless communication technologies, industries automate services like equipment monitoring, automated material handling, and real-time analytics. Periodic maintenance cycles are scheduled using predictive analytics on data collected using NB-IoT-enabled devices. This reduces maintenance costs by approximately 30% and extends equipment lifespan [26]. Industries minimize inventory losses and optimize warehouse layouts using LoRaWAN-based asset tracking [51]. The robotics of 5G-powered improves throughput by empowering synchronized multi-robot operations [52]. Wi-Fi 6E–enabled HD video inspection system boosts defect detection accuracy [53]. BLE-enabled tool tracking systems, along with predictive analytics and AI, are helpful in raising worker productivity [47] (see Table 9.11).

Table 9.11 Industrial IoT (IIoT) applications

Application	*Wireless tech*	*Function*	*Impact*
Hazard warnings	C-V2X (5G/LTE-M)	V2V/V2I safety messaging	Fewer collisions [52]
Cooperative merging	DSRC	Low-latency vehicle coordination	Fewer panic brakes [53]
Fleet telematics	NB-IoT	Location, fuel, driver behavior tracking	Lower fuel cost, faster deliveries [26]
Adaptive traffic control	LoRaWAN	Congestion data to signal controllers	Shorter commutes [54]
EV charging station services	Wi-Fi 6	Firmware updates, payments	Higher station uptime [55]

9.4.5 Smart transportation and connected vehicles

Smart transportation capitalizes on wireless IoT networks to improve traffic flow, safety, and vehicle efficiency. This is achieved by connecting vehicles, infrastructure, and cloud platforms in real time [56]. Various communication techniques like vehicle-to-everything (V2X), -to-vehicle (V2V), -to-road infrastructure (V2I), and -to-networks (V2N) facilitate coordinated driving, reduce congestion, and enhance safety [57]. The networks of connected vehicles demand ultra-low latency, high reliability, and broad coverage for collision avoidance and autonomous navigation [51].

Wireless Technologies and Use Cases: Cellular V2X (C-V2X) uses LTE-M and 5G NR to exchange safety messages between vehicles and roadside units. This reduces accident rates by signaling early hazard warnings [52]. In highway scenarios, dedicated short-range communications (DSRC) operating on the 5.9-GHz band support low-latency braking alerts and cooperative merging to ensure safety and efficient traffic management [53]. NB-IoT trackers embedded in logistics fleets continuously monitor vehicle location, fuel usage, and driver behavior to cut route deviations and lower fuel costs [26]. LoRaWAN-enabled smart traffic sensors in urban grids feed congestion data for adaptive control of signals, which achieves faster green-wave coordination and better traffic management [54]. Wi-Fi 6 hotspots at various charging stations enable high-speed firmware updates and secure payment processing for electric vehicles [55]. Figure 9.8 shows a wireless IoT deployment architecture for smart transportation and connected vehicles. It also depicts that wireless technologies can be used for intelligent traffic management.

Operational Benefits: By integrating various wireless technologies into transportation systems, we are able to deliver safer journeys, ensure lower vehicular emissions, and also optimize logistics. C-V2X real-time brief messaging services send warning signals to vehicles about potential hazards. This approach has been effective in decreasing collisions by nearly 35% [52]. DSRC cooperative driving enables vehicles to communicate with each other in real-time, resulting in drastically reducing the emergency braking events [53]. Fleet telematics via NB-IoT cuts operational costs by constant monitoring of vehicle health. Using predictive analytics, it identifies inappropriate driving styles, which help in better resource utilization. Many manual activities like driver details

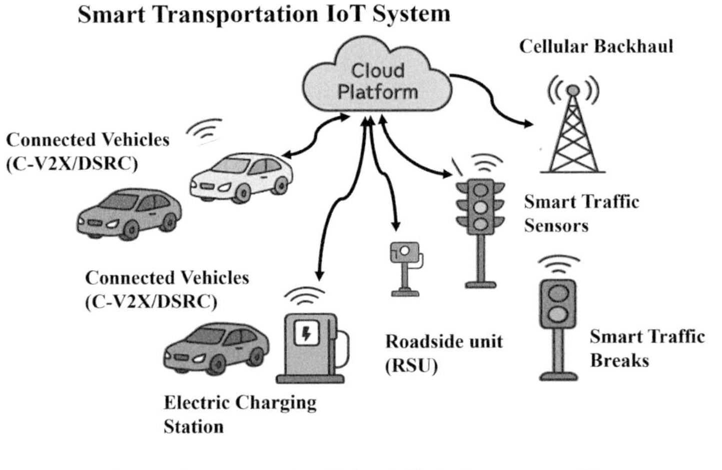

Figure 9.8 Smart transportation and connected vehicles IoT deployment architecture.

lodging, mileage tracking, and maintenance schedules can now be automated using cost-effective NB-IoT devices. LoRaWAN-based intelligent traffic management systems have proven to shorten average commute times by nearly 10–15% by predictively adjusting the signal timings based on real-time traffic flow [54]. Wi-Fi 6–enabled EV charging stations provide opportunities to EV owners to remotely manage their charging sessions [55].

9.4.6 Agriculture and environment monitoring

Agriculture and environmental monitoring are two of the most impactful IoT applications. These require wireless sensor networks to optimize resource usage, enhance crop yields, and protect ecosystems [49]. Smart farming incorporates precision agriculture techniques for real-time monitoring of soil conditions, weather patterns, plant health, etc. On the other side, environmental systems track air quality, water contamination, and wildlife conservation across a vast rural landscape. These applications require long-range, low-power wireless technologies that operate reliably in remote locations and with minimal maintenance for years [55].

Precision Agriculture Applications: LoRaWAN dominates agricultural IoT due to its exceptional range and ultra-low-power consumption. Sensors can therefore operate on a single battery for up to ten years and can cover distances of more than 15 km [57]. Soil moisture sensor grids in the field use LoRaWAN to transmit real-time data on water content, nutrient levels, and pH balance. This allows farmers to optimize irrigation schedules, leading to efficient water management [26]. Weather stations equipped with LoRaWAN connectivity monitor temperature, humidity, wind speed, and rainfall. Using predictive forecasting, it guides farmers to timely schedule planting and harvesting activities. Studies have shown that IoT-based predictive analytics services are able to support smart farming techniques with nearly 90% accuracy under normal scenarios [53].

Cellular network coverage enables NB-IoT devices to transmit data related to livestock tracking in remote pastures. The audio and video data thus transmitted is used to identify distress calls, unusual clustering, or movement to mitigate risks and ensure overall nature well-being. Smart collars monitor cattle location, feeding patterns, and other vital signs to alert farmers. This helps farmers to focus on herds that need attention, leading to their better health maintenance and providing required support within minutes, thereby reducing livestock mortality [54]. GPS-enabled NB-IoT trackers prevent farm equipment from theft. These are also helpful in optimizing field operations by ensuring their proper movement through route optimization across the farm land to cut fuel costs. Wi-Fi 6 provides support for high-bandwidth applications in smart greenhouses and indoor farming facilities. Camera-based systems stream HD videos, which are required for automated crop inspection, detecting diseases and pests. Furthermore, data analytics techniques ensure timely preventive measures against crop infections to increase productivity. Automatic environmental controls based on sensor data analysis are helpful in maintaining ambient lighting, temperature, and CO_2 levels to increase greenhouse yields [52].

Environmental Monitoring Networks: LoRaWAN air-quality sensors are deployed across rural and urban areas to monitor particulate matter (PM2.5, PM10), ozone, carbon dioxide, and nitrogen dioxide levels. These are helpful in drawing real-time pollution maps, which help authorities to identify emission sources and implement corrective measures. Lakes and rivers are equipped with water quality monitoring

Table 9.12 Environmental monitoring IoT applications

Application	*Wireless tech*	*Function*	*Impact*
Soil monitoring sensors	LoRaWAN	Moisture, nutrients, pH tracking	Water reduction, optimized irrigation [26]
Weather station networks	LoRaWAN	Hyperlocal climate data	Forecast accuracy for farm planning [53]
Livestock health tracking	NB-IoT	Animal monitoring and alerts	Reduction in mortality rates [54]
Equipment fleet management	NB-IoT	GPS tracking and optimization	Fuel cost savings
Greenhouse automation	Wi-Fi 6	Vision systems and controls	Yield increase, disease detection [52]
Air quality monitoring	LoRaWAN	Pollution tracking networks	Real-time emission source identification
Water quality surveillance	LoRaWAN	River and lake contamination	Hours vs. days pollution detection
Wildlife conservation	NB-IoT	Species tracking and research	Protected population growth

systems using LoRaWAN. Data related to water quality measures like pH levels, dissolved oxygen, temperature, and chemical contamination is periodically transmitted to clouds and servers. Data-processing techniques are helpful in detecting pollution events and indicating authorities for corrective actions within hours rather than days. Wildlife conservation benefits from NB-IoT tracking collars and tags, which monitor endangered species, their movement patterns, habitat use, and population dynamics. These systems enable researchers to study animal behavior without human intervention. As a result, conservationists and authorities are able to provide necessary support to protect endangered species and prevent human-wildlife conflicts (see Table 9.12).

9.5 CHALLENGES AND FUTURE SCOPE

The wireless communication standards for IoT provide transformative changes, yet faced several challenges to realize its full potential [9]. The main key challenges are security issues in huge device interconnections, interoperability among heterogeneous communication protocols, energy efficiency for battery-operated sensors, limited spectrum availability, and compliance with different regulatory frameworks [38]. It remains critical to ensure data privacy and integrity in distributed networks, particularly in highly sensitive areas like healthcare. Scalability-related challenges appear when billions of devices are connected to the network, which demand robust network management, dynamic resource allocation, and seamless updates without any service disruption [58]. Interference and congestion in bandwidth employ significant obstacles for both short-range and long-range technologies. Different protocols such as Wi-Fi, BLE, Zigbee, and cellular IoT coexist within shared bands. To avoid performance degradation, it needs advanced coexistence mechanisms and cognitive radio capabilities [59]. Energy efficiency is the key requirement for remote deployments. LPWANs extend battery life but often at the cost of data rate and latency [6].

The future of wireless IoT systems relies on emerging trends in wireless communication. Currently, research targeted towards AI-driven network optimization techniques based on predictive analytics leads to self-organized networks capable of dynamically adapting in real-time. Overall, these approaches try to enhance network reliability and performance

[60]. It is the advent of edge computing, which processes data at the local server closer to devices. This plays a key role in reducing latency and bandwidth demands, especially involving mission-critical applications like autonomous vehicles and remote surgery [61]. Currently under development and in its naïve stage, 6G communication technologies are expected to launch in the 2030s. The technology of 6G promises operation in the terahertz frequency band, guaranteeing ultra-high data rates, integration of sensing and communication capabilities, and native support for holographic IoT applications. Overall, research in wireless-based IoT networks is based on the expansion of high-throughput networks offering diverse services [62]. Furthermore, security features in IoT networks are expected to evolve with the growth of blockchain and distributed ledger technologies. These technologies offer decentralized security frameworks for device authentication, data integrity, and mitigate single-point-of-failure risks in networks [63]. Furthermore, digital-twin models of IoT physical systems will enable virtual testing, scenario forecasting, and resource optimization across smart cities, factories, and agricultural ecosystems [64].

IEEE, 3GPP, and ITU have put in continuous efforts for standardization of protocol specifications, regulatory policies, and security frameworks, and will continue to do so to foster global interoperability and trust. Techniques for energy harvesting—such as solar panels, RF, and kinetic scavenging—have started to complement battery-powered devices. This has led to the deployment of self-sustaining IoT nodes under challenging environments [65]. All these technological advances will collectively ensure that IoT-enabled wireless communication networks achieve unparalleled scalability, resilience, and intelligence across different sectors.

9.6 CONCLUSION

IoT and wireless communication technologies have together catalyzed a revolutionary era of innovation across multiple sectors like smart cities, homes, healthcare, industry, transportation, agriculture, and environmental monitoring. Diverse wireless technologies ranging from low-power protocols like BLE, Zigbee, NB-IoT, and LoRaWAN to high-capacity networks such as Wi-Fi, LTE, and 5G have enabled IoT systems to deliver tailored solutions that enhance efficiency, safety, reliability, and sustainability for a better quality of life. In spite of that, IoT networks still continue to face persistent challenges across domains like security, interoperability, energy management, and spectrum allocation. But with ongoing technological advances in areas like AI-driven networking, edge computing, blockchain security, and forthcoming 6G cellular networks, IoT systems show considerable promise to overcome these barriers and deliver unprecedented levels of performance and scalability.

This chapter presented foundational concepts, wireless communication standards, different IoT architectural layers, various smart application areas, and recent emerging trends in both IoT and wireless communication technologies. It also showcases the immense transformative potential inherent in IoT-enabled wireless communication systems and its effect on daily living. As these technologies continue to mature, emerging digital twin architectures, energy-harvesting mechanisms, and AI-powered networks will further elevate the performance of IoT networks and lead to greater dependence of humans on machines. This shall foster intelligent ecosystems that will respond dynamically to the changing needs of humans as well as the environment. Collaborative standardization efforts and cross-domain convergence of IoT devices with wireless communication technologies will remain essential for seamless interoperability and global deployment of IoT-enabled services. Eventually, their fusion will continue to reshape industries and societies. It has and will continue to ensure smarter, safer, and more sustainable solutions in pursuit of a connected world of tomorrow.

REFERENCES

1. Al-Fuqaha, A., Guizani, M., Mohammadi, M., Aledhari, M., & Ayyash, M. (2015). "Internet of Things: A Survey on Enabling Technologies, Protocols, and Applications." *IEEE Communications Surveys & Tutorials*, 17(4), 2347–2376.
2. Adelantado, F., Vilajosana, X., Tuset-Peiró, P., Martínez, B., Melia-Segui, J., & Watteyne, T. (2017). "Understanding the Limits of LoRaWAN." *IEEE Communications Magazine*, 55(9), 34–40.
3. Palattella, M. R., et al. (2016). "Internet of Things in the 5G Era: Enablers, Architecture, and Business Models." *IEEE Journal on Selected Areas in Communications*, 34(3), 510–527.
4. Saad, W., Bennis, M., & Chen, M. (2021). "A Vision of 6G Wireless Systems: Applications, Trends, Technologies, and Open Research Problems." *IEEE Network*, 34(3), 134–142.
5. Gómez, C., Oller, J., & Paradells, J. (2012). "Overview and Evaluation of Bluetooth Low Energy." *Sensors*, 12(9), 11734–11753.
6. Raza, U., Kulkarni, P., & Sooriyabandara, M. (2017). "Low Power Wide Area Networks: An Overview." *IEEE Communications Surveys & Tutorials*, 19(2), 855–873.
7. Gubbi, J., Buyya, R., Marusic, S., & Palaniswami, M. (2013). "Internet of Things (IoT): A Vision, Architectural Elements, and Future Directions." *Future Generation Computer Systems*, 29(7), 1645–1660.
8. Silva, B. N., Khan, M., & Han, K. (2018). "Towards Sustainable Smart Cities: A Review of Trends, Architectures, and Future Research." *Sustainable Cities and Society*, 38, 697–713.
9. Zanella, A., Bui, N., Castellani, A., Vangelista, L., & Zorzi, M. (2014). "Internet of Things for Smart Cities." *IEEE Internet of Things Journal*, 1(1), 22–32.
10. Yaqoob, I., et al. (2017). "Internet of Things Architecture: Recent Advances, Taxonomy, Requirements, and Open Challenges." *IEEE Wireless Communications*, 24(3), 10–16.
11. Chen, L., et al. (2022). "Bluetooth Low Energy 5.x: Performance Evaluation and Energy Optimization." *Sensors*, 22(18), 6934.
12. Kumar, A., et al. (2023). "Large-Scale BLE Mesh Networks: Deployment Strategies and Performance Analysis." *Computer Networks*, 218, 109366.
13. Wang, Y., et al. (2024). "Wi-Fi 6E for IoT: Latency Reduction and Power Efficiency in the 6 GHz Band." *IEEE Transactions on Mobile Computing*, 23(5), 2234–2247.
14. Brown, S., et al. (2023). "Wi-Fi 6E Performance Analysis for Industrial IoT Applications." *IEEE Transactions on Industrial Informatics*, 19(8), 8901–8910.
15. Hassan, M., et al. (2024). "Energy-Efficient Zigbee 3.0 Protocol Stack for Smart Building Applications." *IEEE Internet of Things Journal*, 11(4), 6723–6735.
16. Thompson, E., et al. (2024). "Global LoRaWAN Deployment Trends and Network Performance Analysis." *Computer Communications*, 198, 145–158.
17. Park, J., et al. (2023). "Advanced ADR Algorithms for LoRaWAN: Energy Efficiency and Network Optimization." *IEEE Internet of Things Journal*, 10(12), 10567–10578.
18. Liu, X., et al. (2024). "NB-IoT Release 17: Enhanced Coverage and Capacity for Massive IoT Deployments." *IEEE Wireless Communications*, 31(2), 88–95.
19. Singh, D., et al. (2023). "Deep Indoor Penetration Analysis of NB-IoT for Underground IoT Applications." *IEEE Access*, 11, 87432–87445.
20. Anderson, K., et al. (2024). "5G NR-IoT for Ultra-Reliable Low-Latency Communications." *IEEE Communications Magazine*, 62(3), 45–52.
21. Johnson, R., et al. (2023). "Massive Machine-Type Communications in 5G Networks: Challenges and Solutions." *IEEE Network*, 37(6), 134–142.
22. Martinez, P., et al. (2024). "Hybrid Protocol Deployments in IoT: Performance Benefits and Implementation Challenges." *IEEE Communications Surveys & Tutorials*, 26(1), 456–478.
23. Darroudi, S., & Gomez, C. (2017). "Bluetooth Low Energy Mesh Networks: A Survey." *Sensors*, 17(6), 1306.
24. Smith, L., & Jones, R. (2023). "IoT Parking Solutions." *Sensors*, 24(7), 3017.
25. Balaji, D., & Srinivasan, D. (2023). "Adaptive Lighting Using IoT." *International Journal of Smart Home*, 12(3), 15–26.
26. Patel, R., & Singh, P. (2024). "NB-IoT for Fleet Management." *Sensors*, 24(12), 5678.
27. Ramirez, A., & Liu, Y. (2023). "Energy Savings from Occupancy-Based Lighting Control." *Building Automation Review*, 18(1), 75–84.
28. Zhang, H., et al. (2023). "AI-Enabled Smart Thermostats: Learning and Energy Savings." *IEEE Transactions on Consumer Electronics*, 69(3), 210–217.
29. Nguyen, P., & Patel, S. (2024). "Secure Wireless Video Surveillance." *International Journal of Network Security*, 22(4), 350–359.

30. Chen, J., & Kumar, R. (2024). "Voice Assistants in Smart Buildings." *ACM Computing Surveys*, 56(7), 112–129.
31. Silva, M., & Gomez, C. (2024). "LoRaWAN for Building Safety Applications." *Sensors*, 24(5), 1524.
32. Johnson, M., et al. (2024). "Bluetooth Low Energy in Wearable Health Devices." *IEEE Transactions on Biomedical Engineering*, 71(4), 892–901.
33. Patel, R., & Kumar, S. (2024). "Smart Glucose Monitoring with BLE Connectivity." *Diabetes Technology & Therapeutics*, 26(3), 180–188.
34. Williams, A., et al. (2024). "AI-Enhanced Blood Pressure Monitoring via BLE." *Journal of Medical Internet Research*, 26(1), e45123.
35. Chen, L., & Zhang, H. (2024). "Continuous ECG Monitoring Using BLE Patches." *Circulation: Arrhythmia and Electrophysiology*, 17(2), 78–85.
36. Rodriguez, P., et al. (2024). "Real-Time Cardiac Event Detection with Wireless Monitoring." *New England Journal of Medicine*, 390(8), 712–720.
37. Garcia, M., & Liu, Y. (2024). "LoRaWAN-Based Home Health Networks." *IEEE Transactions on Network Science and Engineering*, 11(2), 1456–1467.
38. Roman, R., et al. (2015). "Security and Privacy in the Internet of Things: Challenges and Solutions." *IEEE Communications Surveys & Tutorials*, 17(3), 1294–1312.
39. Lee, S., et al. (2024). "Remote ICU Monitoring via 5G Networks." *Critical Care Medicine*, 52(4), 445–453.
40. Smith, D., & Taylor, J. (2024). "NB-IoT Applications in Remote Health Monitoring." *IEEE Internet of Things Journal*, 11(6), 10254–10263.
41. Davis, J., & Wilson, P. (2024). "5G Emergency Response Wearables for Elderly Care." *Journal of Medical Systems*, 48(2), 67.
42. Roberts, A., et al. (2024). "Impact of 5G on Emergency Medical Response Times." *Prehospital Emergency Care*, 28(3), 312–319.
43. Patel, R., & Kumar, S. (2024). "NB-IoT-Enabled Predictive Maintenance." *Sensors*, 24(8), 4157.
44. Chen, L., & Garcia, M. (2024). "LoRaWAN in Smart Factories." *IEEE Communications Surveys & Tutorials*, 26(3), 512–536.
45. Smith, D., & Allen, R. (2024). "Private 5G Networks for Industrial Automation." *IEEE Transactions on Industrial Informatics*, 20(1), 45–53.
46. Williams, A., et al. (2024). "Wi-Fi 6E in Industrial Vision Systems." *Journal of Industrial Wireless*, 9(4), 210–219.
47. Davis, J., & Wilson, P. (2023). "BLE Beacons for Asset Tracking." *Sensors*, 23(11), 6784.
48. Zhou, Y., et al. (2024). "Wireless IoT for Industrial Automation." *IEEE Internet of Things Journal*, 11(5), 10345–10358.
49. Kumar, A., & Singh, P. (2024). "Real-Time Monitoring in Industrial IoT." *Journal of Industrial Informatics*, 14(2), 200–212.
50. Lee, J., et al. (2024). "Integrating Legacy Systems with Industrial IoT." *IEEE Communications Surveys & Tutorials*, 26(2), 324–348.
51. Chen, L., et al. (2024). "Ultra-Reliable Low-Latency Communications for V2X." *IEEE Journal on Selected Areas in Communications*, 39(10), 2987–2998.
52. Smith, D., & Allen, R. (2024). "C-V2X Safety Performance Study." *IEEE Transactions on Vehicular Technology*, 72(2), 456–467.
53. Williams, A., et al. (2023). "DSRC-Based Cooperative Driving." *IEEE Access*, 11, 7890–7901.
54. Garcia, M., & Liu, Y. (2024). "LoRaWAN in Smart Traffic Systems." *IEEE Transactions on Network Science and Engineering*, 11(3), 1678–1689.
55. Martinez, C., & Brown, R. (2024). "Wi-Fi 6 for Electric Vehicle Infrastructure." *Journal of Wireless Networks*, 15(1), 45–53.
56. Li, X., et al. (2024). "Wireless IoT for Intelligent Transportation Systems." *IEEE Transactions on Intelligent Transportation Systems*, 25(4), 1234–1247.
57. Zhang, H., & Kumar, R. (2024). "V2X Communications: Standards and Challenges." *IEEE Communications Magazine*, 62(5), 88–95.
58. Borgia, E. (2014). "The Internet of Things Vision: Key Features, Applications, and Open Issues." *Computer Communications*, 54, 1–31.
59. Bhushan, N., et al. (2019). "Spectrum Sharing in 5G Networks." In *5G System Design: Architectural and Functional Considerations*. Wiley.
60. Taleb, T., et al. (2020). "AI-Driven Self-Organizing Networks: Challenges and Perspectives." *IEEE Network*, 34(5), 168–174.

61. Shi, W., et al. (2016). "Edge Computing: Vision and Challenges." *IEEE Internet of Things Journal*, 3(5), 637–646.
62. Saad, W., et al. (2021). "6G: The Next Frontier—From Holographic Communications to Integrated Sensing and Communication." *IEEE Communications Magazine*, 59(12), 65–71.
63. Dorri, A., et al. (2018). "Blockchain for IoT Security and Privacy: The Case Study of a Smart Home." *IEEE Communications Magazine*, 56(11), 23–29.
64. Tao, F., & Qi, Q. (2020). "Digital Twins and Cyber-Physical Systems Toward Smart Manufacturing and Industry 4.0: Correlation and Comparison." *Engineering Science*, 5, 2–11.
65. Paradiso, J. A., & Starner, T. (2005). "Energy Scavenging for Mobile and Wireless Electronics." *IEEE Pervasive Computing*, 4(1), 18–27.

Chapter 10

UAV and drone communication

Suvra Nayak, Gyana Ranjan Nanda, Dilip Senapati, and Sasmita Dash

10.1 INTRODUCTION

The growing demand for drone-based technologies and unmanned aerial vehicles (UAVs) is driven by their affordability, versatility, cost-effectiveness, and ability to perform tasks that pose risks, require extensive time, or are logistically impractical for humans. UAVs are revolutionizing both commercial and defense sectors, from surveillance, disaster management, and precision agriculture to logistics, urban monitoring, and infrastructure inspection. In today's high-tech landscape, with the ability to provide detailed imagery, real-time situational awareness, and function in extreme or inaccessible areas, UAVs are no longer just futuristic gadgets; they have become mission-critical tools. The worldwide demand for UAVs is driven by industries adopting drone-based solutions for automation, smart cities, and Industry 4.0 initiatives. As a result, UAVs have become critical enablers of economic growth and technological innovation, and businesses are also adopting aerial systems in wide applications. Advancements in artificial intelligence (AI), autonomy, and high-speed connectivity, viz., 5G and satellite links, are empowering drones with beyond-line-of-sight and swarm capabilities. The driving progress of UAVs in both the public and private spheres makes resilient communication systems essential for their expanding influence. In an effort to provide better and faster facilities in the context of future traffic and energy demands and the luxury of automated home deliveries, drones are viewed as a connecting link, and automation leads to less human intervention and faster services.

The question itself seems straight out of science fiction. No matter how futuristic it may sound, it is certainly possible with modern technological advances. With the advent of internet of vehicles (IoV) and the wireless sensor network and long-range communication facilities, the data transmission has become easier and faster than ever before and the smart cities have the ambition to equip themselves with all amenities and luxuries; it's evident that the area is full of new research aspects and efforts are being put relentlessly to bring about new innovations.

Drones or UAVs are facilitated with a number of sensors, i.e., Global Positioning System (GPS) and gyroscopes, for smooth transportation and communication. Recently, it has proved to be beneficial in reducing time and cost of human interventions. These are controlled remotely and instrumental in the advancement of critical areas like military recon and telemedicine. They perform a variety of tasks in air efficiently because of the equipped sensors and built-in intelligent functionalities. Its numerous variants depend on the motor and payload-carrying capacity and the type of operations they perform under different terrain and hostile weather conditions. The multi-propeller system that is specially designed for carrying significant loads in disaster response missions and glide control. Drones with an advanced camera are used in aerial photography of remote and unreachable terrain. Drones with proper mechanical support are used in agriculture to spread insecticides, pesticides, and irrigation facilities on crop fields. These are used to provide internet facilities in remote

DOI: 10.1201/9781003637721-10

areas, surveillance in public gatherings, and 3D mapping of infrastructure project planning. Drones in smart cities are used to monitor traffic through the aerial views for better path planning and congestion control. It has been widely used to enforce awareness on the social distancing during the Covid-19 pandemic.

Drone and UAV communication is built on several fundamental principles that ensure efficient, secure, and adaptive operation in various applications. These principles form the backbone of autonomous drone networks and real-time data exchange.

- Communication Architectures: Drones operate within different communication frameworks, including centralized communication (ground station/cloud server), decentralized mesh networks (direct drone-to-drone communication), and edge computing (local data processing to reduce latency).
- Wireless Communication Technologies: UAVs utilize multiple wireless protocols for data transmission, such as Wi-Fi and LTE for remote control, 5G networks for ultra-low latency, and satellite communication for long-range operations.
- Signal Transmission and Processing: Drone networks rely on secure and adaptive transmission techniques, including encryption protocols to prevent unauthorized access, bandwidth optimization for efficient data management, and interference mitigation for stable communication.
- Autonomous Decision-Making: Artificial intelligence improves UAV operations through collision avoidance using LiDAR and sensor fusion, path optimization algorithms for efficient navigation, and real-time data processing for dynamic adaptation.
- Regulatory and Safety Considerations: Global UAV communication must comply with international standards, including FAA and ICAO guidelines on frequency usage, cybersecurity measures to protect communication networks, and interoperability standards that ensure cross-network compatibility.

  This section lays the groundwork for understanding UAV communication principles, providing essential insights for designing efficient drone-based networks.

Smart cities enhanced with drones: The following is a collection of the changes that would happen if the drones were put in charge of the on-road traffic.

- Humans will be replaced by UAVs: The aim of smart cities is to modernize all aspects and make the facilities sophisticated and provide its citizens with the utmost luxuries, and this glamor draws a lot of population to settle in the city and the roads will eventually become more busy. This scenario requires a new and well-organized traffic monitoring and regulation system. Human speed and perception might not be sufficient to meet demands in order to smoothly run the traffic, and more and more traffic jams may arise. The entry of drones would make it a lot easier to keep track of the vehicles and to monitor them with the help of technologies, viz., GPS and the IoV communication paradigm. This would lead to the scene where the cars are directed by the drones directly on their dashboard with routes and the possible traffic ahead in real time, faster than any other existing framework.
- Fewer traffic jams: With the battery-powered drones hovering continuously over roads and directly gathering information relayed by vehicles, the geographic location and velocity-related data of all vehicles running will be stored in the database of the drone network, and the routing will be based on these data, leading to an uninterrupted flow of traffic and less traffic jams.
- Enhanced communication: Current communication channels in the IoV paradigm involve a remote base station that analyzes data and responds, which involves a longer waiting time and the transmission mode is not line of sight, which makes it even more inefficient. This issue is directly handled by the drones as they are

hovering directly overhead the road in the city's airspace, which makes the data transmission in line of sight and the latency is further reduced to avail better services to the client vehicles.

- Safer roads: Faster service rate, enhanced communication, and better routing lead to fewer traffic jams and, consequentially, lead to much lesser road accidents as the vehicles are monitored continuously based on their speed and location, and all the information of the nearby vehicles and the route intersection data is relayed to the service-availing vehicle.

Autonomous drone deliveries: The standard of living has been changed, and especially in urban areas in terms of the services like smooth home delivery systems. Many firms have expanded their service initiatives for autonomous drones in new possibilities and application areas [1]. This chapter brings the details of the automation of the drone delivery network systems using self-piloted drones, and an authentic architecture is proposed to support the system, making it capable of payload pickup and delivery with optimal use of the hangars in the shortest possible path plannings.

Drone edge computing: The futuristic idea discusses here are possible and require for local data processing and computing to support the backhaul communication facilities. The major advantage of drone-based services is that they operate at the edge of the processing environment, which makes them faster, reliable, and more efficient. In this chapter, a novel queueing theory-based local processing paradigm is discussed, which enhances the concept of drone edge computing. In addition, the incorporation of edge computing with the drone framework reduces the waiting time for the service and reduces the cost.

The chapter explains and deduces mathematical evidence and portrays corresponding results in support of findings.

10.2 LITERATURE REVIEW

UAVs, commonly known as drones, have made significant advances in recent years, positioning them as crucial assets in modern wireless communication systems. Their versatility has enabled a wide spectrum of applications, ranging from environmental monitoring and emergency response to precision agriculture and infrastructure inspection. The increasing integration of UAVs across diverse sectors has underscored the critical need for communication systems that are not only reliable and resilient but also capable of intelligent adaptation to dynamic operational environments.

This chapter highlights the distinguished work and innovation brought about in the field of drone computing and the use of drones in the area of traffic monitoring and regulation and the last-mile delivery made by drones in the urban environment. In this section, a brief description of some relevant work focusing on the IoV, 5G communication, and drone technology is provided. The rapidly growing transportation environment and the increasing frequency of on-road vehicles require smooth channelization and cost optimization poses a challenge. In this context, the incorporation of drone technology into the field of IoV is of great help when it comes to wireless communication. The drone acts as a base station and assists the ground base station and enhances communication. The paper [2] provides optimization of the algorithm-based heuristic approach to obtain the 3D positioning of the drone with respect to the on-road vehicles to enhance communication and minimize fading between the base station and the vehicles where the drone acts as an intermediate aerial base station in challenging terrain. The paper [3] presents the mathematical model for the routing infrastructure in the drone-based delivery architecture and promotes green delivery and pollution reduction. The work also addresses the optimization of fuel cost and the minimization of traffic congestion [4].

In this chapter, the work is focused on the clustering of vehicles exhibiting correlated data, and an algorithm has been proposed on the basis of a self-organizing communication system in the context of network stability and minimization of end-to-end communication delay. The article [5] deals with the benefits of drone-assisted communication in the vehicle-to-vehicle network and introduces drone-assisted vehicular networks (DAVNs), highlighting their advantages and providing a study on the network throughput. The paper [6] highlights the application and advantages of 5G communication in the field of V2E and drones. It also provides detailed features on various cellular technologies in support of drone architectures. Furthermore, this chapter [7] discusses the application of mission control machine type communication (mcMTC) in the context of improvements in link availability and reliability. It was observed to have a strong effect in the IoT business areas. The paper [8] characterizes IoV channels in the context of scalability, storage, transmission, and computing in VANETs and discusses the issues related to big data in the field of IoV. Gharibi et al. [9] demonstrate the area of application of the internet of drones (IoD) and identifies the key features of its various implementations, such as access of UAVs to the controlled airspace and wireless connectivity among drones.

As the performance of wireless communication devices is highly dependent on signal strength and relay medium, factors such as fading play a pivotal role in facilitating good communication. In this context, the authors [10, 11] characterize various wireless channels considering the phenomenon of fading. The work provides excellent agreed results with the help of non-extensive parameter q [12]. In the scope of IoT, all wireless devices are controlled by a centralized system, which in turn leads to high latency, so in the context of real-time applications, mobile edge computing reduces the processing time at the level of data acquisition and analysis and improves latency [13]. In [14], the work is focused on the relay positioning of the drone using a multi-arm bandit algorithm, which treats a predefined location as an action and accordingly determines the reward based on the best location that helps the drone-to-drone and the drone-to-vehicle communication. The work [15] presents a mixed integer programming (MIP) algorithm for the purpose of optimizing delivery in terms of the time taken and the distance covered in the context of dynamic pickup and delivery locations for the drone. In this work [16] a mathematical solution is provided for the multi-truck and multi-drone problem in the delivery context. The vehicle routing problem with drones is solved using a metaheuristic approach. This paper [17] specialized in the formulation of a multi-arm bandit algorithm-based optimization procedure for UAV path planning to serve a disaster-hit location where it acts as a mobile base station to provide the necessary communication facilities [18].

Furthermore, a taxonomy of some relevant works in the field of drone technology and its application in the delivery service is presented. Drones being one of the most studied and researched topics in logistics in recent years have excelled in their application in various environments, and the number of research works has also grown [19–27]. The paper [28] studies the energy consumption of the drone-enabled and the conventional payload delivery systems and the energy consumption model for the drone parcel delivery system was provided given the constraints like the flight pattern. There has been significant progress in the application of the drones into the delivery service; the paper [29, 30] provides a model of the drone delivery service that assists the truck delivery system and thus a heuristic approach is used to provide a solution to the multiple flying sidekick traveling salesman problem [11, 31–35]. Moreover, the papers [15, 36, 37] provide another novel approach to solve the traveling salesman problem to minimize the operational and the overall service cost of the delivery system. The next paper works on a dynamic programming approach to solve the traveling salesman problem with drones, and the routing restrictions for the delivery trucks are discussed to minimize the time for recovering the assistant drone [38].

The paper [39] uses a heuristic approach to solve the traveling salesman problem (TSP) and then applies a variable neighborhood problem to find the routes for the truck and the drone operating together to make the last-mile delivery. The work [40] focuses on a two-tier robot-based delivery system and demonstrates its advantages over the conventional truck delivery system, the paper uses a p-median problem to propose the solution to the problem. In [41], a vehicle routing based on energy optimization and a heuristic approach model is proposed to optimize the energy consumption. The next work [42] focuses on formulating a hybrid genetic algorithm-based solution for the TSP with drones to optimize either the operational cost or the service time. A metaheuristic approach is used to solve the traveling salesman problem for the drones in paper [1] and the approach proves to be working well with the small- and medium-level instances. In paper [43], the authors use a two-layer model to optimize and conserve the energy of the drones by using the transit vehicles in the route of delivery of the drones to the destination and the algorithms, and the mathematical solutions are proposed to support the idea. Furthermore, paper [44] works on different algorithms and the mathematical formulations for solving and optimizing the results of the flying sidekick TSP.

After the TSP, the vehicle routing problem for the drones has also been one of the major areas of interests; the paper [45] models two heuristic algorithms for solving the problem. Paper [46] models the drone supply to the delivery vehicles such that the vehicle does not have to return to the depots to take the supplies; thereby, a routing for facilitating the drone resupply for the same-day delivery is formulated. There is a huge revenue generated by the drone service platforms, and the paper [47] provides a scheduling algorithm for the drone delivery service systems to maximize the revenue based on the time constraint [48–51]. As the delivery networks using the drones prosper, the security issues also seem to amplify the risk. Thereby, the paper [52] works on the concept of providing the necessary security to the drone delivery network right from the vendor to the customer using a zero-sum interdiction game. Apart from all the necessary optimizations based on the delivery time and other operational constraints, another major area of focus is the payload management during the airborne maneuvers, which is considered, and the design is proposed to hold the packages till the delivery [53]. The paper [54] provides an analytical examination of the emissions by the drone delivery network and deals with a genetic algorithm-based approach to solve the model and facilitate a sustainable and green aspect of the last-mile parcel delivery through the drones. In [55, 56] a continuous wireless charging environment is proposed to provide a seamless drone delivery service.

10.3 TECHNOLOGIES IN DRONE AND UAV COMMUNICATION

Drone and UAV communication systems have advanced significantly due to innovations in AI and next-generation wireless technologies. These developments enable seamless data exchange, autonomous decision-making, and enhanced operational efficiency, making drones integral to modern smart infrastructure.

10.3.1 AI and machine learning

AI and machine learning (ML) serve as a key catalyst in advancing UAV communication. AI-powered drones optimize flight paths, process data efficiently, and detect obstacles through real-time computations. Reinforcement learning techniques allow UAVs to exhibit agile adaptability to environmental conditions by continuously improving their decision-making capabilities.

Computer vision improves drone operations by enabling autonomous object recognition, tracking, and terrain mapping. Advanced image processing algorithms help UAVs analyze the environment with precision, supporting applications in disaster response, surveillance, and environmental monitoring. AI-driven predictive analytics also improves fault detection by monitoring sensor data to identify potential failures before they occur, reducing maintenance costs and operational downtime.

AI in multi-drone coordination allows UAVs to work collectively in swarm intelligence systems. By communicating data with neighboring drones, AI-powered UAVs optimize aerial formations, improving efficiency in large-scale mapping, delivery services, and industrial inspections. Furthermore, AI aids in secure communication protocols, detecting cybersecurity threats, and ensuring encrypted data transmission among UAV networks.

10.3.2 5G and beyond

The integration of 5G networks has significantly improved UAV communication by providing high-speed connectivity, ultra-low latency, and enhanced reliability. Drones operating on 5G infrastructure benefit from faster data transmission, enabling real-time video streaming, autonomous navigation, and AI-powered analytics. Beyond 5G, next-generation networks like 6G and satellite communication offer promising advancements. 6G is expected to introduce AI-driven connectivity for self-optimizing UAV networks, ensuring seamless adaptability in dynamic environments. Meanwhile, satellite-based communication enables drones to function efficiently in remote locations where ground-based networks are unavailable.

10.3.3 Integration with Smart Cities

The role of UAVs in smart cities is rapidly evolving and provides intelligent solutions for traffic monitoring, infrastructure inspection, and environmental tracking. Drones equipped with 5G connectivity streamline urban planning efforts by collecting data on congestion patterns, air quality, and real-time events. AI-enabled UAVs improve public safety by assisting in automated emergency response, providing medical supplies, and supporting crowd management surveillance systems. 5G-connected drones can also optimize delivery networks, facilitating efficient logistics in densely populated areas.

Challenges in Network Infrastructure: Despite technological advancements, UAV communication systems face infrastructure-related challenges. Signal interference in urban environments can disrupt data transmission and requires advanced frequency management techniques to maintain stable connections. Latency concerns in mission-critical operations necessitate the adoption of edge computing, minimizing delays in drone response times.

Cybersecurity threats pose additional risks to UAV networks, requiring robust encryption protocols and AI-based security mechanisms to protect sensitive drone communication channels. Limitations in network infrastructure, such as base station availability and regulatory compliance, also impact drone scalability, emphasizing the need for innovative solutions in future UAV deployments. The continuous evolution of UAV communication technologies ensures smarter, safer, and more efficient drone operations. As AI, 5G, and beyond reshape aerial networks, UAVs will become increasingly integrated into smart cities, industrial automation, and advanced surveillance systems.

10.4 DYNAMIC TRAFFIC CHANNELIZATION

The advancement of technology is highly dependent on the data acquisition from the sensory devices and the wireless channels. The management and analysis of this huge generated data is an integral part toward providing standard service. In particular, the IoV paradigm plays an important role when it comes to data generation supported by wireless communication, and a major concern in the field of IoV is traffic channelization [57]. The present static traffic channelization systems are not enough to handle the extensive growth of traffic, so dynamic traffic channelization is inevitable [4]. Dynamic channelization can be achieved with the assistance of the drone technology framework that has significant advantages over the existing static framework [3]. For instance, fading in wireless communication is minimized, and the communication delay in drone-to-drone and drone-to-vehicle channels is reduced as well. Moreover, the drone facilitates the real-time routing of the vehicles, which reduces the chances of a bottleneck situation in the traffic flow. This routing is achieved by continuously monitoring the vehicles and collecting the sensory and location-related data and, accordingly, suggesting the best route considering the traffic intensity, thus effectively channelizing the congestion. Following the mammoth task of providing the service to vehicles, a large amount of data is generated at each moment. Moreover, as the concept of edge computing comes into play, each drone works as a base station for processing the collected data. The processing paradigm comes closer to the vehicle, and the vehicle does not have to wait for the processing in any remote base station, which in turn minimizes the waiting time index for the vehicles as the data are transmitted from the drone directly to the vehicle without any fading and interference.

In this chapter, the real-time 5G traffic scenario is generated using the SUMO network simulator. It takes the desired open-street data and mimics a particular location to support 5G V2V communication in the context of IoV. A simulation representing an urban traffic jam scenario is demonstrated, which is created if congestion is not channelized.

10.4.1 Framework for traffic optimization

In this section, an effective model for traffic congestion channelization in future smart cities is provided. Figure 10.1 demonstrates a pictorial representation of a portion of a busy road network in a modern city [58]. The framework highlights the deployment of drones in the controlled airspace of cities to monitor vehicles [18] and route vehicles taking the distance from source to destination and dynamic traffic intensity in real time. The drones are placed above the roads to facilitate line-of-sight communication between the vehicles and the drones, which minimizes fading and facilitates better communication [10–12, 26]. The optimal 3D positioning of the drones above the road is made for improving communication [2, 14]. In this architecture, drones act as an airborne base station that continuously collects location data and sensor data and processes the data simultaneously to provide service at an exponential rate.

The drones create a cellular network and each vehicle in the city is connected to that network, enabling the transmission of data to and from the drone network, but with a huge number of vehicles continuously accessing the drone network, it creates a bottleneck situation over the 5G communication channel, leading to low latency and the network traffic can lead to low service performance. In order to counter the latency issues and to provide faster processing, the drone edge computing is incorporated.

10.4.2 Stochastic routing protocol (SRP)

This is an extension of the static Dijkstra's algorithm that takes the real-time traffic intensity as a random Poisson input based on which the algorithm determines the route. With each

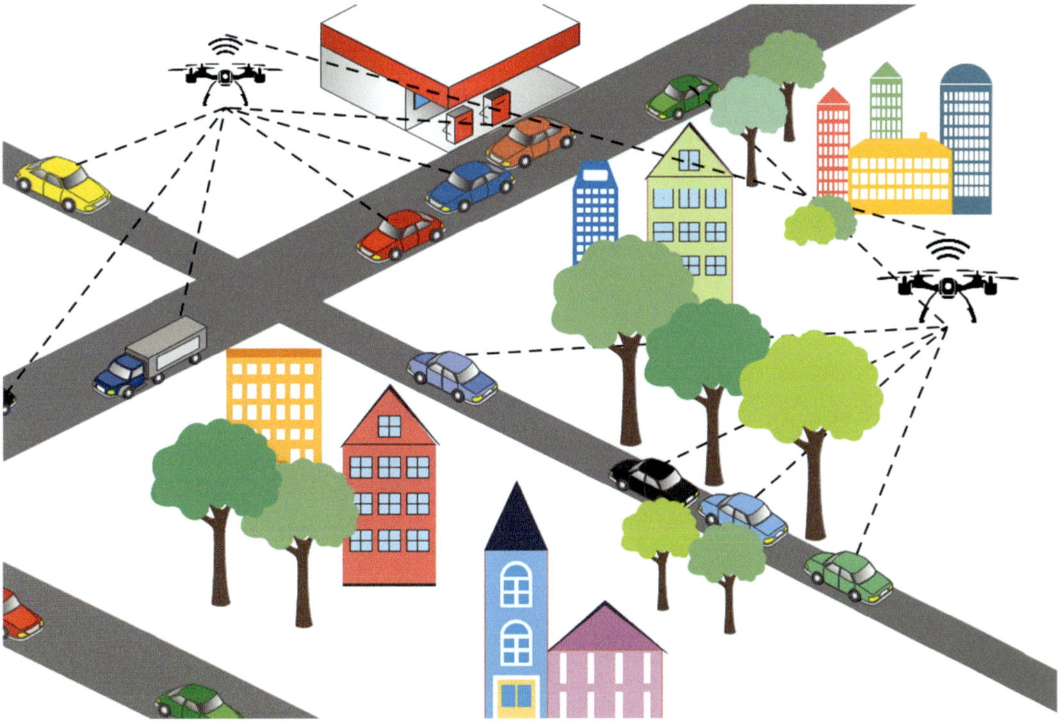

Figure 10.1 Drone architecture in an urban scenario.

change in traffic intensity, the algorithm recursively routes the vehicle in the path at the minimum cost, leading to low congestion on the streets. The simulated scenario of traditional static routing is shown in Figure 10.2, which does not consider traffic intensity and fails to optimize the route and congestion arises, whereas the SRP generates the route in real time that continues to update based on the change in the traffic intensity, which is demonstrated in Figure 10.3 at time (t) and in Figure 10.4 at time (t+1).

Stochastic Routing Protocol (SRP)

```
G (Graph), θ (Edge weights), R (Source node), F (Final node)
foreach v ∈ V[G] do
    D[v] ← ∞ ;                                  // Set initial distance
    Parent[v] ← ∅ ;                             // No parent yet
D[R] ← 0 ;                                      // Source distance is zero
Visited ← ∅ ;                                   // Initialize visited set
Q ← V[G] ;                                      // Initialize queue
while Q[F] ≠ u do
    while Q ≠ ∅ do
        Generate random λ
        u ← min{Q, λ} ;                         // Select node with min weight
        Visited ← Visited ∪ {u} ;               // Mark as visited
        foreach v ∈ Adj[u] do
            if D[v] > D[u] + θ[u, v] then
                D[v] ← D[u] + θ[u, v] ;         // Update distance
                Parent[v] ← u ;                 // Update path
```

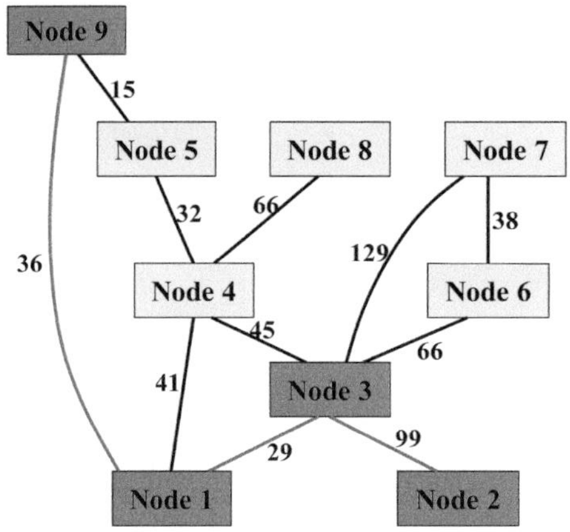

Figure 10.2 Shortest route under static scenario.

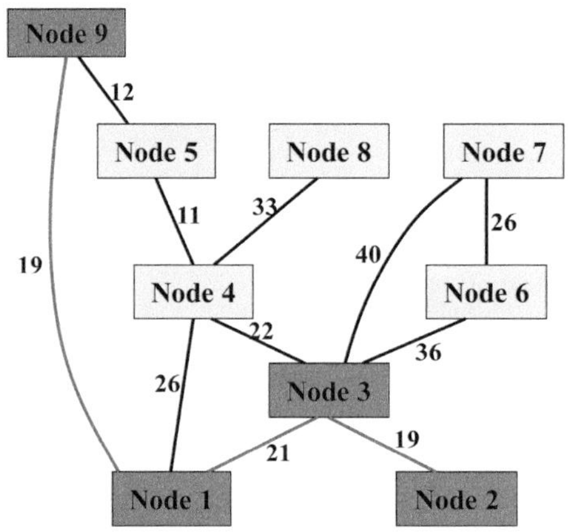

Figure 10.3 Shortest route with varying traffic intensity at time t.

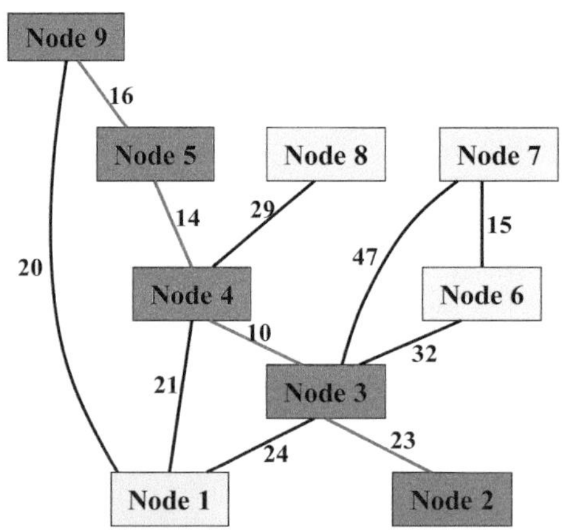

Figure 10.4 Shortest route with varying traffic intensity at time t+1.

Figure 10.5 The local drone processing to convert the arrival Poisson.

10.4.3 Role of drone edge computing

The advent of edge computing has opened new opportunities for faster data analysis, thereby improving the service. In the drone routing context, edge computing plays a major role when it comes to the QoS of the drone architecture, as it improves the data processing as the processing paradigm moves closer to the data source, i.e., the vehicles. The voluminous access of the drone network by the vehicles creates huge traffic over the 5G communication channel that might affect the service rate, and the performance of the drone network is hampered. Moreover, delayed service leads to an increase in waiting time for the vehicles in the queue for the service that causes congestion over the network due to a bottleneck situation created by the slow processing rate. The traffic in the 5G communication channel is created by the vehicles that access the drone network for the service. The arrival of the number of service requests to the drone processing environment is treated as a non-Poisson (Figure 10.5) and the drone processes the requests and converts the non-Poisson arrival into a Poisson arrival which is fed as the input into the queueing model (Figure 10.6). The traffic over the network leads to the delay in processing time, which is addressed in this section considering two scenarios. Thereafter, the drone processing efficiency is calculated in a framework constrained with traffic and without edge computing, and then the same is calculated with the drone edge computing. First, the waiting time of the vehicles in a traffic jam where the PDF of the waiting time of the vehicles is calculated considering the processing intensity of the drone, which depends on the number of vehicles requesting the service. Secondly, the PDF of the waiting time of the vehicle in-line for service is calculated with the drone edge computing.

The QoS of any drone service paradigm is a major concern and is largely dependent on the processing cost, and edge computing contributes greatly to minimizing the cost of the service. The cost is minimized when the data is processed near the source of its generation, thus the service provided is faster with lower latency. The comparative study of the service cost of drone processing using edge computing and without using is given below.

Figure 10.6 The local drone processing and exponential service.

10.4.4 Problem formulation

In this section, a problem is considered based on a hypothetical scenario with numerous possible constraints, the solution to the problem is devised, and the results are portrayed.

Waiting Time Under Traffic Jam:

Let W denotes the waiting time for vehicles. The PDF of waiting time is given by,

$$\psi(\omega\,|\,\mu)=\sum_{n=0}^{\infty}\psi(\omega\,|\,n)P_n, \tag{10.1}$$

where $\psi(\omega\,|\,n)$ be the probability density function of waiting time of vehicles given that the system has n number of vehicles with a service rate μ. It can be stated as,

$$\psi(\omega\,|\,n)=\frac{\mu^{n+1}}{n!}e^{-\mu\omega}\omega^n, \tag{10.2}$$

and P_n be the probability of n vehicles in the system, given as,

$$P_n=\rho^n(1-\rho), \tag{10.3}$$

where ρ denotes the processing intensity corresponding to the n vehicles in the system.

From Eqs. (10.1) and (10.2), we get,

$$\psi(\omega)=\sum_{n=0}^{\infty}\frac{\mu^{n+1}}{n!}e^{-\mu\omega}\omega^n\rho^n(1-\rho), \tag{10.4}$$

or,

$$\psi(\omega)=\mu(1-\rho)e^{-\mu\omega}\sum_{n=0}^{\infty}\frac{(\mu\omega\rho)^n}{n}, \tag{10.5}$$

$$\psi=\mu(1-\rho)e^{-\mu\omega}e^{-\mu\omega\rho}. \tag{10.6}$$

Therefore, the PDF of the waiting time of vehicles in the system is given as,

$$\psi(\omega)=\mu(1-\rho)e^{-\mu\omega(1-\rho)}. \tag{10.7}$$

Waiting Time Under Edge Computing: It is well-known that the inter-arrival of vehicles follows the exponential distribution. Therefore, the PDF of waiting time of vehicles under service of drone edge computing can be defined as,

$$\phi(\omega)=\int_0^{\infty}f(\omega\,|\,\mu)S(\mu)\,d\mu. \tag{10.8}$$

The inter-arrival rate of vehicles can be characterized by the exponential distribution and is given by,

$$S(\mu)=\mu e^{-\mu}. \tag{10.9}$$

From Eqs. (10.8) and (10.9), we obtain,

$$\phi(\omega)=\int_0^{\infty}(\mu-\lambda)e^{-\omega(\mu-\lambda)}\mu e^{-\mu}\,d\mu. \tag{10.10}$$

Hence,

$$\phi(\omega)=\int_0^\infty (\mu^2-\lambda\mu)e^{-(\omega)\mu}d\mu. \tag{10.11}$$

Therefore, the PDF of waiting time distribution under drone edge computing can be obtained as,

$$\phi(\omega)=\frac{1}{C}\frac{(2-\lambda)e^{\lambda\omega}}{\omega}, \tag{10.12}$$

where C is the normalization constant.

10.4.5 Service cost estimation

With Drone Edge Computing: The cost for transmitting the data packets from moving vehicles to drone edge is given by,

$$C(\mu)=\frac{C_1\mu}{k}+C_2L_1, \tag{10.13}$$

where μ is the processing rate of the data packets, C_1 is the processing cost per data packet, C_2 is the cost to wait for a single data packet, and L_1 denotes the count of distinct data packets at the drone edge.

From the conventional queueing model, $(M/M/1):(\infty/FCFS)$, we have,

$$C(\mu)=C_1\mu+C_2\left(\frac{\lambda}{\mu-\lambda}\right). \tag{10.14}$$

where λ is the rate at which data packets arrive at the drone edge.

Without Drone Edge Computing: The service cost estimation without using drone edge computing is given as,

$$C(\mu)=C_1\mu+C_2L_2, \tag{10.15}$$

where L_2 denotes the number of data packets on the drone edge, and using $(M/E_k/1):(\infty/FCFS)$ we have,

$$L_2=\frac{k+1}{2k}\frac{\lambda^2}{\mu(\mu-\lambda)}+\frac{\lambda}{\mu}. \tag{10.16}$$

From Eqs. (10.15) and (10.16), we get

$$C(\mu)=C_1\mu+C_2\left[\frac{k+1}{2k}\frac{\lambda^2}{\mu(\mu-\lambda)}+\frac{\lambda}{\mu}\right]. \tag{10.17}$$

Figure 10.7(a) and (b) illustrate a comparative study between the conventional service mechanism, and our proposed edge computing framework using the well-known queueing model comparing the service cost with varying arrival rate (λ) and service rate (μ). It is observed that the proposed framework outperforms the conventional mechanism and provides a better agreement for the support of 5G traffic. An interesting result is observed in Figure 10.7(c) and (d) that both the proposed and conventional models demonstrate similar behavior with the unit flocking coefficient, i.e., $k=1$.

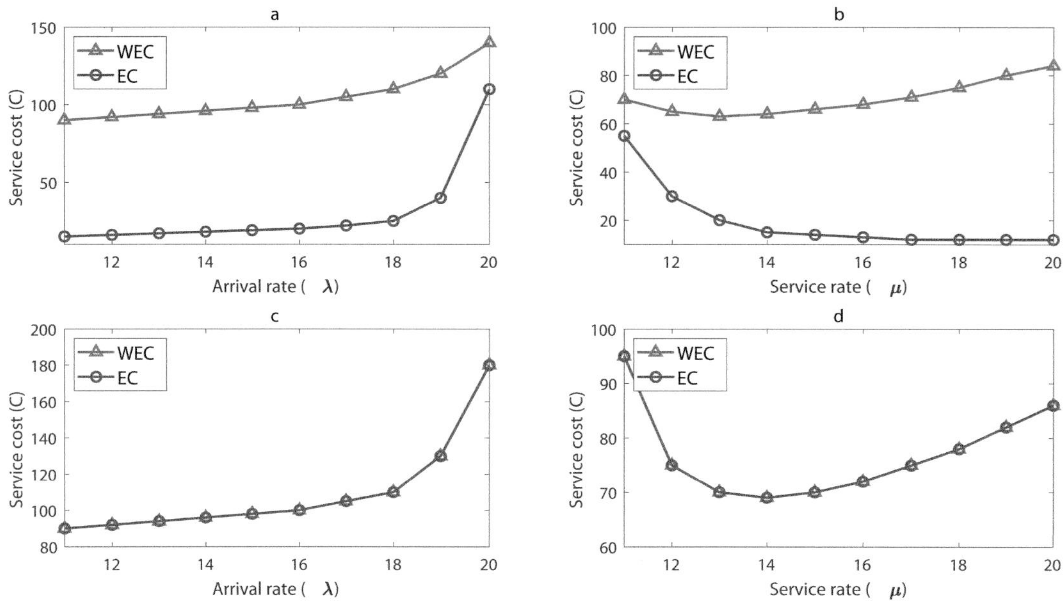

Figure 10.7 Parts (a) and (b) show the comparison of the service cost in EC and WEC models corresponding to different arrival rate and service rate. Parts (c) and (d) illustrate the service cost of EC and WEC models when $k = 1$.

10.5 CONCLUSION

In this chapter, a cognitive dynamic routing framework for monitoring and channelizing the traffic in real time is proposed, which optimizes the parameters like resources, and time. For this context, this chapter has generated a random 5G traffic scenario using the SUMO network simulator and provided the cost-optimized route considering the geographic distance and the real-time traffic intensity. The solution was provided by the SRP algorithm, which continuously generates a shortest route in each unit of time, taking the varying traffic intensity between the nodes as input. A comparative study between the conventional service mechanism and our drone edge computing framework using the well-known queueing model was provided. It is observed that the proposed framework operates well and provides better agreement for the support of 5G traffic. Although rapid technological growth has led to the invention of several dynamic techniques for the management of traffic and control of congestion in smart cities and optimization of the service. The traffic over the communication line is also addressed continuously with repeated innovations. As the civilization grows, and new vehicles and governing technologies come into existence, there is always room for further research to provide a better-than-ever-before service. The chapter focuses on highlighting the use of drones in the urban scene to facilitate technological innovations, and the arguments were supported by mathematical and visual results.

Future research in UAV communication should focus on several emerging areas to enhance drone efficiency, security, and integration within smart networks. Advancements in AI can enable autonomous UAV decision-making, allowing drones to optimize flight paths, perform real-time environmental analysis, and collaborate in swarm operations. Next-generation communication technologies, including 6G and quantum communication, promise ultra-fast and secure data exchange, which will be crucial for seamless connectivity in complex urban environments. Edge computing offers opportunities to reduce latency and optimize real-time mission execution by decentralizing data processing closer to UAVs. Multi-agent systems and swarm coordination research can refine algorithms for large-scale drone collaboration in surveillance, mapping, and logistics. Strengthening cybersecurity measures for UAV networks is essential to mitigate threats, which requires advanced encryption techniques and AI-driven security protocols. Regulatory and ethical considerations must also be addressed to ensure responsible deployment of UAVs, covering issues such as airspace governance, AI ethics, and compliance with smart city infrastructure. Additionally, UAVs can contribute significantly to environmental monitoring by providing real-time data for pollution detection, disaster management, and climate change analysis. Future research must continue to push the boundaries of UAV technology to ensure safe, intelligent, and efficient drone ecosystems that support a wide range of applications in both commercial and governmental sectors.

REFERENCES

1. Mauro Dell'Amico, Roberto Montemanni, and Stefano Novellani. Matheuristic algorithms for the parallel drone scheduling traveling salesman problem. *Annals of Operations Research*, 1–16, 2020.
2. Elham Kalantari, Halim Yanikomeroglu, and Abbas Yongacoglu. On the number and 3d placement of drone base stations in wireless cellular networks. In *2016 IEEE 84th Vehicular Technology Conference (VTC-Fall)*, pages 1–6. IEEE, 2016.
3. Sinem Han, Begüm Özer, Bars Alioglu, Özgür Polat, and Ayse Tülin Aktin. A mathematical model for the delivery routing problem via drones. *Pamukkale Üniversitesi Mühendislik Bilimleri Dergisi*, 25(1):89–97, 2019.
4. Kai Lin, Fuzhen Xia, and Giancarlo Fortino. Data-driven clustering for multimedia communication in internet of vehicles. *Future Generation Computer Systems*, 94:610–619, 2019.
5. Weisen Shi, Haibo Zhou, Junling Li, Wenchao Xu, Ning Zhang, and Xuemin Shen. Drone assisted vehicular networks: Architecture, challenges and opportunities. *IEEE Network*, 32(3):130–137, 2018.
6. Hanif Ullah, Nithya Gopalakrishnan Nair, Adrian Moore, Chris Nugent, Paul Muschamp, and Maria Cuevas. 5G communication: An overview of vehicle-to-everything, drones, and healthcare use-cases. *IEEE Access*, 7:37251–37268, 2019.
7. Antonino Orsino, Aleksandr Ometov, Gabor Fodor, Dmitri Moltchanov, Leonardo Militano, Sergey Andreev, Osman NC Yilmaz, Tuomas Tirronen, Johan Torsner, Giuseppe Araniti, et al. Effects of heterogeneous mobility on D2D-and drone-assisted mission-critical MTC in 5G. *IEEE Communications Magazine*, 55(2):79–87, 2017.
8. Wenchao Xu, Haibo Zhou, Nan Cheng, Feng Lyu, Weisen Shi, Jiayin Chen, and Xuemin Shen. Internet of vehicles in big data era. *IEEE/CAA Journal of Automatica Sinica*, 5(1):19–35, 2017.
9. Mirmojtaba Gharibi, Raouf Boutaba, and Steven L Waslander. Internet of drones. *IEEE Access*, 4:1148–1162, 2016.
10. Sujit Bebortta, Dilip Senapati, Nikhil Kumar Rajput, Amit Kumar Singh, Vipin Kumar Rathi, Hari Mohan Pandey, Amit Kumar Jaiswal, Jia Qian, and Prayag Tiwari. Evidence of power-law behavior in cognitive IoT applications. *Neural Computing and Applications*, 32:1–13, 2020.
11. Tanmay Mukherjee, Amit Kumar Singh, and Dilip Senapati. Performance evaluation of wireless communication systems over weibull/q-lognormal shadowed fading using tsallis entropy framework. *Wireless Personal Communications*, 106(2):789–803, 2019.
12. Dilip Senapati et al. Generation of cubic power-law for high frequency intra-day returns: Maximum tsallis entropy framework. *Digital Signal Processing*, 48:276–284, 2016.
13. Hesham El-Sayed, Sharmi Sankar, Mukesh Prasad, Deepak Puthal, Akshansh Gupta, Manoranjan Mohanty, and Chin-Teng Lin. Edge of things: the big picture on the integration of edge, IoT and the cloud in a distributed computing environment. *IEEE Access*, 6:1706–1717, 2017.
14. Pouya Pourbaba, Samad Ali, KB Shashika Manosha, and Nandana Rajatheva. Multi-armed bandit learning for full-duplex uav relay positioning for vehicular communications. In *2019 16th International Symposium on Wireless Communication Systems (ISWCS)*, pages 188–192. IEEE, 2019.

15. Patchara Kitjacharoenchai, Mario Ventresca, Mohammad Moshref-Javadi, Seokcheon Lee, Jose MA Tanchoco, and Patrick A Brunese. Multiple traveling salesman problem with drones: Mathematical model and heuristic approach. *Computers & Industrial Engineering*, 129:14–30, 2019.
16. David Sacramento, David Pisinger, and Stefan Ropke. An adaptive large neighborhood search metaheuristic for the vehicle routing problem with drones. *Transportation Research Part C: Emerging Technologies*, 102:289–315, 2019.
17. Yu Lin, Tianyu Wang, and Shaowei Wang. UAV-assisted emergency communications: An extended multi-armed bandit perspective. *IEEE Communications Letters*, 23(5):938–941, 2019.
18. Emmanouil Barmpounakis and Nikolas Geroliminis. On the new era of urban traffic monitoring with massive drone data: The pneuma large-scale field experiment. *Transportation Research Part C: Emerging Technologies*, 111:50–71, 2020.
19. Sujit Bebortta, Aditya Ranjan Dalabehera, Bibudhendu Pati, Chhabi Rani Panigrahi, Gyana Ranjan Nanda, Biswajit Sahu, and Dilip Senapati. An intelligent spatial stream processing framework for digital forensics amid the Covid-19 outbreak. *Smart Health*, 26:100308, 2022.
20. Sujit Bebortta, Bibudhendu Pati, Chhabi Rani Panigrahi, and Dilip Senapati. Dynamic performance modeling framework for QoS-aware 5g-based IoT-edge systems. *International Journal of Information Technology*, 16(4):2149–2160, 2024.
21. Sujit Bebortta, Nikhil Kumar Rajput, Bibudhendu Pati, and Dilip Senapati. A real-time smart waste management based on cognitive IoT framework. In *Advances in Electrical and Computer Technologies: Select Proceedings of ICAECT 2019*, pages 407–414. Springer, 2020.
22. Sujit Bebortta, Dilip Senapati, Nikhil Kumar Rajput, Amit Kumar Singh, Vipin Kumar Rathi, Hari Mohan Pandey, Amit Kumar Jaiswal, Jia Qian, and Prayag Tiwari. Evidence of power-law behavior in cognitive IoT applications. *Neural Computing and Applications*, 32(20): 16043–16055, 2020.
23. Sujit Bebortta and Dilip Senapati. Empirical characterization of network traffic for reliable communication in IoT devices. In *Security in Cyber-Physical Systems: Foundations and Applications*, pages 67–90. Springer, 2021.
24. Sujit Bebortta and Dilip Senapati. A secure blockchain-based solution for harnessing the future of smart healthcare. In *IoT-Based Data Analytics for the Healthcare Industry*, pages 167–191. Elsevier, 2021.
25. Sujit Bebortta and Dilip Senapati. Characterizing the epidemiological dynamics of Covid-19 using a non-parametric framework. *Current Science*, 122(7):790, 2022.
26. Sujit Bebortta, Amit Kumar Singh, Surajit Mohanty, and Dilip Senapati. Characterization of range for smart home sensors using tsallis' entropy framework. In *Advanced Computing and Intelligent Engineering*, pages 265–276. Springer, 2020.
27. Sujit Bebortta, Amit Kumar Singh, and Dilip Senapati. Performance analysis of multi-access edge computing networks for heterogeneous IoT systems. *Materials Today: Proceedings*, 58:267–272, 2022.
28. Thomas Kirschstein. Comparison of energy demands of drone-based and ground-based parcel delivery services. *Transportation Research Part D: Transport and Environment*, 78:102209, 2020.
29. Chase C Murray and Amanda G Chu. The flying sidekick traveling salesman problem: Optimization of drone-assisted parcel delivery. *Transportation Research Part C: Emerging Technologies*, 54:86–109, 2015.
30. Chase C Murray and Ritwik Raj. The multiple flying sidekicks traveling salesman problem: Parcel delivery with multiple drones. *Transportation Research Part C: Emerging Technologies*, 110:368–398, 2020.
31. Amartya Mukherjee, Nilanjan Dey, and Debashis De. Edgedrone: QoS aware MQTT middleware for mobile edge computing in opportunistic internet of drone things. *Computer Communications*, 152:93–108, 2020.
32. Tanmay Mukherjee, Sujit Bebortta, and Dilip Senapati. Stochastic modeling of q-lognormal fading channels over tsallis' entropy: Evaluation of channel capacity and higher order moments. *Digital Signal Processing*, 133:103856, 2023.
33. Tanmay Mukherjee, Gangadhar Nayak, and Dilip Senapati. Evaluation of symbol error probability using a new tight gaussian q approximation. *International Journal of Systems, Control and Communications*, 12(1):60–71, 2021.
34. Tanmay Mukherjee, Bibudhendu Pati, and Dilip Senapati. Performance evaluation of composite fading channels using q-weibull distribution. In *Progress in Advanced Computing and Intelligent Engineering: Proceedings of ICACIE 2019, Volume 1*, pages 317–324. Springer, 2021.

35. Suvra Nayak, Chhabi Rani Panigrahi, Bibudhendu Pati, Sarmistha Nanda, and Meng-Yen Hsieh. Comparative analysis of HAR datasets using classification algorithms. *Computer Science and Information Systems*, 19(1):47–63, 2022.
36. Quang Minh Ha, Yves Deville, Quang Dung Pham, and Minh Hoàng Hà. On the min-cost traveling salesman problem with drone. *Transportation Research Part C: Emerging Technologies*, 86:597–621, 2018.
37. Shen Lin and Brian W Kernighan. An effective heuristic algorithm for the traveling-salesman problem. *Operations Research*, 21(2):498–516, 1973.
38. Paul Bouman, Niels Agatz, and Marie Schmidt. Dynamic programming approaches for the traveling salesman problem with drone. *Networks*, 72(4):528–542, 2018.
39. Júlia Cária de Freitas and Puca Huachi Vaz Penna. A variable neighborhood search for flying sidekick traveling salesman problem. *International Transactions in Operational Research*, 27(1):267–290, 2020.
40. Iurii Bakach, Ann Melissa Campbell, and Jan Fabian Ehmke. A two-tier urban delivery network with robot-based deliveries. *Working Paper Series*, 78: 461–483, 2020.
41. Kevin Dorling, Jordan Heinrichs, Geoffrey G Messier, and Sebastian Magierowski. Vehicle routing problems for drone delivery. *IEEE Transactions on Systems, Man, and Cybernetics: Systems*, 47(1):70–85, 2016.
42. Quang Minh Ha, Yves Deville, Quang Dung Pham, and Minh Hoàng Hà. A hybrid genetic algorithm for the traveling salesman problem with drone. *Journal of Heuristics*, 26(2):219–247, 2020.
43. Shushman Choudhury, Kiril Solovey, Mykel J Kochenderfer, and Marco Pavone. Efficient large-scale multi-drone delivery using transit networks. *arXiv preprint arXiv:1909.11840*, 2019.
44. Mauro Dell'Amico, Roberto Montemanni, and Stefano Novellani. Models and algorithms for the flying sidekick traveling salesman problem. *arXiv preprint arXiv:1910.02559*, 2019.
45. Daniel Schermer, Mahdi Moeini, and Oliver Wendt. Algorithms for solving the vehicle routing problem with drones. In *Asian Conference on Intelligent Information and Database Systems*, pages 352–361. Springer, 2018.
46. Iman Dayarian, Martin Savelsbergh, and John-Paul Clarke. Same-day delivery with drone resupply. *Transportation Science*, 54:229–249, 2020.
47. Shireen Seakhoa-King, Paul Balaji, Nicolas Trama Alvarez, and William J Knottenbelt. Revenue-driven scheduling in drone delivery networks with time-sensitive service level agreements. In *Proceedings of the 12th EAI International Conference on Performance Evaluation Methodologies and Tools*, pages 183–186, 2019.
48. Gangadhar Nayak, Amit Kumar Singh, Subarna Bhattacharjee, and Dilip Senapati. A new tight approximation towards the computation of option price. *International Journal of Information Technology*, 14(3):1295–1303, 2022.
49. Gangadhar Nayak, Amit Kumar Singh, and Dilip Senapati. Computational modeling of non-gaussian option price using non-extensive tsallis' entropy framework. *Computational Economics*, 57(4):1353–1371, 2021.
50. Amit Kumar Singh, Dilip Senapati, Sujit Bebortta, and Nikhil Kumar Rajput. A non-stationary analysis of erlang loss model. In *Progress in Advanced Computing and Intelligent Engineering: Proceedings of ICACIE 2019, Volume 1*, pages 286–294. Springer, 2021.
51. Amit Kumar Singh, Dilip Senapati, Tanmay Mukherjee, and Nikhil Kumar Rajput. Adaptive applications of maximum entropy principle. In *Progress in Advanced Computing and Intelligent Engineering: Proceedings of ICACIE 2019, Volume 1*, pages 373–379. Springer, 2020.
52. Anibal Sanjab, Walid Saad, and Tamer Basar. Prospect theory for enhanced cyber-physical security of drone delivery systems: A network interdiction game. In *2017 IEEE International Conference on Communications (ICC)*, pages 1–6. IEEE, 2017.
53. Christopher Burke, Hung Nguyen, Mary Magilligan, and Rafiqul Noorani. Study of a drone's payload delivery capabilities utilizing rotational movement. In *2019 International Conference on Robotics, Electrical and Signal Processing Techniques (ICREST)*, pages 672–675. IEEE, 2019.
54. Wen-Chyuan Chiang, Yuyu Li, Jennifer Shang, and Timothy L Urban. Impact of drone delivery on sustainability and cost: Realizing the uav potential through vehicle routing optimization. *Applied energy*, 242:1164–1175, 2019.
55. Chung Hoon Choi, Hyeon Jun Jang, Seong Gyu Lim, Hyun Chul Lim, Sung Ho Cho, and Igor Gaponov. Automatic wireless drone charging station creating essential environment for continuous drone operation. In *2016 International Conference on Control, Automation and Information Sciences (ICCAIS)*, pages 132–136. IEEE, 2016.

56. Aurélien Froger, Jorge E Mendoza, Ola Jabali, and Gilbert Laporte. Improved formulations and algorithmic components for the electric vehicle routing problem with nonlinear charging functions. *Computers & Operations Research*, 104:256–294, 2019.
57. Omprakash Kaiwartya, Sushil Kumar, DK Lobiyal, Pawan Kumar Tiwari, Abdul Hanan Abdullah, and Ahmed Nazar Hassan. Multiobjective dynamic vehicle routing problem and time seed based solution using particle swarm optimization. *Journal of Sensors*, 2015, 2015.
58. Giuseppe Piro, Ilaria Cianci, Luigi Alfredo Grieco, Gennaro Boggia, and Pietro Camarda. Information centric services in smart cities. *Journal of Systems and Software*, 88:169–188, 2014.

Chapter 11

Telemedicine and remote healthcare in wireless communication

Satwik Chatterjee, Ashni Chatterjee, and Rashmi R.

11.1 INTRODUCTION

11.1.1 Introduction to telemedicine

Telemedicine is a cutting-edge innovation in healthcare, overcoming the spatial divide between medical practitioners and patients [1]. Telemedicine is the communication of medical information and medical services with the help of telecommunication technology over distances. Telemedicine was initially designed for use in geographically distant or rural areas, but it now includes a host of modalities such as synchronous virtual consultations, mobile health (mHealth) apps, and systems of patient monitoring remotely [2]. Better digital infrastructure and internet coverage globally are driving telemedicine to take center stage in new healthcare provision, translating into improved clinical outcomes, increased efficiency, and reduced healthcare expenditure [3].

Telemedicine, in the health dynamics of the world, has been a necessary tool for ensuring continuity of care, especially in emergencies. Telemedicine proved priceless during the COVID-19 pandemic as it eased the pressure on hospitals and allowed real-time interaction with physical distancing being a must. Telemedicine is scalable and sustainable in providing timely medical intervention in conflict-affected areas, natural disaster-struck areas, or in poorly developed health systems. Remote healthcare systems are transforming patient accessibility in the following ways: reducing treatment and diagnosis delays, facilitating long-term management of chronic diseases, and reducing the rate of hospital readmission in real time.

11.1.2 Wireless communication application in healthcare

Wireless communications are the technological basis upon which telemedicine is constructed. From patient data transmission to real-time diagnosis, the systems allow dynamic, secure, high-rate medicine interaction [4]. The prominent wireless technologies involved with telemedicine are often employed in hospitals as a way of connecting local equipment, electronic medical records (EMRs), and internal systems.

Bluetooth: Facilitates the use of medical wearable devices and distant health monitoring.

Fifth Generation (5G): Enables ultra-low latency and high bandwidth that is required for high-risk applications such as remote surgery and real-time video consults [5].

6G: The technology of 6G will transform medicine with immersive experiences such as augmented reality diagnosis and digital twins [6].

Web/Internet-Based Platforms: Serve as the universal platform for cloud-based health management, electronic prescriptions, and digital diagnosis.

DOI: 10.1201/9781003637721-11

The principal applications of these wireless technologies are teleconsultation and electronic prescriptions, wearable and implantable medical devices, mHealth platforms and real-time alerts, remote diagnostic imaging and teleradiology, Tele-Intensive Care Units (ICUs) and remote emergency care [7]. The rising rates of chronic diseases, aging populations, healthcare worker shortages, and pandemics all necessitate new approaches in healthcare. Telemedicine can resolve all of these issues by decentralizing the treatment and improving effectiveness. The significant world drivers are: *pandemics*, enabling safe, socially distanced clinical contact; *aging populations*, long-term care and home-based services; and *workforce shortages*, maximizing resource efficiency and reducing burnout for clinicians.

Telemedicine is a significant contributor to the following objectives: Universal Health Coverage (UHC) by extending coverage for underserved groups, financial robustness by reducing transport and hospital infrastructure costs, and sustainable development goal: assuring health and well-being and quality health care for all [8].

Prospects that dictate telemedicine's evolution include Artificial Intelligence (AI) integration: maximizing clinical decision support, cybersecurity, and ethical frameworks: addressing data privacy issues and digital literacy: facilitating both health consumers and health professionals in adopting and adjusting accordingly [9].

11.2 TELEMEDIA AND WIRELESS COMMUNICATIONS

11.2.1 5G in healthcare

Figure 11.1 shows the general applications of 5G network technology. With the evolution of the 4G network, the 5G network came into play, taking part in smart healthcare. The only motive for this is to increase the people's quality of life. The place we are surviving has more medical emergencies than doctors or medical facilities. To overcome this, the 4G network and other wireless communications were used, but these technologies proved to be incapable of meeting the dynamic and complex demands of smart healthcare applications [10]. The leakage of various information and communications changed radically and created an ecosystem for new opportunities in every sector and industry, resulting in digital transformation [11]. For the further expansion of telehealth, the 5G networks, along with AI, have been developed, creating a digital innovation. One of the important

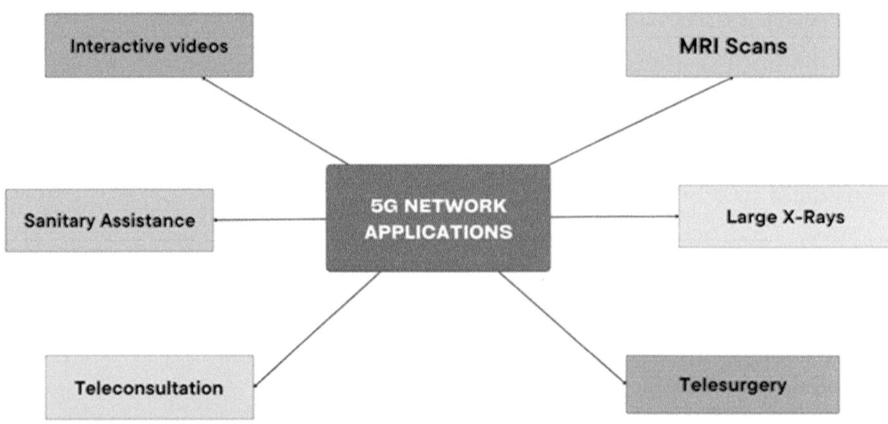

Figure 11.1 General applications of 5G network technology.

applications of 5G networks is *smart healthcare*. Their applications have increased due to the advancements of the Internet of Things (IoT) and AI, making smart healthcare into a new ecosystem [12].

5G is the interface that is both unified and powerful, and built with extra capacity to support next-generation user experiences [13]. The 5G technology improves the provision of health services, including monitoring, care, and remote surgeries. For example, Google has created so many smart devices for health monitoring, and one of those is Google's Fitbit tracker, which measures oxygen level in the blood, pulse rate, and sleep cycle. The features of 5G network, including eMBB (enhanced mobile broadband) and URLLC (ultra-reliable low-latency communications), generate healthcare systems that minimize the face-to-face assistance of patients. Although 4G has been used in all wireless communication devices, it has faced many challenges, especially when the number of device connections increases. But 5G can transfer and manage data in bulk, which makes it more efficient for usage. The other applications are listed in Figure 11.1.

Clinical Applications of 5G Network: The capacity of 5G telecommunications creates an ecosystem that will be helpful in re-assessing the care delivery by incorporating AI and additional wireless technologies. There are examples of using remote clinics and ambulances where HD consultation was successfully performed without the presence of the patient. It is also mentioned that remote surgeries were carried out using a surgical robot [14]. For example, remote laparoscopic surgery was carried out using the robot called 'Micro hand' on a pig, which was 3000 km away from the surgeon. These applications do not end here but also include a remote ultrasound system, which was developed during COVID-19 to examine the heart, lungs, blood vessels, and other organs [14]. This comes in two environments, where the first one has a workstation and a doctor, and the other environment is where the patient will be located, along with the ultrasound system. This also holds a special place in other domains like Pulmonology, Imaging, Ultrasonography, Ophthalmology, Cardiology, Emergency Medicine, Obstetrics, and Surgery. The block diagram of this is given below in Figure 11.2.

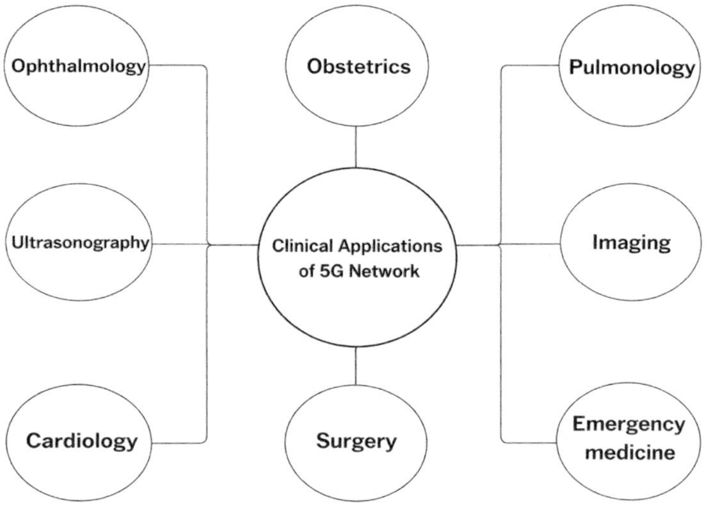

Figure 11.2 Clinical application of 5G wireless networking.

Figure 11.2 explores the clinical application of 5G wireless networking. The data will be streamed on any of the electronic devices that are connected to the 5G router and the 5G infrastructure. The responsible person can access it through a dedicated dashboard [15]. Various technologies like threshold algorithms, machine learning (ML), deep learning, and clustering methods are used for analyzing and diagnosing the massive number of medical images, which takes a lot of time for manual segmentation. There is a recent introduction to dynamic telepathology where, through telecommunication, the transmission of the microscopic slide images to the recipient is done. It has its advantages for its usage in remote ultrasonography, as the examiners are protected against cross-infection, they can perform real-time scans remotely on patients, and this method can be performed anywhere, even in the patient's home. These technologies are even used in the ophthalmology domain as they provide data-rich images that are essential for them. The robotic telestenting, in which the 5G wireless technology is used, allows for where percutaneous coronary intervention (PCI) to be performed on a patient who is located remotely. The telemedicine worked best for the emergency medicine department. This allowed patients to get treated anywhere rather than coming to the ED. The chatbots help in the consultation of patients through video consultation. For obstetric care, this technology can be incorporated for the monitoring of the fetus at home. This is also used in robotic surgeries, including pre-operative, intraoperative, and postoperative procedures. One of the important surgical applications is trauma and accident injury. Through this technology, it is now easy for a first responder to virtually connect with a healthcare professional to diagnose the patient anywhere. Telesurgery or remote surgery provides high-quality care in most surgeries. The example for this is the surgery that is currently used, i.e., Robot-assisted Minimally Invasive Surgery (RAMIS). It is also used in robot-assisted spinal surgery, which is a reliable surgical procedure. To conclude, it can be accepted that the use of 5G remote ultrasound will be a good choice for monitoring and diagnosing during the period of pandemics.

Figure 11.3 explains the 5G features involving D2D communication where direct communication is done to exchange data, software defined network which integrates many network technologies to make a network more flexible, advanced massive MIMO which increases the healthcare network capacity by increasing the number of sending and receiving antennas, and NOMA which allows flexibility between strong and weak users and mitigates the effects of multiple access interference.

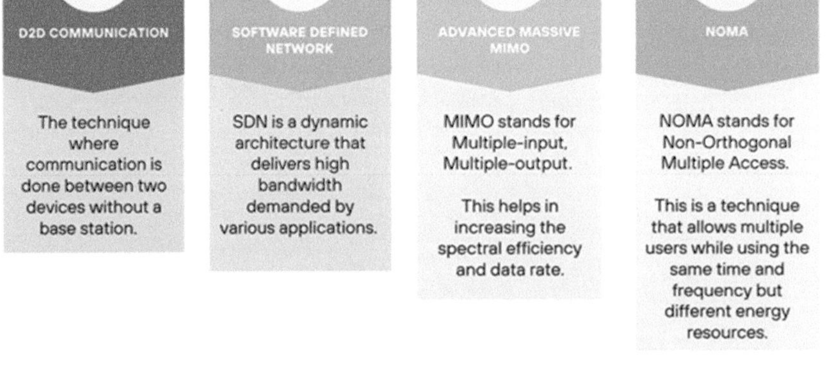

Figure 11.3 5G features and enabled technologies.

11.2.2 Bluetooth and Wi-Fi in healthcare

Wireless communication technologies, such as Bluetooth and Wi-Fi, have become core components of any healthcare system, such as remote patient monitoring (RPM) and telemedicine, and they allow for the near-instantaneous and unbroken transmission of medical data from a patient to a provider [16]. The cuts to nursing care capacity and efficiency that are now so prevalent will no longer exist. Bluetooth is primarily used for short-range range, low-power devices, such as wearable glucose monitors, pulse oximeters, and ECG sensors. Wi-Fi permits much higher data usage, but battery longevity and range are shorter. The Bluetooth and Wi-Fi technology in telehealth systems are shown in Figures 11.4 and 11.5, respectively. The comparison between Bluetooth and Wi-Fi technologies is stated in Table 11.1. The advantages in clinical and home settings are represented in Table 11.2.

Modern healthcare ecosystems successfully deploy both Bluetooth and Wi-Fi technologies. Bluetooth can be leveraged to connect a limited-distance wearable device – typically

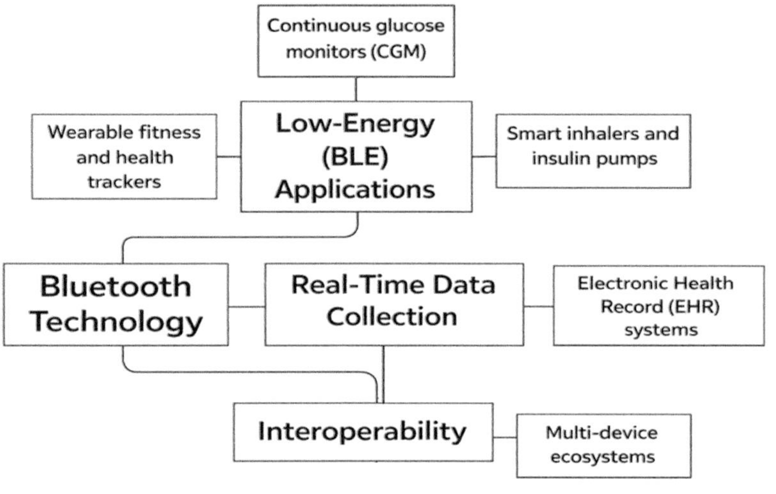

Figure 11.4 Bluetooth technology in healthcare.

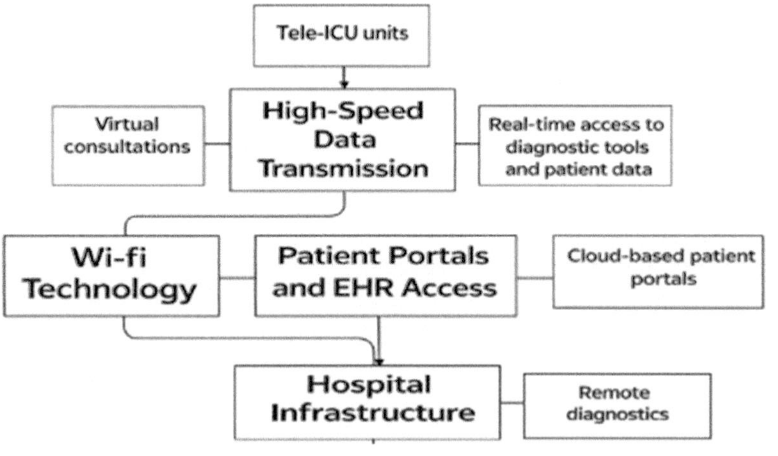

Figure 11.5 Wi-Fi in telehealth systems.

Table 11.1 Comparison between Bluetooth and Wi-Fi

Features	*Bluetooth*	*Wi-Fi*
Range	Short range	Long range
Power use	Low energy	High energy
Ideal use	Wearables, mobile health devices	Video consultations, large data transfer
Setup	Easy pairing	Requires network infrastructure

Table 11.2 Advantages in clinical and home settings

Clinical benefits	*Homecare benefits*
Continuous monitoring and alerts in ICUs	Elderly care with fall detection
Enhanced workflow with reduced manual entries	Chronic disease management (e.g., diabetes, hypertension)
Real-time decision-making based on live data	Mental health monitoring through biofeedback wearables

used to acquire data such as heart rate or activity level. Wi-Fi can connect cloud services, transmit information over longer distances, and provide clinicians with dashboards [17].

For instance, a patient could wear a Bluetooth-enabled ECG patch, the information then relayed via smartphone with Wi-Fi to a cloud server where a cardiologist can see the results in real time. Both of these technologies are needed and also require the same security protocols that include encryption, authentication protocols, and network monitoring to protect sensitive health data. Future innovations include: Bluetooth mesh networking to provide connectivity in multi-device environments, Wi-Fi 6/6E developed for faster and more secure healthcare applications, and AI/ML technologies as collaborative platforms or predictive diagnostics. Using Bluetooth and Wi-Fi alongside AI-enabled analytics will give both patients and physicians the tools to jointly, actively co-manage their health conditions by providing timely alerts, comparing data trends, and treating accordingly.

11.2.3 Internet of Things (IoT) in healthcare

As a frontier of innovation in modern healthcare infrastructure, healthcare IoT helps to facilitate RPM, diagnostics, and clinical decision support [18]. Healthcare IoT integrates well across digital healthcare platforms and facilitates smart data flow, improved diagnostics, and proactive intervention, leading to improved patient outcomes and effective healthcare provision. Real-world applications of IoT in healthcare are discussed below:

RPM: The system consists of multiple connected elements that use Internet protocols to handle health information exchange through real-time data collection, transmission, and analysis. The modern healthcare infrastructure includes healthcare IoT as an innovative field that enables remote patient care. Healthcare IoT integrates seamlessly between digital healthcare platforms, helping facilitate smarter data flow, better diagnostics, and proactive interventions that can lead to better patient outcomes and the provision of effective healthcare.

In-Hospital Monitoring Systems: In hospitals, staffing also coordinates a multitude of items catapulted by the increase in IoT devices. Smart beds measure movement to detect when a patient attempts to get out of bed. Smart beds may reduce fall opportunities while allowing the staff to respond sooner [19]. Asset tracking systems use sensors to recover equipment such as ventilators, infusion pumps, or heart monitors to minimizing the time lost due to consistently lost items. Temperature sensors are important in areas such as ICUs that measure air flow, temperature, humidity, and lighting. Temperature systems provide clinical staff with the accurate, real-time data required to provide quality evidence-based safety measures. Air quality and temperature will possibly unite covertly to prevent infections and offer the best healing for a patient.

Medication Management: The technology of smart pill bottles can remind users to take medicine at an appropriate time. It also alerts the patients or the users if the course of the medicine is missed. In this way, the IoT helps patients adhere to the medication. A dispenser with tracking abilities can record when a patient took medications. It also allows caregivers to monitor medication routines from a distance [20]. This will lessen the probability of a mistake, omitted dose, or adverse drug interaction. Overall, smart medication tools allow patients to be more proactive in managing their own health and improve their daily routines.

Emergency Care: IoT technology allows improvement in emergency treatment, which is important even though the emergency service is influenced by speed. For example, currently in an ambulance, devices like portable ECG's and ventilators integrate wirelessly together inside the ambulance. Therefore, paramedics can actually, while en route to the hospital in an ambulance, check the patient's heart rhythm or status of breath. The ambulance also tracks the actual location to assist in very timely delivery to the hospital, and it sends the patient's vital signs to the ER for timely delivery of treatment. When the patient arrives at the hospital with chest pain, the emergency department doctors see what is happening in real time prior to giving the patient any treatment. The benefit is that it takes less time for the emergency team to work and, with the best possible coordination, increases the patient survival rate. Emergency teams with access to information through IoT-enabled ambulances mean access to information toward the faster provision of care in urgent situations.

The benefits of IoT in healthcare are shown in Table 11.3. The inclusion of IoT in healthcare is revolutionizing the sector by providing a vibrant, real-time interpretation of patients' health and enabling users and professionals with actionable information. From chronic care to preventive health approaches, IoT is central to digital health ecosystems (next generation). Although it comes with issues regarding data privacy, interoperability, and infrastructure, there are continued advances in AI, 5G, and edge computing that will ensure a connected, responsive, personal future for healthcare.

Table 11.3 Benefits of IoT in healthcare

Benefit	*Description*
Continuous Monitoring	Enables 24/7 health data collection and timely interventions
Enhanced Patient Engagement	Empowers individuals to manage their health proactively
Cost Efficiency	Reduces hospital readmissions and in-person visits, optimizing resource allocation
Data-Driven Clinical Decisions	Providers receive more accurate, real-time insights leading to better outcomes
Personalized Healthcare	Facilitates tailored treatment plans based on patient-specific data
Improved Access in Remote Areas	IoT bridges geographical barriers, bringing quality care to underserved regions

11.3 WIRELESS COMMUNICATION APPLICATIONS IN TELEMEDICINE

Telehealth stands for the integration of information plus communication. The concept of wireless communications or telehealth transformed the way the medical field has been accessed. It exposes the wide spectrum applications ranging from inpatient consultations to remote consultations, from clinical interventions to health education, and many more [21]. The crucial component of telehealth is telemedicine, which involves the delivery of medical services, either emergency or clinical, using telecommunication systems, including both real-time interactions and stored-and-forwarded consultations. Since the pandemic, telemedicine has become an important healthcare practice that works for both emergency and routine checkups. The potential of remote healthcare systems gained its potential due to the advancements in wireless technology, including AI, IoT, or cloud computing [22]. These innovations made RPM an easy task by using sensors, data aggregation units, and many more. These devices detect early signs of disease and also empower patients to manage their routine healthcare at home. It explores RPM, key technological innovations, system components, and ongoing research. This unit has the objective to highlight the impact of telemedicine on global healthcare systems. The key parameters monitored through wireless communication are shown in Figure 11.6.

11.3.1 Technologies enabling remote patient monitoring (RPM)

Wireless Sensors and Wearables: Now, contemporary RPM systems utilize small wearable biosensors that can record data in real time. Gadgets such as smartwatches, adhesive patches, and smart rings collect ongoing physiological information [23].

Internet of Things (IoT): IoT links these devices to each other through a network, which allows the flow of data between patients and healthcare providers over Wi-Fi, Bluetooth, or cellular networks.

Cloud Computing and Data Collection: Data is transmitted to secure cloud systems where it is stored, churned, and made available to medical professionals. Aggregate profiles support trend monitoring and trend analysis of the user's health.

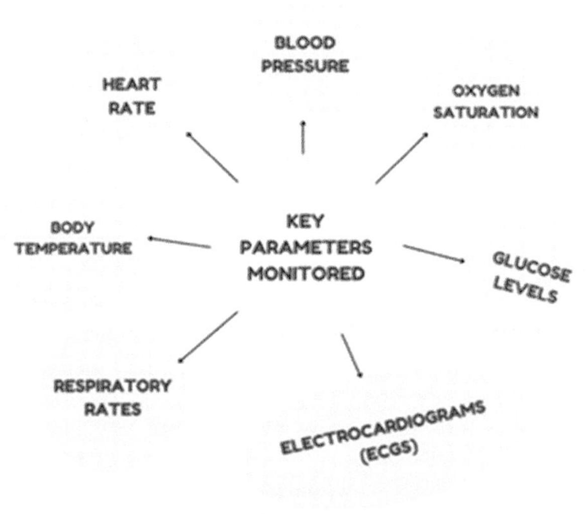

Figure 11.6 Key parameters monitored through wireless communication.

AI and Analytics: Algorithms in AI sort through patient data to identify trends and anomalies.
Predictive Modeling Using ML: To predict adverse events and aid in clinical decision making.
Mobile Apps and User Interfaces: Through intuitive mobile apps and dashboards, patients and healthcare providers engage with RPM systems, improving communication.

Chronic disease management is central to RPM. The approach allows providers to keep an eye on chronic patients, flag any issues early, and avoid high-frequency visits to the hospital. RPM is beneficial for managing diseases such as diabetes, high blood pressure, heart issues, respiratory conditions, and for elderly or post-surgery patient care [24]. For diabetes, one of the most popular is the continuous glucose monitor (CGM): a patch on your skin that measures your blood sugar throughout the day, so you don't have to prick your finger all the time. The devices transmit real-time data to healthcare teams, which can adjust therapies accordingly. This prevents occurrences of dangerously high or low blood sugar and results in better control over the disease. Those with hypertension use wireless blood pressure cuffs that automatically track and send readings. These are tools that enable doctors to see how well medications are working, and to gauge whether other things, such as stress or diet, may be influencing blood pressure. For heart patients, a wearable ECG patch that reads data or a remote, data-connected monitor installed on a pacemaker can continually monitor heart rhythm. The device identifies arrhythmia and irregular heartbeats, which indicate potential stroke or heart attack risks. RPM allows patients to stay at home instead of visiting clinics repeatedly.

Patients who have respiratory diseases, including asthma and COPD, use pulse oximeters together with spirometers. The device tracks blood oxygen levels and sends alerts to caregivers when oxygen levels become dangerous. Spirometers enable doctors to assess lung function, which helps them create individualized treatment plans based on patients' breathing abilities. The early intervention allows us to prevent patients from developing more severe conditions that require hospitalization. RPM is also important in the care of elderly and postoperative patients. Older people often have a number of health problems and need to be carefully and constantly monitored. RPM targets early detection of health decline (due to dehydration, infection, or heart issues, for example) so that they can continue living at home. This helps avoid hospital readmissions and keeps seniors comfortable [25]. Post-surgery patients would use remote checks to make sure they're healing properly, to catch any infections early, and to avoid unnecessary trips back to the hospital.

For maternal health, there are also great benefits of RPM. Pregnant aging women can receive remote fetal heart rate, uterine contractions, and signs of gestational diabetes monitoring. This is particularly beneficial in rural or disadvantaged areas, as there may be limited healthcare access. The remote corners of the world can be monitored through connected devices, which include neonatal monitors to track infant health during their first essential days. The advancement of mental health treatment occurs through the implementation of wearable technology. The combination of wearable tools enables users to track their sleep activities and their stress levels [25]. The collected data enables healthcare teams to identify symptoms of depression, anxiety, and post-traumatic stress disorder. The system provides patients with immediate feedback through its real-time monitoring capabilities. The system enables users to recognize their personal patterns and develop strategies to handle their triggers. The growing implementation of remote monitoring systems enables patients to achieve greater control over their medical conditions. It reduces the burden of frequent clinic visits and allows early detection of problems. With the right devices and technology, long-term health conditions become more manageable, safer, and less disruptive to daily life. To summarize, RPM is changing the methods of healthcare delivery by allowing the elderly and the physically disabled to receive accurate and real-time monitoring through RPM systems, which extends the hands of caregivers farther, improving the hospital dependency ratio while increasing quality of care.

11.3.2 Telehealth services and virtual consultations

Telehealth relates to the comprehensive and increasing use of electronic communication methods for service exchange in health care and sharing information about one's health. It covers a range of services facilitating interaction between users and healthcare service providers without requiring physical consultations. Video conferencing for consultation and diagnostics remotely so that physicians can review the data from the devices, education of patients via the internet, and use of Electronic Health Files or Records (EHRs) are some of the examples. All these systems are interconnected to enhance the convenience and efficiency of healthcare services. The enhanced connectivity of the internet, the increase in the use of mobile phones, and the creation of intelligent medical equipment have significantly impacted the delivery of health care services [26]. Patients can now attend virtual consultations, and the visits are now standard for routine checkups, chronic illness management, and mental healthcare in primary care settings. Patients can now access an attending physician or a subspecialist through Zoom, Doxy.me, and secure hospital portals from their homes or workplaces. The means enable patients to receive care in a qualified manner and with more relative ease compared to having to wait many days or weeks. Further, more patients, including those living in rural and underprivileged regions, can be reached by healthcare providers than was formally possible. The components and modes of virtual healthcare are shown in Table 11.4.

Virtual consultations have benefits for both patients and providers. Patients from remote areas or places with restricted healthcare options have easier access to specialists who otherwise would have taken them hours to reach. Access to healthcare becomes less complicated for patients as the long travel hours and costs associated with it are eliminated. For working patients, virtual consultations provide unrivaled convenience as the time away from work or childcare that needs to be arranged becomes minimal. Telehealth has further expanded the already limited mental health services available. Patients undergoing therapy no longer need to attend in-person sessions and can now meet with their therapists and psychiatrists through video calls, which makes seeking help a lot easier for a lot of patients. This is particularly helpful in reducing stigma and local barriers where there are limited mental health resources. For postoperative care, virtual follow-up consultations have also been a success. Postoperative patients don't have to visit the hospital in person but can show their incision sites, explain their pain levels, and go over postoperative instructions over the devices.

The implementation of telehealth provides vital advantages for both pediatric and elderly patient populations. Through telehealth services, parents receive convenient access to pediatricians from their own homes, which reduces their time commitments and travel inconveniences. Telehealth services benefit senior citizens with physical limitations because they eliminate the need to face the physical and mental obstacles of traveling long distances. Virtual appointments enable caregivers to perform regular health checks without

Table 11.4 Components and modes of virtual healthcare

Components and modes of virtual healthcare	*Key technologies involved*
Live video (Synchronous)	High-speed internet
Store-and-forward (Asynchronous)	Secure video conferencing platforms
Remote patient monitoring (RPM)	Wearable devices and mobile health apps
Mobile health (mHealth)	EHR and data integration tools

interruptions. Yet, some specific barriers remain. Telehealth services do not reach all geographic areas equally [27]. The combination of rural settings and low-income populations in some areas results in either weak or nonexistent internet connections. The digital tools required for telehealth services pose a challenge for most elderly people, thus preventing them from using this healthcare option. The digital storage and transmission of sensitive health data through online platforms creates major concerns regarding privacy and data security because it increases the probability of information breaches. Remember that healthcare systems operate under strict privacy regulations, which include the Health Insurance Portability and Accountability Act (HIPAA) in the United States and the GDPR in Europe, to protect patient information. Patients can experience difficulty in receiving medical treatment across different regions because licensing requirements vary according to the specific situation. The quality of a session can deteriorate due to technical issues like poor internet connections and software malfunctions.

Telehealth services provide major benefits for caregivers who need to support children and elderly people. Through virtual medical consultations, parents have an easy way to discuss their children's health with doctors while avoiding travel and time-consuming tasks. Elderly patients and those with movement limitations find telehealth particularly beneficial because it removes the physical and psychological burdens of lengthy healthcare visits. Caregivers can monitor health conditions more consistently through virtual appointments. The process of accessing telehealth services continues to face various obstacles. Telehealth services remain inaccessible to certain geographic regions. The combination of low-income areas and rural environments results in slow internet.

11.3.3 Electronic Health Files or Records (EHRs) and seamless data exchange

EHR systems function as the electronic equivalent of traditional patient paper charts. Health professionals can easily access and update medical data because electronic health records store essential information in digital form. Electronic health records enable real-time updates while traditional paper charts remain stationary in one location. Whenever healthcare staff members enter new information into the system, it becomes instantly visible to authorized viewers. Electronic health records serve as patient-centered information systems that present full medical backgrounds. The electronic health record system contains specific patient-related information, including treatment histories and current medical conditions, along with vaccination records, allergy information, and diagnostic imaging data. Diagnostic imaging data alongside laboratory test results are also available through the electronic health record. Electronic health records serve as essential tools for maintaining accurate, unbroken medical care. Healthcare operations require a single system to store all collected data. A complete patient medical record enables healthcare providers to stop dangerous medication errors through allergy information access. The full availability of health information enables providers to identify patient health patterns and changes at higher speeds. Doctors who treat patients with fever and cough can access recent test results and prior diagnostic information simultaneously to provide prompt medical intervention [28].

During medical emergencies, EHRs enable doctors to retrieve vital patient information rapidly. EHRs contribute to better communication between healthcare providers. Specialists receive complete access to patient health records through their primary care doctors in real time when patients visit them. The usefulness of these records depends on wireless communication technology. Secure wireless networks enable healthcare workers to exchange information rapidly without using traditional paper files or telephone conversations. The ability to transfer information between doctors and clinics benefits patients who receive treatment from multiple healthcare providers. Emergency room doctors, local clinic staff, and remote hospital personnel utilize wireless data exchange to receive identical real-time medical information [29]. The use of wireless data exchange enhances both the rapid and effective delivery of healthcare, especially in situations that need immediate attention. Emergency room staff can use wireless systems to obtain patient allergy information and recent scan results if the patient is unconscious. This technology can help medical professionals save lives through faster emergency treatment.

Telemedicine receives support from wireless data exchange systems. The practice of consulting doctors through video calls from remote or rural areas has become common for many patients. Through secure and fast wireless connections, medical professionals can access EHRs when they conduct virtual patient visits. As technology advances, more healthcare facilities are adopting wireless systems. The goal is to make patient data available instantly and securely, no matter where care takes place. Experts believe that these wireless systems will continue to grow and become a standard part of healthcare.

11.4 WIRELESS COMMUNICATION AND TELEMEDICINE'S IMPLICATIONS IN HEALTHCARE

During medical emergencies, EHRs enable doctors to retrieve vital patient information rapidly. EHRs contribute to better communication between healthcare providers. Specialists receive complete access to patient health records through their primary care doctors in real time when patients visit them. The usefulness of these records depends on wireless communication technology. Secure wireless networks enable healthcare workers to exchange information rapidly without using traditional paper files or telephone conversations. The ability to transfer information between doctors and clinics benefits patients who receive treatment from multiple healthcare providers. Emergency room doctors, local clinic staff, and remote hospital personnel utilize wireless data exchange to receive identical real-time medical information [30].

The use of wireless data exchange enhances both the rapid and effective delivery of healthcare, especially in situations that need immediate attention. Emergency room staff can use wireless systems to obtain patient allergy information and recent scan results if the patient is unconscious. This technology can help medical professionals save lives through faster emergency treatment. Telemedicine receives support from wireless data exchange systems. The practice of consulting doctors through video calls from remote or rural areas has become common for many patients. Through secure and fast wireless connections, medical professionals can access EHRs when they conduct virtual patient visits. With real-time monitoring devices plus those wearable health-tracking devices and remote diagnostic tools, clinicians can intervene much earlier, plan treatments more fully aligned to their patients, and always monitor the progress of their patients. Hence, it always facilitates disease management, which in turn, keeps the patient from going back to the hospital, thus improving eventual welfare [31]. Going forward, the chapter will discuss each of these implications; hence, it will deal with how wireless communication and telemedicine enhance accessibility, improve efficiency, and foster better patient outcomes worldwide.

11.4.1 Expanded accessibility

Figure 11.7 shows the impact of telemedicine on health outcome metrics. Access to quality healthcare remains a challenge faced by all humans across the globe, especially in the remote and rural areas [32]. Remote areas in particular suffer from a shortage of trained medical professionals, transportation problems, with poor cooperating infrastructure, and an

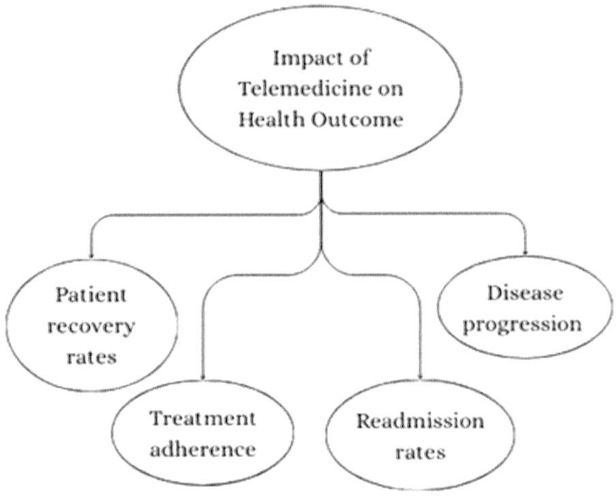

Figure 11.7 Impact of telemedicine on health outcome metrics.

overall low level of awareness about health services, thereby creating a stagnant situation. In consequence, millions remain underserved leading to delayed diagnosis, treatable diseases, and poor health outcomes [33]. The establishment of wireless communication technologies to further telemedicine has carved a revolutionary solution to this problem. Telemedicine strives to lift medical services above spatial distances so that those from far away can seek specialist consultation and timely intervention without having to take extremely long journeys to the treatment centers.

The concept of expanded accessibility goes beyond geographical boundaries and includes social and economic restrictions as well as physical limitations. The specific groups that benefit from telemedicine include the following: elderly people who have difficulty moving around, women living in traditional communities who experience cultural limitations when going to medical facilities, people with disabilities who encounter problems with reaching hospital facilities, and families with low income who lack the resources to cover transportation expenses and time away from work [34].

Multilingual and Mobile Solutions: One of the biggest challenges in remote areas is language and digital literacy. But modern telemedicine solutions are addressing this with: multilingual interfaces (Hindi, Kannada, Marathi, etc.), voice enabled health apps for illiterate or visually impaired users, and mHealth vans with wireless diagnostics and internet. Through user-friendly platforms even patients who are not tech savvy can get health services with the help of local coordinators or health aides [35]. For example, a mobile telehealth van in rural Maharashtra can use satellite internet and AI powered tools to connect a patient with a pediatrician in Mumbai so that the child can get the right treatment on time. The impact on emergency and chronic care is shown in Table 11.5.

Table 11.5 Impact on emergency and chronic care

Impact on emergency care	*Impact on chronic care*
Pre-hospital triage through mobile apps	Regular virtual follow ups
Remote monitoring of vitals by emergency medical teams	Remote prescription renewals
Coordination with tertiary care centers before patient arrival	Health monitoring through wearables

11.4.2 Efficiency and expense reduction

Telemedicine, powered by wireless technology, is the backbone of modern healthcare. One of the biggest benefits is the economic gains [36]. By cutting down on extra costs, making smarter use of resources, and running things more smoothly, telemedicine proves to be a sustainable approach for patients, healthcare providers, and governments alike.

Overhead and Operational Costs: Telemedicine provides healthcare organizations with substantial cost savings because it reduces their operational expenses. Healthcare delivery through traditional methods requires organizations to spend large amounts on building facilities, paying for utilities, and maintaining staff who work directly at the location. Telemedicine platforms provide users with several advantages including: the virtual nature of consultations through telemedicine enables healthcare providers to minimize their need for physical facilities and reduce their requirement for clinic expansion, the implementation of digital tools for appointment scheduling and record management and follow-up care reduces the need for administrative staff, healthcare facilities require less energy consumption and maintenance work because they receive fewer patients through physical visits. The 2022 AMA study demonstrated that hospitals which implemented telehealth solutions achieved annual savings between 15% and 20% for their non-clinical expenses.

Cost Savings for Patients and Insurance Providers: Telemedicine provides patients with simplified healthcare access at reduced expenses. People who reside in distant locations no longer need to spend money on doctor visits because telemedicine services are available. The need to purchase fuel for cars and public transportation and taxi services has become unnecessary. Patients who require hospital stay for medical treatment can remain at the hospital without needing to pay for accommodations. Virtual doctor visits eliminate the need for patients to miss work hours. The flexible appointment scheduling through telemedicine enables patients to maintain their work schedule while reducing healthcare expenses. Telemedicine technology decreases the number of emergency room visits that patients need to make. And on top of that, telemedicine reduces ER trips. Your doctor can monitor your health through tracking devices and scheduled check-ins which enables early detection of medical issues before they develop into major problems.

Early disease detection at an early stage leads to reduced complications which results in decreased emergency room visits that save money while extending patient health duration. Patients with diabetes and heart problems experience a breakthrough when they receive immediate medical guidance and small treatment adjustments without needing to visit the emergency room. Insurance companies receive financial advantages because patients who receive proper care for their conditions require less expensive large-scale medical payments. The combination of preventive care with ongoing monitoring helps decrease both hospital expenses and expensive medical procedures which results in reduced total healthcare expenses. The US healthcare system will achieve $250 billion annual savings through telehealth chronic care services according to McKinsey & Company research. The system generates financial benefits for all users.

Enhancing Workforce Efficiency: Healthcare providers have completely adopted the telehealth revolution which has evolved into a new medical practice. Telehealth technology gives medical staff expanded scheduling options for their work activities. Physicians now have the ability to meet their patients at home during evening and weekend hours. The flexible nature of telehealth practice allows doctors to achieve better work–life balance, which decreases their risk of burnout. Telehealth systems automate various tasks which healthcare providers need to perform repeatedly. The integration of telemedicine systems with electronic health record systems enables healthcare providers to complete paperwork and billing tasks more efficiently and accurately. The system enables healthcare providers to dedicate their time to patient care instead of administrative work which decreases the risk of medical mistakes. Healthcare organizations have developed advanced scheduling systems for their operations. Telehealth teams use staggered shift schedules to increase patient access without requiring health systems to add more staff members. Some medical facilities have extended their operating hours to include weekend shifts, which better serves patients who need care during their preferred time. The system will enhance healthcare service availability for patients who live in areas with restricted medical facilities.

11.4.3 Improved healthcare outcomes

Telemedicine has brought healthcare into a new era, which delivers more than basic convenience and cost reduction because it produces better medical results. Telemedicine delivers healthcare services through wireless technology, which enables real-time communication and remote diagnostics and personalized data-driven therapy to provide patients with timely and dependable care. The following section explains how telemedicine enhances patient health results through ongoing tracking and immediate medical responses, and forecasting healthcare needs for enhanced treatment results, superior life quality, and equal healthcare access.

Real-Time and Timely Intervention: One of the most advantageous aspects of telemedicine is the ability to provide real-time interventions, particularly in the event of an emergency or a deteriorating chronic condition: the system provides continuous remote monitoring through devices which include wearable heart monitors and glucose sensors that detect patient health status changes to alert both patients and their healthcare providers instantly. The system enables healthcare providers to make fast medical choices through immediate data access, which allows them to activate remote specialist support when needed.

Less Time from Symptoms to Diagnosis: Digital health consults, a telemedicine service, minimize the time from symptoms to diagnosis and thereby reduce the risk of disease progression. A 2023 World Health Organization (WHO) report noted that timely interventions via telemedicine for stroke and cardiac cases produced 20%–30% lower mortality rates at pilot hospitals in Asia and Europe.

Personalized Care Plans: Telehealth systems leverage AI, ML, and big data analytics to deliver personalized care plans, incorporating data from genomics and lifestyle.

Post-Acute Care and Rehabilitation: The patients usually experience challenges during the recovery process when they end up in Intensive care units as extended stay, or recovering from surgical procedures. Virtual follow-up and rehab have greatly benefited these patients during their recovery. Taking the process home allows the physical and occupational therapist to lead the patient's therapy with distant interventions. Virtual physical rehabilitation minimizes patients' travelling to a facility or a clinic to receive one-on-one, in-person therapy; it saves time and effort. Most patients feel more comfortable doing their exercises at home, and completing the rehab opens the door for more consistent therapy. Remote guidance also allows caregivers to assist and record the progress. Additionally, the ability to reduce hospital readmissions is a significant advantage of virtual care. When the level of care is constant and provided through remote oversight, health care professionals are likely to identify complications at their earliest onset [37].

Healthcare wearables (or even just routine virtual check-in) promote an early detection of infections, potential blood clots, or other problem signs or symptoms, which can result in a readmission. Early detection limits these smaller issues from becoming larger ones and reduces the chances of a readmission. Surveillance studies determined that remote monitoring significantly reduces the incidence of postoperative complications and reduces hospital admissions. Patients recovering from orthopedic surgeries and cardiac surgeries saw the best results. Telemedicine has an important place within recovery care. It helps to make rehabilitation accessible, adaptable, and timely. Telecare allows patients to rehabilitate safely while still being able to draw on expertise and support. This provides less physical pressure and emotional burden for patients and can facilitate timely healing on their way to recovery.

Equity in Health Outcomes: Telemedicine also changes how healthcare comes to different communities. It helps to reduce barriers of distance, income, and language. Many rural or remote communities do not have access to specialists due to a history of denying them access to care. People traveled distances for routine care. Telemedicine connects these communities and allows patients access to healthcare. Virtual visits within telemedicine allow for the populations in rural or remote areas to access expert advice simply and quickly without travel. Language and culture can also create limits to care accessibility. Telehealth platforms provide interpreter services and culturally appropriate services. For instance, a Spanish-speaking patient can access a bilingual doctor or interpreter through their appointment. This improves communication between the patient and provider and builds trust. Another benefit is cost savings for patients. Regular care via telehealth does reduce the cost of travel, including child care and time missed from work. Patients who may have avoided care due to these costs can pursue care more easily. The greater access means the health outcomes do not depend mainly on a person's income or geography. It creates a more equal opportunity for every person to receive care.

11.5 CHALLENGES AND LIMITATIONS TOWARD WIDESPREAD ADOPTION

The implementation of telemedicine and wireless communication in healthcare faces numerous obstacles that prevent its universal adoption and complete integration into healthcare practice. Telehealth systems promise cost savings, enhanced access, and better results, but these advantages face substantial resistance from technological, regulatory, and social factors. The protection of patient information privacy stands as the main security concern. The implementation of wire networks and cloud systems for healthcare delivery created an urgent need to protect electronic health records from unauthorized access and system breaches. Service providers and patients require proof that their submitted information will stay confidential and fully compliant with security standards. The technical obstacles related to interoperability represent major challenges for healthcare systems. Standardized communication protocols, compatibility of the software, and integration building blocks for medical devices and platforms are still absent, leading to inefficiency and ineffectiveness of telemedicine systems. In the absence of effortless data passing between different healthcare systems, care continuity and joined-up responses are always limited.

In addition, the digital divide also widens the differences with regard to accessing telehealth services. Lack of internet access, digital literacy, and smart devices are disparately experienced by poor populations, especially in rural and remote areas. By doing so, the very promise of equitable and inclusive healthcare is at risk of being undermined.

Under this section, we explore these fundamental limits to this technology – security and privacy risks, interoperability voids, and digital inequality – which require robust approaches, accommodative policies, and modern strategies to unlock the telemedicine potential for all individuals.

11.5.1 Concerns regarding security and privacy

With the emergence of telemedicine and wireless health systems, transmission and storage of patient data have been moved from physical files to digitized records and cloud storage and wireless networks. This change has made efficient and real-time delivery of care possible while introducing massive vulnerabilities. Keeping confidential medical information safe is non-negotiable anymore [38]. It is a necessity that is both legal and ethical as well as functional. Healthcare providers need to tackle the data breaches and unauthorized access, as well as issues with data integrity and regulatory compliance, in a world where cyberattacks are becoming more and more complex. In telemedicine, a wide range of patient information is collected, processed, and transmitted, often over open or semi-secure networks. The data types at risk in telemedicine include personal identifiers, medical history and diagnostic results, prescriptions and treatment plans, billing and insurance information, and real-time sensor data. Figure 11.8 represents the key threats to security and privacy. It explains about different types of threats that affect people's privacy.

In order to preserve data privacy, various regulations and frameworks are applied in several parts of the world.

i. *HIPAA (USA)*: The HIPAA enforces the use of administrative, physical, and technical safeguards by healthcare providers.
ii. *GDPR (EU)*: GDPR calls for severe data protection rules, including those involving cross-border data transfer.
iii. *DISHA (India – Draft)*: The Digital Information Security in Healthcare Act is the act that intends to control the digital health data and secure it in terms of its confidentiality, integrity, and availability.

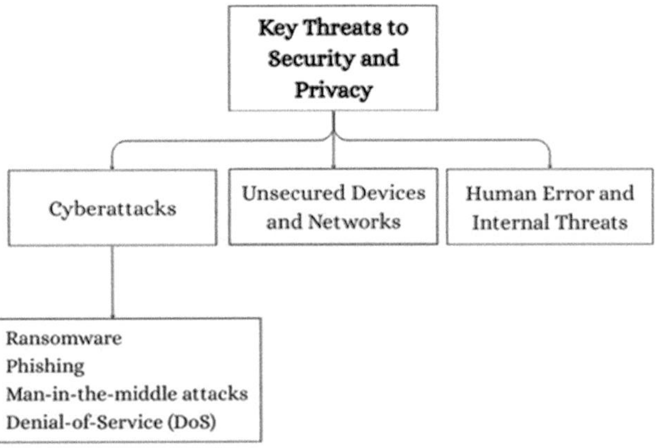

Figure 11.8 Key threats to security and privacy.

Although these laws do provide a legal structure, it is still difficult to enforce and align technology in the real world. To curb risks and promote a safe implementation of wireless communication in healthcare, the following practices are important:

i. Data Encryption: Measures such as implementing end-to-end encryption will make sure the patient data is encrypted in transit as well as in storage, even if intercepted.
ii. Multi-Factor Authentication (MFA): Adding layers of identity verification increases the access controls significantly, particularly for remote access by healthcare providers.
iii. Secure APIs and Protocols: Telehealth platforms must rely on standardized secure APIs and protocols for communication (e.g., HTTPS, TLS, HL7 FHIR).
iv. Scheduled Security Audits and Updates: Regular vulnerability assessment and penetration testing, and software updates, help to uncover loopholes in the system and take corrective measures.
v. User Education and Training: Both the suppliers and the patients should be educated on the best practices, such as strong passwords, identification of phishing emails, and safety of unverified platforms [39].

The fact that telemedicine and wireless health systems have become a common reality promises enormous potential – which can only be realized if patient data becomes secure. The protection of private data and cybersecurity requires multiple stakeholders to take responsibility because it involves more than technological solutions. Into the future, there has to be a balancing act between innovation and protection, where convenience and accessibility will not trade off ethics and patient rights. Funding security-by-design frameworks, fostering the world's cooperation, and stimulating interdisciplinary research would be imperative to create a future in which telehealth is helpful and credible.

11.5.2 Interoperability issues

Interoperability also defines the capability of different information-based devices, systems, and applications to access, exchange, coordinate, and integrate pieces of data between and within the boundaries of an organization. In the field of healthcare, this ability is extremely important for the effective and timely delivery of seamless care, especially against the background of telemedicine and wireless communication. But since there are no common standards and procedures in platforms, fragmentation, data silos, and communication impasses occur.

The lack of interoperability causes a number of negative consequences, including sysops plagiarizing sysops and sysops stealing sysops [40]. Delayed diagnoses: failing to get a patient's full history compromises on clinical decision, redundant tests: frequent repeating of the tests because of inaccessible prior results increases costs and burden to patients, fragmented care: there is no continuity of care when the patients transfer between different providers or health care systems, inefficiency: time and effort wasted in reconciliation or re-entry of data [41]. In situations that are remote and when emergencies arise, where prompt access to detailed information is essential, these interoperability problems may cause a threat to life. The core challenges of interoperability are a lack of standardized protocols, incompatible medical devices, data silos, and vendor lock-in. Case examples are discussed below:

EHR Integration Failure: In a multi-hospital network, patient records were not accessible in the main urban hospital from the rural clinic because they were working with different EHR vendors. Therefore, allergies to medication were overlooked during the emergency operation: the complication ensued.

Device Data Incompatibility: When a diabetic patient tried to sync glucose readings from a home glucose monitor with their doctor's dashboard – they found they could not do so, since the device lacked API compatibility. This led to the manual input of data, which created more chances of mistakes.

Strategies for Improvement [42] Are Adoption of Global Standards: (a) promoting the standard adoption of HL7 FHIR, SNOMED CT, LOINC, and ICD-10 for all exchanges of data, (b) APIs and open source platforms: use of open APIs for third-party tools and apps to be able to connect to EHRs and devices securely, (c) health information exchanges (HIEs): development of regional or national health information exchanges to allow for effortless data movement, (d) government regulation and incentives: the drafting of the regulatory guidelines, audit of compliance and financial incentives to put into use interoperable solutions, (e) vendor collaboration: creating collaboration and interoperability commitment incentive to medical technology vendors.

AI and middleware platforms can make data formats normal, translate systems to each other, and cover the semantic gaps. These tools are bridges, allowing an automated integration process between disparate systems while simplifying the manual workload. For example, AI-based clinical decision-making systems that extract structured data from non-coherent EHRs and middleware that combines sensor data of various wearables and normalizes it into a universal form. In order for wireless communication and telemedicine to achieve their optimal potential, interoperability is a given. The solution to these problems needs more than technological advancements because it requires new policies and better cross-sectoral collaboration and healthcare services that focus on patient needs and patients. A truly interoperable health system may be able to facilitate: continuity of care across providers, personalized and data-driven interventions, real-time emergency response, patients empowered and active health management, and investments in interoperability – now – will decide the quality, availability, and sustainability of healthcare systems of tomorrow.

11.5.3 Digital divide and access to technology

Digital divide is the term that describes how differently equipped individuals or communities are with modern ICT (information and communication technology). In the world of telemedicine and wireless healthcare, this separation is capable of creating enormous disparities in healthcare distribution, reach, and results. This challenge looms larger when tech-dependent health services are championed without resolving the infrastructural, economic, and digital literacy inequality. Telemedicine has the ability to revolutionize access only if the fundamentals of access – the internet, devices, power, and literacy – are built [43].

Issues of the digital divide are geographical disparities, socioeconomic inequality, educational and digital literacy gaps, languages, and cultural barriers. Sometimes, rural and remote areas lack proper services for the internet. Inadequate coverage offered by cellular networks and high latency affect the use of telehealth services. Erratic power supply limits the devices' use and charging capacity even more. Low-income groups find it difficult to pay for smartphones, tablets, etc., or for data plans. The families could prioritize the need for essentials over technology, thus having minimal gadget ownership. Poor financial status also hinders access to digital health insurance platforms or subscription-based health apps. Several individuals do not know how to use a telehealth platform, an online consultation tool, or a digital health record. Older patients or patients who haven't completed their formal education may not use or misuse telemedicine services. Lack of digital confidence results in decreased adoption and engagement. Interfaces or health instructions in a non-native language make accessibility hard. Use is restricted by cultural taboos or a lack of awareness about the remote healthcare options.

The digital divide, or the gap in access to digital technology, could be addressed by a great number of strategies. Improving infrastructure is one of the significant techniques developed. This includes collaborations between governments and private firms that aim at bringing high-speed internet – including broadband and 5G, into areas that do not have such services. The other effective approach is the provision of subsidies toward affordable technology. This is offering low-priced or no charges on smartphones, tablets, and internet programs for poor families. Cards like the Lifeline Program in the US assist in defraying the costs for other people to be able to connect to online health resources, as well as telemedicine services.

Digital literacy programs are also very important. Such initiatives teach patients how to use their devices and become familiar with about navigating health apps, telemedicine systems, and electronic health records (EHR). For example, various community centers organize classes that teach how to order prescription refills online or schedule virtual doctor visits for older people. When an individual learns how to use such tools confidently, he/she becomes more self-managing for the purpose of health. It is also important to develop simple platforms for all groups to use. The culturally adaptive apps take into consideration the language, norms of a given culture, and literacy levels. Many health apps come along with several language choices, audio instructions, big icons, and easy instructions. This facilitates the process of acquiring healthcare for users with low literacy or technology inexperience.

Other international organizations, such as the WHO and the UNICEF, are also supporters of the projects related to community-based digital health. Such efforts target countries where access is poor as a result of bad infrastructure or high levels of poverty. For instance, air access for all people regardless of UNICEF has been funding telehealth programs in Africa and enabling rural clinics to reach out to doctors overseas. This wide support is the key to designing fair access for all people regardless of the place where they live. All these strategies are aimed at establishing that digital health services can be accessed, used, and affordable to any person anywhere. The authors recognize that technology by itself does not resolve the existing digital gap. The combination of support services with training programs and culturally appropriate tools creates meaningful value. The combination of these elements enables more people to access healthcare services through digital platforms. The digital health divide requires more than device and internet distribution because it needs systematic solutions to create equal healthcare access for all. A health system needs digital equality, which protects all people from exclusion because of their location, financial status, reading abilities, or age. Telemedicine presents an attractive future, but its success depends on creating equal access for everyone to use this technology. The system will achieve democratic, efficient, and sustainable healthcare transformation through wireless communication only when all people have equal access.

11.6 EXPECTATIONS AND TRENDS FOR THE NEXT GENERATION

The combination of new technologies with telecommunication progressions creates conditions for healthcare to undergo both development and transformation. The current telemedicine extensions will not satisfy the complete potential of future telemedicine systems. The future of healthcare delivery will transform globally through telemedicine, which will create individualized care access for all patients. The combination of 6G networks with neural networks and ML algorithms will enhance remote healthcare system performance and speed, which will create an efficient and natural healthcare experience. The transforming healthcare environment provides us with unrestricted medical knowledge and services through enhanced data transmission and unified healthcare systems. The digital health infrastructure development enables healthcare access for all patients regardless of their location, between rural areas and urban centers. Telemedicine evolves from its current reactive function into a system that delivers preventive, predictive, and personalized healthcare services through data-based healthcare systems. The healthcare system of tomorrow will deliver individualized care through real-time data analysis and AI technology and wearable devices, which will result in improved diagnostic accuracy, earlier treatment, and superior patient outcomes. This section examines three active trends that describe wireless technology advancement and healthcare information system development, and individualized care models that transform telehealth into a core healthcare service for modern, equitable healthcare delivery.

11.6.1 Improvements in wireless technology

Telemedicine has grown with each generation of wireless communication, which saw the transition from 3G, which enabled basic remote consultations, to 5G, which supported real-time diagnostics and high-resolution medical imaging [44]. As we move into the 6G era, and we incorporate neural networks and ML, we see the growth of telemedicine at an exponential rate. In this new age of tech, which telemedicine is entering, we will see very precise, intelligent, and responsive health care services, which in turn will do away with the gap between patient and health care provider.

Soaring on the Promise of 6G – for Telemedicine is expected to appear commercially around 2030 and is meant to be about more than just a faster network. It promises 1000 Gbps of data transmission speed, enabling ultra-definition remote surgeries and diagnosis. Less than 1 ms latency providing real-time decision-making in life-threatening situations. Edge AI incorporation, putting AI closer to the edge of the network for real-time diagnosis [45]. For telemedicine, that means the potential for more seamless robotic-based surgeries from a distance, for high-definition video consultations with no lag, and for the possibility of real-time patient monitoring even in data-rich settings such as ICUs.

ML is an important mechanism to bring intelligence and improvement in telemedicine. It provides an essential layer of intelligence that enables providers to better understand the needs and respond to the needs of patients. One component of ML is predictive analytics. These algorithms sift through huge quantities of past and current health data to predict how a disease might progress. The system uses ML to analyze patient breathing data and other signals, which enables it to forecast asthma attacks before they happen. The early detection of medical problems enables doctors to send alerts to patients while they make necessary changes to their treatment plans. The preventive approach would protect patients from fatal outcomes while preventing them from needing emergency medical care. The process of risk-based patient classification plays a crucial role in medical practice. The system uses ML to identify patients who require immediate medical attention based on their treatment urgency. The combination of medical history data with laboratory test results and personal lifestyle information helps doctors identify patients who need urgent medical care. Doctors would dedicate additional time to patients who show high-risk factors, which would prevent any patient from receiving inadequate care. For example, an ML system would identify candidate patients with diabetes who have impending signs of complications so that intervention could be made at an early stage. In this way, targeted care would provide for better utility of scarce healthcare resources, both in crammed clinics and while serving huge populations.

Automation can be developed for clinical rehabilitation and referrals. ML may attend to routine tasks such as appointment booking, prescription ordering, and inputting data into patient records. Hence, many administrative duties would be taken off the physicians' and nurses' hands, thereby freeing up precious time for pure patient care. To give an example, the ML application would send refill reminders and confirm appointment times without any human intervention [46]. These systems are literally never asleep and can gladly take thousands of requests in parallel, thus accomplishing the administrative processes much faster with fewer mistakes to be accounted for. Millions of smart features keep functioning on wireless-enabled wearables and cloud-based platforms! For instance, these wearables continuously collect a person's vital signs and health status 24×7. This data from the device is instantaneously processed by the ML algorithms in the cloud platforms. The system observes each patient's unique pattern of health and adjusts its own responses as more health information comes into the picture. The system operates through an endless learning process that delivers personalized alerts and recommendations to users. The healthcare system becomes more digital through these features, which deliver personalized care to patients, resulting in enhanced medical outcomes and better quality of life.

The proposed system faces multiple challenges during its implementation process. The system requires a powerful network infrastructure to transmit high-definition diagnostic data and continuous critical signs information at high bandwidth rates. The system requires advanced encryption methods to protect sensitive health data because wireless systems need to prevent security breaches. The solutions work to achieve better integration while providing enhanced security measures. The upcoming 6G network will unite with neural networks and ML to create autonomous telehealth systems, which will function as intelligent healthcare platforms. The devices will develop self-learning abilities to enhance their performance through automated learning processes without requiring human supervision. The wireless technology sector will focus on creating standardized systems, which will enable seamless telemedicine services between different countries for international patient care. Healthcare providers will work alongside AI systems to create better treatment plans and delivery methods through collaborative partnerships. Wireless communication development will enhance telemedicine services while transforming global healthcare delivery through connected and intelligent medical care systems.

11.6.2 Global healthcare integration

Health networks enable doctors to access patient records through secure online storage systems, which provide continuous medical care when patients relocate or travel between different countries. Wireless networks enable emergency response teams to establish immediate communication with worldwide teams during pandemics and natural disasters. The development of new healthcare technologies enables high-quality medical services to reach all parts of the world. Cloud computing and global health databases operate as a transformative system for medical information management of patient information. Medical staff worldwide can access cloud-based Electronic Health Files or Records (EHRs) through authorized access, which provides patients with continuous access to their medical information. Medical AI receives better training through global datasets because these databases contain diverse symptoms, languages, and medical conditions. Medical researchers from various countries can work together to conduct extensive clinical research and monitor emerging diseases through real-time monitoring. Cloud service integration enables healthcare organizations to expand their services while maintaining high reliability for delivering better medical care to all patients.

Healthcare operations today require maximum data protection because everything exists in a highly interconnected system. Blockchain technology enables secure protection of medical records through tamper-proof authentication, which allows patients to manage their personal data access permissions regardless of their location or treatment organization. The system protects against fraud because it prevents unauthorized changes to medical records, which is crucial for international health insurance claim processing. The development of new technology creates conditions for establishing a unified worldwide healthcare system with seamless operations. A worldwide health ID system combined with AI and mobile data monitoring enables borderless disease outbreak detection. Virtual hospitals operate continuously online through virtual platforms that connect doctors and specialists from worldwide locations. The combination of international cooperation and technological progress could lead to a future healthcare system that delivers equal access to medical services for all people across the globe.

Health organizations and medical facilities rapidly increased their service capacity to handle the rising number of patients during the pandemic. The pandemic created a surge in telehealth service usage because more people began to use this technology. The adoption rate of telemedicine services differs between locations because certain areas adopted the technology at different speeds. The adoption of telemedicine depends on multiple factors, which include internet speed, government policies, and patient comfort with digital health tools, thus creating significant differences between different regions [46].

High Adoption Regions: The pandemic brought telemedicine to become a major healthcare trend throughout North America, especially in the United States and Canada. The number of doctors in the United States who provide virtual care has reached 86% since 2019, when only 15% of doctors offered this service. The rapid growth of remote healthcare services demonstrates how essential virtual care became during the pandemic years. Teladoc and Amwell became leading platforms while numerous clinics started using video calls and online chats as standard healthcare options for patient care. The CDC reported that telemedicine services became standard practice for specific healthcare services because of their rapid adoption during the pandemic. The UK and Germany, along with other European countries, implemented substantial changes to their healthcare systems.

Moderate Adoption Regions: Telehealth services received expanded support from European countries, which integrated them into their national healthcare infrastructure. The UK NHS expanded its virtual appointment services and introduced digital support resources, while Germany established new laws to support digital healthcare development. The European regions adopted telemedicine at different rates because urban areas with reliable internet access adapted quickly, but rural communities struggled with slow internet speeds and technological barriers among their elderly population.

In South America, for example, Brazil and Chile launched new telehealth programs to streamline access to care. Brazil's government expanded online health services to reach the slums, and particularly Brazil's isolated indigenous population, and Chile used telemedicine to keep treating patients even as the country was under lockdown. Even so, a gulf exists between rich, urban areas with fast internet, and rural and poor areas, for whom internet is a luxury. Telemedicine becomes increasingly less accessible as the distances and lack of infrastructure increase.

Low Adoption Regions: The slowest uptick in accepting telemedicine has been in Africa. There are many countries where even the most basic internet – most notably in rural zones. Without access to consistent internet or adequate numbers of smartphones and computers, telehealth is out of reach for many people. So far, there are relatively few laws governing the way in which telemedicine should and should not work, which creates legal roadblocks. In many countries, there are just a few pockets of telemedicine, typically in large cities or special programs. Wide use is still challenging.

Parts of Asia, including Pakistan and Bangladesh, grapple with similar challenges. Internet access is spotty, and many people do not have the devices required for remote health visits. These countries' telemedicine utilization is estimated to be at the extremely low value of 0.1%. Researchers suggest that healthcare professionals in these three countries have so far been using telemedicine hesitantly. Unless connectivity is drastically improved in these countries and supportive laws are in place, telehealth will remain hard to roll out in large swathes of these nations. Like a patchwork quilt, the global pattern of telemedicine has seen rapid rise during the pandemic. More affluent and better-connected areas raced to embrace digital health care. Others, particularly poorer countries, still have a long way to go. The future of telehealth is in developing infrastructure, setting up enabling policy initiatives, and showing its true value through the benefits that it could generate. This would mean even greater access to remote care on a global scale.

11.6.3 The potential of data-driven, individualized healthcare

Generalized treatment outlines have been heavily relied upon by traditional health care approaches. These plans are also made based on what seems to help the majority of new patients with a common condition. The researchers studied general responses instead of individual variations between participants. A doctor uses the same prescription for treating high blood pressure patients as if it would work for everyone. The method produces occasional success, but it fails to recognize essential details about each individual candidate.

Modern technology has started to transform this approach. Smartwatches and sensors operating through wireless technology enable doctors to track patient health conditions in real time. The devices track various health indicators, which include body temperature, movement patterns, sleep patterns, heart rate, and breathing rate. The advanced data processing systems operate at high speed when handling extensive amounts of information. Doctors obtain complete patient health information through this system at any point in time. The healthcare industry faces an unprecedented transformation because big data has taken over its operations. The system enables healthcare professionals to obtain health information from millions of people [46]. Wearable devices enable researchers to identify health patterns that were previously undetectable through data analysis. The analyzed data undergoes processing through AI and ML systems. The tools identify subtle patterns in patient data to forecast their reactions to different medical treatments.

Wearable sensors serve as essential components for this system. The sensors deliver ongoing live feedback, which enables doctors and patients to track their health status. The system detects warning signs of medical issues through its ability to identify irregular heartbeats and increasing blood sugar levels. Patients can track their health through mHealth applications, which run on their smartphones. The information collected from these tools enables developers to create new algorithms that identify health risks before patients show any symptoms. The healthcare industry has transitioned to individualized care delivery because recent medical discoveries support this approach. Research indicates that medical treatment which matches individual patient requirements produces superior therapeutic results. The method eliminates the extensive testing process that medical professionals currently perform. Medical professionals create individualized treatment plans for their patients through their unique patient profiles. Most experts agree this is a big deal. The combination of data analysis with personalized care approaches enables healthcare providers to deliver better results while reducing costs. Patients achieve enhanced control over their medical care through this system. Patients receive individualized solutions that help them understand their specific treatment needs. The transition from standardized care to individualized treatment represents the beginning of an exciting new era. The advancement of wireless technology, data analytics, and emerging tools drives this transformation. The approach delivers improved medical results together with reduced side effects and individualized treatments for each patient, making this an achievement in healthcare services.

Data-driven healthcare means the system collects continuous data through wearables, remote monitoring systems, mHealth applications, and Electronic Health Files or Records (EHRs). The system uses AI/ML to analyze integrated data for pattern detection, anomaly identification, and trend recognition. The system generates personalized recommendations through analysis of patient medical records, life patterns, and current biological measurements. The system generates continuous feedback because increased data collection produces better insights, which result in enhanced medical diagnosis, prevention, and treatment methods.

Healthcare organizations can transition from treating patients to preventing diseases through personalized data analysis. The system includes three main features. The system uses personal risk factors to schedule preventive screenings for patients. The system provides individualized lifestyle guidance, which helps patients enhance their eating habits, physical activity, and stress control abilities. The system delivers specific vaccination and prophylaxis treatments to patients based on their exposure risks and immune system status. The healthcare system experiences decreased workload because patients receive preventive care that stops them from needing hospitalization and developing chronic diseases.

Genomic data integration into personal healthcare plans enables precision medicine to treat oncology patients and rare disease cases. Wireless platforms provide users with immediate access to genomic information while allowing them to consult genetic counselors remotely and receive drug treatment plans based on their pharmacogenetic profile. The decreasing cost of sequencing technology allows medical professionals to use DNA data for routine patient care, which results in personalized medical treatment. The combination of predictive analytics with telemedicine enables healthcare providers to create individualized treatment plans while performing continuous patient monitoring through a complete proactive healthcare system. Patients will transition from receiving passive medical care to taking an active role in their health through intelligent systems that deliver customized treatment approaches. The combination of wireless communication advancements with universal healthcare access and customized medical care will revolutionize the entire healthcare industry. The upcoming telemedicine system will focus on delivering accessible, efficient, patient-centered care to all parts of the world.

11.7 CONCLUSION

The telemedicine revolution represents a major achievement in healthcare delivery because it unites advanced wireless communication systems with the global requirement for accessible healthcare services at affordable prices. The chapter demonstrates that telemedicine functions as an essential component for building enduring healthcare systems instead of serving as a temporary solution for emergencies. Telemedicine started in areas with limited access to healthcare before the COVID-19 pandemic, and it was used to establish fast diagnosis times, improve chronic disease management, and create enduring doctor–patient relationships. The foundation of telehealth infrastructure depends on wireless communication systems, which include Wi-Fi and Bluetooth, and 5G and 6G networks for conducting real-time consultations and monitoring patient health through wearable devices. The practice of medicine undergoes a cultural transformation through wireless communication technology integration because healthcare providers must adapt to new medical practices. Telemedicine functions as a permanent solution that scales up to meet the needs of healthcare systems that face staff shortages and treat increasing numbers of chronic disease patients and aging populations. The global health initiative for UHC and SDG 3 recognizes digital health as a fundamental tool to achieve health equity and improve population well-being. The introduction of new technologies requires the development of solutions to protect personal data privacy and ensure digital literacy and proper regulatory frameworks for safe and equitable practice. Telehealth solutions will achieve expanded reach and better effectiveness through the integration of AI and blockchain security and immersive virtual technology, including augmented reality and digital twins, during the upcoming decades. Telemedicine represents a permanent healthcare delivery system that requires sustained research and innovative approaches and international governance to achieve worldwide implementation.

REFERENCES

1. Adeghe, E. P., Okolo, C. A., and Ojeyinka, O. T. (2024). "A Review of Emerging Trends in Telemedicine: Healthcare Delivery Transformations." *International Journal of Life Science Research Archive*, 6(1), 137–147.
2. Ahad, A., Ali, Z., Mateen, A., Tahir, M., Hannan, A., Garcia, N. M., and Pires, I. M. (2023). "A Comprehensive Review on 5G-Based Smart Healthcare Network Security: Taxonomy, Issues, Solutions and Future Research Directions." *Array*, 18, 100290. https://doi.org/10.1016/j.array.2023.100290
3. Ahad, A., Tahir, M., and Yau, K.-L. A. (2019). "5G-Based Smart Healthcare Network: Architecture, Taxonomy, Challenges and Future Research Directions." *IEEE Access*, 7, 100747–100762. https://doi.org/10.1109/ACCESS.2019.2930628
4. Nawaz, N. A., et al. (2022). "Impact of Telecommunication Network on Future of Telemedicine in Healthcare: A Systematic Literature Review." *International Journal of Advanced and Applied Sciences*, 9(7), 122–138. https://doi.org/10.21833/ijaas.2022.07.013
5. Alemdar, H., and Ersoy, C. (2010). "Wireless Sensor Networks for Healthcare: A Survey." *Computer Networks*, 54(15), 2688–2710. https://doi.org/10.1016/j.comnet.2010.05.003
6. Ayo-Farai, O., Ogundairo, O., Maduka, C. P., Okongwu, C. C., Babarinde, A. O., and Sodamade, O. T. (2023). "Telemedicine in Health Care: A Review of Progress and Challenges in Africa." *Matrix Science Pharma*, 7(4), 124–132. https://doi.org/10.4103/mtsp.mtsp_24_23
7. Ayyappan, V., and Coffin, J. (2024). "Telemedicine Is Reshaping the Healthcare Landscape." *Healthcare Administration Leadership & Management Journal*, 2(3), 123–126. https://doi.org/10.55834/halmj.7598913119
8. Bashshur, R. L., Howell, J. D., Krupinski, E. A., Harms, K. M., Bashshur, N., and Doarn, C. R. (2016). "The Empirical Foundations of Telemedicine Interventions in Primary Care." *Telemedicine Journal and e-Health*, 22(5), 342–375. https://doi.org/10.1089/tmj.2016.0045
9. Canipel, L. (2016). "La Télémédecine, Ciment de l'équipe de Soins." *Soins*, 61(810), 35–37. https://doi.org/10.1016/j.soin.2016.09.006
10. Damaševičius, R., and Abayomi-Alli, O. O. (2023). "The Future of Telemedicine." *Advances in Medical Technologies and Clinical Practice*, 306–338. https://doi.org/10.4018/978-1-6684-9823-1.ch010
11. David, Y. (2019). "Telehealth, Telemedicine, and Telecare." In E. Iadanza (Ed.), *Clinical Engineering Handbook* (Second Edition), Academic Press, Elsevier eBooks, 550–555. https://doi.org/10.1016/b978-0-12-813467-2.00083-3
12. Devi, D., Duraisamy, K., Armghan, A., Alsharari, M., Aliqab, K., Sorathiya, V., Das, S., and Rashid, N. (2023). "5G Technology in Healthcare and Wearable Devices: A Review." *Sensors*, 23, 2519.
13. Elendu, C., Egbunu, E. O., Opashola, K. A., Afuh, R. N., and Adebambo, S. A. (2023). "The Role of Telemedicine in Improving Healthcare Outcome: A Review." *Advances in Research*, 24(5), 55–59. https://doi.org/10.9734/air/2023/v24i5958
14. Elgharably, R., Marzban, E., Belal, S., Ahmad, B., AbdElLatif, I., Atef, R., and ElBabli, I. (2008). "Wireless-Enabled Telemedicine System for Remote Monitoring." *Cairo International Biomedical Engineering Conference*, 1–4. https://doi.org/10.1109/cibec.2008.4786070
15. Gadish, M. A., and Denko, M. K. (2008). "Challenges, Systems and Applications of Wireless and Mobile Telemedicine." In N. Wickramasinghe & E. Geisler (Eds.), *Encyclopedia of Healthcare Information Systems*, IGI Global, 201–209. https://doi.org/10.4018/978-1-59904-889-5.ch028

16. Georgiou, K. E., Georgiou, E., and Satava, R. M. (2021). "5G Use in Healthcare: The Future Is Present." *JSLS*, 25(4). https://doi.org/10.4293/JSLS.2021.00064
17. Guerrazzi, C., and Feldman, S. S. (2020). "Health Information Exchange: What Matters at the Organizational Level?" *Journal of Biomedical Informatics*, 102, 103375. https://doi.org/10.1016/j.jbi.2020.103375
18. Gupta, R., Gamad, R. S., and Bansod, P. (2014). "Telemedicine: A Brief Analysis." *Cogent Engineering*, 1(1), 966459. https://doi.org/10.1080/23311916.2014.966459
19. Islam, M. Z., Islam, M. M., and Asraf, A. (2020). "A Combined Deep CNN-LSTM Network for the Detection of COVID-19 Using X-Ray Images." *Informatics in Medicine Unlocked*, 20, 100412. https://doi.org/10.1016/j.imu.2020.100412
20. Javaid, M., Haleem, A., Singh, R. P., and Suman, R. (2023). "5G Technology for Healthcare: Features, Serviceable Pillars, and Applications." *Intelligent Pharmacy*. https://doi.org/10.1016/j.ipha.2023.04.001
21. Kailas, A., and Ingram, M. A. (2009). "Wireless Communications Technology in Telehealth Systems." *IEEE Wireless VITAE*, 926–930. https://doi.org/10.1109/wirelessvitae.2009.5112574
22. Kailas, A., and Ingram, M. A. (2009). "Wireless Aspects of Telehealth." *Wireless Personal Communications*, 51(4), 673–686. https://doi.org/10.1007/s11277-009-9763-7
23. Kelly, J. T., Campbell, K. L., Gong, E., and Scuffham, P. (2020). "The Internet of Things: Impact and Implications for Health Care Delivery." *Journal of Medical Internet Research*, 22(11), e20135. https://doi.org/10.2196/20135
24. Kruse, C. S., Karem, P., Shifflett, K., Vegi, L., Ravi, K., and Brooks, M. (2016). "Evaluating Barriers to Adopting Telemedicine Worldwide: A Systematic Review." *Journal of Telemedicine and Telecare*, 24(1), 4–12. https://doi.org/10.1177/1357633X16674087
25. Laxminarayan, S., and Istepanian, R. (2000). "UNWIRED E-MED: The Next Generation of Wireless and Internet Telemedicine Systems." *IEEE Transactions on Information Technology in Biomedicine*, 4(3), 189–193. https://doi.org/10.1109/TITB.2000.5956074
26. Matusitz, J., and Breen, G.-M. (2007). "Telemedicine: Its Effects on Health Communication." *Health Communication*, 21(1), 73–83. https://doi.org/10.1080/10410230701283439
27. Institute of Medicine. (1996). *Telemedicine*. Washington, DC: National Academies Press. https://doi.org/10.11226/5296
28. Mohanty, S. N., Anita, V., Baral, S. K., Goel, R., and Singh, T. (2023). "Telemedicine and Healthcare Setting for Remote Care of Patients." *Lecture Notes in Electrical Engineering*, 103–112. https://doi.org/10.1007/978-981-19-9090-8_10
29. Ng, H. S., Sim, M. L., Tan, C. M., and Wong, C. C. (2006). "Wireless Technologies for Telemedicine." *BT Technology Journal*, 24(2), 130–137. https://doi.org/10.1007/s10550-006-0050-9
30. Pattichis, C. S., Kyriacou, E., Voskarides, S., Istepanian, R., and Schizas, C. N. (2002). "Wireless Telemedicine Systems: An Overview." *IEEE Antennas and Propagation Magazine*, 44(2), 143–153. https://doi.org/10.1109/MAP.2002.1003651
31. Pollard, J. K., Fry, M. E., Rohman, S., Santarelli, C., Theodorou, A., and Mohoboob, N. (2002). "Wireless and Web-Based Medical Monitoring in the Home." *Medical Informatics and the Internet in Medicine*, 27(3), 219–227. https://doi.org/10.1080/1463923021000014130
32. Pratama, R. D., Mahmudah, A., Septiana, L., and Rita, R. S. (2024). "Implementation of Telemedicine in Medical Practice." *Global International Journal of Innovative Research*, 1(2), 146–153. https://doi.org/10.59613/global.v1i2.16
33. Rashvand, H. F., Salcedo, V. T., Sanchez, E. M., and Iliescu, D. (2008). "Ubiquitous Wireless Telemedicine." *IET Communications*, 2(2), 237–254. https://doi.org/10.1049/iet-com:20070361
34. Rejeb, A., Rejeb, K., Treiblmaier, H., Appolloni, A., Alghamdi, S., Alhasawi, Y., and Iranmanesh, M. (2023). "The Internet of Things in Healthcare." *Internet of Things*, 22, 100721. https://doi.org/10.1016/j.iot.2023.100721
35. Sadoughi, F., Behmanesh, A., and Sayfouri, N. (2020). "Internet of Things in Medicine: A Systematic Mapping Study." *Journal of Biomedical Informatics*, 103, 103383. https://doi.org/10.1016/j.jbi.2020.103383
36. Shaderkin, I. A. (2022). "Telemedicine Barriers and Ways to Overcome Them." *Russian Journal of Telemedicine and e-Health*, 8(2), 59–76. https://doi.org/10.29188/2712-9211-2022-8-2-59-76
37. Shafik, W. (2024). "Data-Driven Future Trends and Innovation in Telemedicine." In N. Geada (Ed.), *Improving Security, Privacy, and Connectivity Among Telemedicine Platforms*, IGI Global, 93–118. https://doi.org/10.4018/979-8-3693-2141-6.ch005
38. Shen, X., Misic, J., Kato, N., Langenorfer, P., and Lin, X. (2011). "Emerging Technologies and Applications of Wireless Communication in Healthcare." *Journal of Communications and Networks*, 13(2), 81–85. https://doi.org/10.1109/JCN.2011.6157406

39. Shorna, S. A., and Sultana, R. and Hasan, M. (2024). "Big Data Applications In Remote Patient Monitoring And Telemedicine Services: A Review Of Techniques And Tools." *SSRN Electronic Journal*. https://doi.org/10.2139/ssrn.5026572
40. Sri-Ganeshan, M., and Cameron, P. (2024). "Remote Monitoring in Telehealth: Advancements, Feasibility and Implications." In T. F. Heston and C. R. Doarn (Ed.), *A Comprehensive Overview of Telemedicine*, IntechOpen. https://doi.org/10.5772/intechopen.1004661
41. Tachakra, S., Wang, X. H., Istepanian, R. S. H., and Song, Y. H. (2003). "Mobile e-Health." *Telemedicine Journal and e-Health*, 9(3), 247–257. https://doi.org/10.1089/153056203322502632
42. Tian, W. (2020). "Exploration and Prospect of 5G Application in Telemedicine." *Chinese Medical Journal*, 58(1), 1–4. https://doi.org/10.3760/cma.j.issn.0529-5815.2020.01.001
43. Vimarlund, V., and LeRouge, C. M. (2013). "Barriers and Opportunities to Telemedicine Adoption." *Studies in Health Technology and Informatics*. https://doi.org/10.3233/978-1-61499-289-9-933
44. Vudathaneni, V. K. P., et al. (2024). "The Impact of Telemedicine and Remote Patient Monitoring." *Cureus* 16(3), e55534. https://doi.org/10.7759/cureus.55534
45. Wagholikar, S., and Wagholikar, O. (2023). "Recent Trends in Remote Healthcare Applications." *IEEE OTCON Proceedings*, 1–6. https://doi.org/10.1109/otcon56053.2023.10113949
46. Wen, C. L. (2011). "Telemedicine, eHealth and Remote Care Systems." In H. de Fátima Marin, E. Massad, M. A. Gutierrez, R. J. Rodrigues, and D. Sigulem (Eds.), *Global Health Informatics*, Academic Press, 168–194. https://doi.org/10.1016/b978-0-12-804591-6.00009-4

Chapter 12

Visible light communication

Kapila W. S. Palitharathna and Sasmita Dash

12.1 INTRODUCTION

12.1.1 Overview of VLC

The ever-growing requirement of wireless data communication, driven by the proliferation of mobile devices, smart systems, high-definition content, and emerging digital applications, is rapidly exhausting the available radio frequency (RF) spectrum. To address these challenges, researchers, academics, and industry stakeholders are increasingly exploring alternative communication paradigms. One of the most promising among these is visible light communication (VLC), a technology that utilizes the visible light range (380–780 nm wavelength range) of the electromagnetic spectrum to transmit data through modulated light signals, typically emitted by light-emitting diodes (LEDs), and captured by photodiodes (PDs) [1, 2].

VLC offers several compelling advantages such as a vast unregulated spectrum, high spatial reuse, electromagnetic interference immunity, high security, low latency, and dual functionality for lighting and communication [2, 3]. With LEDs already deployed in homes, offices, vehicles, and public spaces, VLC provides a low-cost infrastructure for high-speed, secure, and short-range communication, and existing LED infrastructure can be easily adapted as VLC transmitters [1]. As VLC transmitters exploit the LED infrastructure, which is used for illumination purpose, the communication requires only a minimum additional power, acting as an energy-efficient communication solution. Due to the use of unregulated spectrum, VLC is immune to electromagnetic interference. Moreover, VLC is a promising solution for indoor and underwater setups in which ambient light conditions are minimum [4]. The fact that light waves are confined to a room area allows VLC to act as a highly secure communication method suitable for conference rooms [2]. The negligible multipath signal reception due to directive light patterns and low reflection coefficients provides low inter-symbol interference even under high data rate settings [3].

VLC has evolved beyond its foundational role as an alternative to RF communication and is now being explored for a wide range of innovative wireless communication applications. In smart homes, offices, and industrial settings, VLC can enable ultra-fast, secure data links that integrate seamlessly with lighting systems, creating energy-efficient, robust, and multifunctional environments. In vehicular communication, VLC allows vehicles on the road to exchange critical safety information using their headlights and taillights, facilitating advanced driver assistance systems and opening the opportunity for intelligent transportation systems [5]. In hospitals and aircraft cabins, where RF emissions must be minimized, VLC offers a safe and interference-free communication alternative. VLC also shows a strong potential in hybrid systems, complementing RF, light-fidelity (Li-Fi), or infrared—to support seamless handovers and ensure continuous connectivity in heterogeneous networks [6].

DOI: 10.1201/9781003637721-12

Moreover, the integration of VLC systems with the Internet of Things (IoT), 6G networks, digital economy, and machine learning (ML)-based adaptive system control is accelerating its role in next-generation communication architectures, making it a pivotal enabler of the data-driven, interconnected world.

12.1.2 Motivation for VLC

Despite significant advances, VLC systems face several inherent challenges that require novel and robust system design and optimization compared to traditional RF systems. Key challenges of VLC systems include signal blockages, user mobility, non-linear effects, and random receiver orientation, all of which can result in substantial performance degradation in wireless communication [7, 8]. Additionally, the non-linear characteristics of LED transmitters impose constraints on waveform design and parameter optimization limitations, requiring advanced techniques such as pre-distortion and adaptive modulation to maintain signal fidelity [9]. The highly directional nature of LED transmitters' light emission, while beneficial for spatial reuse, improved channel gains, and physical-layer security, often results in limited coverage and uneven illumination, particularly in indoor VLC environments. Another crucial consideration is improving the energy efficiency of VLC systems. Since VLC transmitters or LEDs also provide illumination, dual use of indoor lighting and communication reduces overall system power consumption. However, VLC system designers must carefully balance brightness/illumination requirements with modulation depth to ensure data integrity and human visual comfort [2]. These challenges are required to be addressed prior to the deployment of practical VLC systems.

At the same time, several novel technologies have arisen under the umbrella of VLC systems. Among them, simultaneous lightwave information and power transfer (SLIPT) protocol has emerged as a promising enhancement to VLC systems, allowing receivers to harvest energy while decoding information [10, 11]. This technology is especially valuable in battery power-constrained IoT devices and low-power sensor networks. In addition, visible light positioning (VLP) technology is an emerging technique that uses lightwave signals to determine the position of the users or objects within an indoor environment [12]. Further, for broader adoption, VLC must also integrate with heterogeneous networks, such as RF and millimeter-wave (mmWave) systems, forming hybrid wireless architectures. This integration allows seamless handovers, extended coverage, and improved reliability, especially in dynamic environments. Addressing these challenges demands a multi-disciplinary approach that includes optical system modeling, signal processing, hardware design, and communication theory.

Therefore, identifications of fundamentals and possible applications are required while careful attention is paid to challenges of VLC and designing, performance evaluations, and optimization of system parameters of VLC systems integrating emerging technologies. To this end, this chapter presents a comprehensive overview of VLC systems, focusing on their principles, architecture, modulation techniques, and multiple access. Moreover, possible application scenarios and challenges are identified. Exploring the emerging advances in wireless communication, computer engineering, and physics, possible future research directions are identified, and integration into existing and future wireless communication frameworks is also presented.

The remaining part of this chapter is organized as follows. Section 12.2 focuses on the detailed system models applicable for VLC systems, transmitter and receiver architectures, channel models that are being used, modulation techniques, and multiple access techniques. Section 12.3 presents the diverse applications of VLC, including indoor wireless communications, vehicular communications, underwater communications, IoT applications, and

UAV-based communication systems. Section 12.4 presents the possible challenges faced by the VLC systems including blockages, user mobility, random receiver orientations, and non-linearities. Possible future research directions applicable for VLC systems integrating modern technologies are discussed in Section 12.5 which covers artificial intelligence, reconfigurable intelligent surfaces (RISs), semantic communication, quantum communication, and liquid lens technology. Finally, the main conclusions are drawn in Section 12.6.

12.2 FUNDAMENTALS OF VLC

VLC systems operate by modulating the light intensity at the transmitter at speeds unidentifiable to the normal human eye. The light source, typically an LED, transmits data via rapid fluctuations in brightness, which are detected by a PD or camera sensor. Unlike RF systems, VLC is inherently directional and often requires a line-of-sight (LoS) connection, making it ideal for secure and localized communication. In the following, the transceiver architecture of VLC systems is presented.

12.2.1 Transceiver architectures

In general, the VLC systems consist of several blocks as shown in Figure 12.1. Specifically, a VLC transmitter includes a data source, modulator, LED driver, LED, and an optical filter. The transmitted light signal after propagating through the VLC channel is captured by the receiver, which includes an optical filter, PD, trans-impedance amplifier (TIA), demodulator, and data sink [2].

- **Modulator**: The modulator is responsible for converting the binary data stream into an analog signal suitable for driving LEDs. VLC systems adhere to intensity modulation (IM) at the transmitter and direct detection (DD) at the receiver, necessitating real-valued and non-negative signals being used. Commonly used modulation schemes include on–off keying (OOK), pulse position modulation (PPM), pulse width modulation (PWM), discrete multi-tone (DMT) modulation, and optical orthogonal frequency division multiplexing (O-OFDM) [2].
- **LED driver**: The LED driver converts the modulated voltage into a current signal suitable for optical transmission. It typically uses BJT/MOSFET-based amplifiers and must account for limitations in peak/average power, signal linearity, and bandwidth [13].
- **LED**: LEDs convert the electrical current signal generated from the LED driver into optical output. Various types of LEDs used in VLC systems are as follows [14].

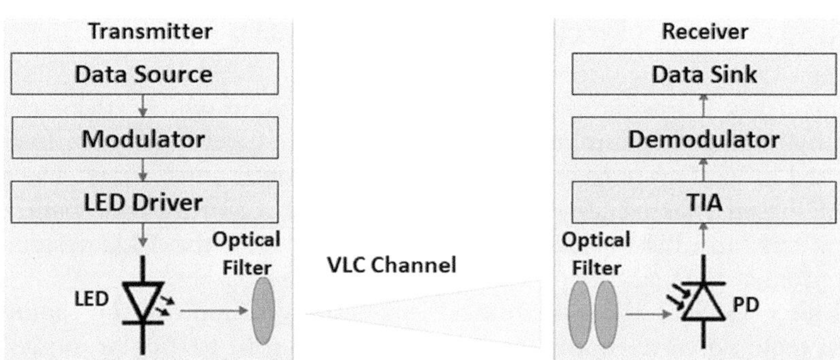

Figure 12.1 Transceiver architecture of VLC systems.

- *White LED*: The most popular kind of LED utilized in VLC systems is the white LED. A blue LED chip covered in yellow phosphor, which absorbs some blue photons and reemits yellow ones, makes up the white LED. White light is created simultaneously when the blue photons that are released directly combine with the yellow photons. As a result, the transmitted light signal appears as white light at the end. However, the modulation bandwidth is restricted to a few MHz due to the phosphor's slow relaxation time and the time required for the aforementioned process, and hence, white LEDs are suitable only for low data rate applications.
- *Red–green–blue (RGB) LED*: These types of LEDs are made of three LEDs allowing more flexible color tuning. Hence, there is a high interest toward using them for VLC applications. In these LEDs, there is no phosphor coating, and hence, RGB LEDs provide a modulation bandwidth of a dozen MHz which is favorable for high-speed data transmission. The concept of modulating the three-color LED sources independently to implement wavelength division multiplexing is commonly studied. Further, several practical high-speed VLC links have been demonstrated using this technique. The challenge arising in these systems is to avoid undesired color patterns for the human eye and propose fair power control strategies. However, for underwater systems this approach is promising.
- *Organic LED (OLED)*: Using organic materials to produce LEDs with a similar functionality to RGB LEDs, OLEDs provide VLC systems the capacity to be deployed in the form of a large area with extremely low cost and high energy efficiency. Nevertheless, this benefit is accompanied by the disadvantage of OLEDs' low modulation bandwidth, which is only a few kHz. They have been utilized for low data rate applications that do not surpass a few Mbps because of their intrinsically high plate capacitance. However, OLEDs are preferred for certain VLC applications, including smart homes, inside vehicles, screen-based VLC, and wearable devices.
- *Micro-LED*: This type of LEDs is manufactured using gallium nitride (GaN) and can achieve up to a bandwidth of hundreds of MHz resulting Gbps data rates which are suitable for most of the modern wireless communication applications. In addition, the small size makes these systems suitable for practical VLC systems that require small transmitter size or need large transmitter arrays. The micro-LED manufacturing process is highly complex and comes with a high cost. The small size and output power limitations of micro-LEDs make them unsuitable for indoor illumination. However, the system design comes with challenges such as heat control and non-linear effects.
- *Laser diode (LD)*: LDs have a higher modulation bandwidth than typical LEDs; specifically LDs have bandwidth in the range of GHz. However, LDs have not been commonly used in VLC systems due to their drawback being hard-to-design VLC systems that concurrently satisfy the illumination constraints, especially in indoor setups. Also, LDs emit extremely narrow light beam which affects the eye safety and homogeneous illumination in the room. However, with a diffuser, LDs can be used in VLC systems with proper transmit power constraints. The deployment of LD-based systems comes with a drawback of requiring wide-band electronics, amplifiers, and PDs, which can increase the cost of the VLC system and could compromise the viability of industrial VLC products.

- **Optical filter:** Optical filters are optical lens designs to improve the channel gain and filter out required wavelengths for data transmission. In particular, optical filters are used in transmitters to filter the required wavelengths as well as to focus the light beams toward desired directions. Similarly, optical filters are used at the receiver end to improve the receiver's field-of-view (FoV), remove unwanted optical wavelengths from the received signal, and minimize noise levels and interference from other sources [15].

- **PD**: In general, the optical receiver of the VLC systems is a PD. The PD converts the received optical signal inside it to a current signal at the receiver. However, in some systems CMOS sensors, photovoltaic (PV) cells, or a camera can be used to detect optical signals. The main types of PDs used are positive–intrinsic–negative (PIN) PD and the avalanche PD (APD) [16].
  - *PIN PD*: PINs can be identified as the most commonly used PD receivers in VLC systems. PIN PDs consist of an undoped semiconductor region interposed between heavily doped n-type and a p-type semiconductor regions. PIN is connected in reverse-biased conditions to increase the depletion region in its operation. Electron-hole pairs are generated when photons of light with the correct wavelength and sufficient energy enter the depletion region. This allows a free current to pass through the apparatus. Photocurrent is generated in proportion to the amount of light that strikes the PD-active surface. In a PIN PD, this idea is used to transform incident light power into a current signal.
  - *APD*: As another common type of PD, APDs are used in modern VLC systems where a high data transmission speed as well as low noise conditions are preferred. APDs are used in reverse-biased operation mode at a high voltage without exceeding their breakdown voltage. This phenomenon creates a large electric field in the avalanche semiconductor region. The impact-ionization process is repeated when carriers produced by the incident light collide with the APD's lattice atoms, creating new electron-hole pairs. This operation leads to a significant improvement in SNR compared to PIN PDs when APDs are used. At the same time, the multiplication process in these APDs generates an excess amount of shot noise. Therefore, APDs are well suited for low-ambient light applications with weak-incident optical power such as closed room or deep underwater applications. Operating APDs in Geiger mode, in which the VLC devices operate at a voltage above the breakdown voltage, can help avoid excess noise of these devices.
  - *Single-photon APD (SPAD)*: SPAD is an APD that operates in Geiger mode, and it uses only a single photon to trigger an avalanche mode by creating a single electron–photon pair. This mode allows PDs to count light photons, opening a new optical communication direction, especially for low power and high precision optical communication systems and also for quantum communication systems. The downside of SPADs is once a photon is detected, there is a dead time, and detection is not possible at that time. Hence, the transient effect is present and requires more advanced system/algorithm design. Also, the output received signal is not proportional to the light intensity, which limits its use in conventional VLC systems.
  - *Silicon Photomultipliers (SiPM)*: To overcome the limitations of single SPAD, SiPM are made by placing SPADs in an array to form silicon photomultiplier (SiPM). They provide very high sensitivity in weak illumination conditions yet are not effective in the cases with high background light as SiPM are shot noise limited.
- **TIA**: In a typical VLC receiver, the current signal generated by the PD or PV cell is sent through a device called TIA to convert it to a voltage signal for further processing. In general, TIA is an operational amplifier (OPAMP)-based closed-loop circuit with a high feedback resistance. It provides high current-to-voltage conversion ratio, where nA current signals are converted to V-range voltage signals. The frequency characteristic of the TIA is a critical parameter that limits the data rate at the hardware level [2].

- **Demodulator:** The demodulator reconstructs the original binary data from the received analog voltage. It compensates for channel impairments, filtering, and noise, and performs synchronization and decoding.

12.2.2 Channel models

The VLC system has a quasi-static, LoS-dominant and intensity-modulated channel. In general, it is assumed to be a linear, time-invariant response for indoor applications, where mobility is limited. The received optical signal at the PD is proportional to the transmitted optical intensity at the LED, with the channel represented as a linear convolution as

$$y(t) = rx(t) * h(t) + n(t), \tag{12.1}$$

where $y(t)$ is the receiver-generated current signal, r is the responsivity of the PD in A/W, $x(t)$ is the LED-transmitted non-negative intensity-modulated optical signal, $h(t)$ is the VLC channel impulse response, $n(t)$ is the additive white Gaussian noise.

12.2.3 LoS channel gain

In most VLC systems, it is assumed LoS DC gain and can be derived using the Lambertian radiation model of the LEDs and can be expressed as [3, 17]

$$h_{LoS}(t) = \frac{(m+1)A}{2\pi d^2}\cos^m(\theta)\cos(\varphi)T_s(\varphi)g(\varphi)\Pi\left(\frac{\varphi}{\Phi}\right), \tag{12.2}$$

where $m = -\ln(2)/\ln(\cos\theta_{1/2})$ denotes the Lambertian order of the LED, A is the PD effective area, d is the distance between LED and PD, θ is the angle of irradiance at the transmitter, φ is the angle of incidence at the receiver, $T_s(\varphi)$ is the optical filter gain, $g(\varphi)$ is the optical concentrator gain, Φ is the transmitter's divergence angle, and $\theta_{1/2}$ is the half power angle of the LED. The concentrator gain $g(\varphi)$ can be further expressed as

$$g(\varphi) = \begin{cases} \dfrac{\rho}{sin^2(\Psi)}, & \varphi \le \Psi, \\ 0, & \text{otherwise}, \end{cases} \tag{12.3}$$

where ρ is the refractive index of the optical lens, Ψ is the FoV of the PD. The LoS components dominate when the receiver is directly within the beam of the transmitter, especially under directional lighting. A figure illustrating critical parameters of VLC channels is shown in Figure 12.2.

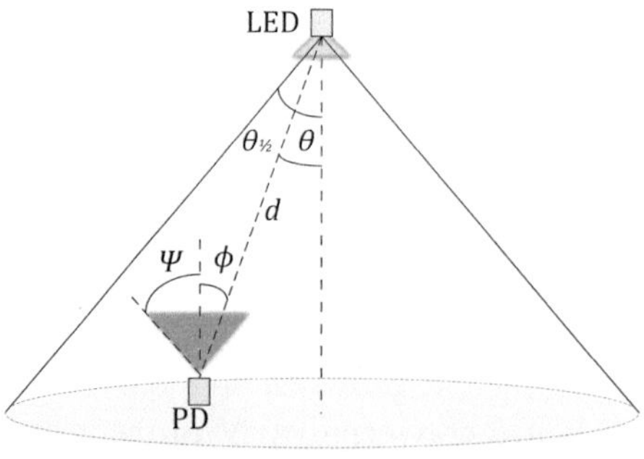

Figure 12.2 VLC channel, including all parameters.

12.2.4 Non-LoS channel gain

Non-LoS (NLoS) components are the direct consequence of surface reflections. In particular, static objects such as walls, ceiling, tables, chairs can contribute toward NLoS component. NLoS is weaker but can affect the overall impulse response in large or complex rooms. A light signal reflected from infinitesimal areas on room surfaces that successively arrive at the receiver can be decomposed into an infinite number of components of the diffusive NLoS component of the VLC channel, $h_N LoS(t)$. Assume that the LED is placed on the ceiling, and that *NLoS* can be expressed as *NLoS* and calculated using an iterative algorithm

$$h_{NLoS}(t) = \sum_{j=1}^{\infty} h^{(j)}_{NLoS,T,R}(t). \tag{12.4}$$

After k reflections ($k \geq 1$), the impulse response of the NLoS channel can be approximated as

$$h^{(j)}_{NLoS,T,R}(t) = \sum_{i=1}^{I} \rho_{\varepsilon_n} d\varepsilon_n h^{(j-1)}_{NLoS,T,\varepsilon_n^r}(t) * h^{(0)}_{NLoS,\varepsilon_n^t,R}(t), \tag{12.5}$$

where I is the number of small reflecting elements of the room surface, ε_n denotes the reflecting elements with the Lambertian radiation pattern, ρ_{ε_n} is the coefficient of reflectivity, $d\varepsilon_n$ is the small surface area of each reflecting element. ε_n^t is the reflecting element when it acts as a transmitting element compared to the PD R, while ε_n^r is the reflecting element when it acts as a receiving element compared to the LED, T. $h^{(0)}_{NLoS,\varepsilon_n^t,R}(t)$ is the LoS impulse response from the element ε_n to the PD R. Due to the directive nature of LED light patterns and LED/PD placement inside a room, h_{NLoS} is negligible in most cases, and hence, only LoS component is considered. However, in conditions such as underwater or small rooms, this effect needs to be considered.

12.2.4.1 Underwater channel

Underwater optical communication is sensitive to several factors, including the attenuation due to water particles and turbulence effect due to oceanic currents. The underwater optical channel gain from an LED to PD, h can be represented using the product of three coefficients given by $h = h_g h_p h_t$, where h_g is the geometric loss coefficient, h_p is the path loss coefficient, and h_t is the weak oceanic turbulence-induced fading [18].

The geometric loss component of the LoS channel between the transmitter and receiver, h_g is the loss due to the geometric placement of the transmitter/receiver as well as the beam-divergent pattern at the LED and h_g can be expressed similar to (12.2).

Optical signal propagation through water is subjected to interactions between each light photon and the seawater particles. These seawater particles can include water, mud, colorings, fish feces, and other sediments. As a result of these particle interactions, absorption and scattering effects can take place which can reduce the mean irradiance of the light beam. In particular, absorption is due to lightwave energy waste due to particle interactions while scattering is the refraction of photons by particles. Beer's law can be used as the simplest and most commonly used mathematical model for representing the absorption and scattering effects and can be expressed as

$$h_p = \exp(-c_a(\lambda)d), \tag{12.6}$$

where $c_a(\lambda)$ is the underwater optical attenuation coefficient which can be expressed as $c(\lambda) = a(\lambda) + b(\lambda)$. Here $a(\lambda)$ and $b(\lambda)$ represent the absorption and scattering coefficients, respectively.

To account for the underwater turbulence conditions, various stochastic distributions are used in the literature. Among them log-normal distribution, *K* distribution, gamma–gamma distributions have been used in common. However, as the most common type of turbulence in the underwater environment, oceanic current-induced weak turbulence conditions better fit the log-normal distribution and are widely used in the literature to model the channel fading. The probability density function (pdf) of the log-normal distributed fading coefficient, h_t, can be expressed as

$$f_{h_t}(h_t)=\frac{1}{2h_t\sqrt{2\pi\sigma_x^2}}\exp\left(-\frac{(\ln(h_t)-2\mu_x)^2}{8\sigma_x^2}\right), \tag{12.7}$$

where μ_x and σ_x^2 are the mean and the variance of the Gaussian-distributed log-amplitude factor given by $X=\frac{1}{2}\ln(h_t)$. The energy of the fading coefficient is preserved by normalizing the fading amplitude using $E\{h_t\}=1$, where $E\{.\}$ is the expectation operator giving $\mu_x=-\sigma_x^2$.

12.2.5 Modulation techniques

VLC systems employ IM at the transmitter and DD at the receiver. Hence, the modulation techniques used in VLC required to generate an LED-driving signal which is real-valued, non-negative, and unipolar. Modulation schemes used in VLC systems can be categorized into two groups: (i) single-carrier and (ii) multi-carrier modulation schemes. The main difference between single- and multi-carrier modulation schemes can be explained as follows. In multi-carrier, the signal bandwidth is separated into smaller sub-bands, and each sub-band transmits independent data symbols, while in single band no such separation exhibits. In the following, we explain different modulation schemes used in VLC systems.

12.2.5.1 Single-carrier modulation schemes

- **OOK**: OOK is the fundamental and most commonly used modulation scheme in VLC systems. This is due to OOK's ease of implementation and its compatibility with intensity IM and DD. In this modulation, the presence of light pulse represents binary data, binary '1' and its absence represents a binary '0'. OOK modulation requires minimal processing and is energy-efficient when transmitting sparsely distributed data. However, OOK is susceptible to ambient light interference and offers limited spectral efficiency. It is most suitable for extremely low data rate applications such as VLC-based IoT and sensor networks in indoor environments.
- **PWM**: This is a modulation scheme in which data is encoded in the width of the light pulse. Here the amplitude of the signal remains constant. PWM is commonly used in VLC systems where both lighting and communication functions are integrated. This allows dimming control without altering the average light intensity of the optical signal. PWM is effective for maintaining illumination quality, but its limited bandwidth and relatively low data rates make it less suitable for high-speed VLC applications. However, it is valuable in hybrid lighting-communication systems where illumination quality and communication coexist.
- **PPM**: PPM conveys data by varying the position of an active light pulse within a fixed time slot. PPM offers high energy efficiency and is resilient to background noise. This makes PPM well suited for low-power VLC applications. However, PPM requires a precise synchronization mechanism and it suffers from bandwidth inefficiency, as higher-order PPM increases symbol duration. In addition, variants of PPM like

differential PPM, and multiple PPM have been proposed to improve spectral efficiency while preserving low power consumption.

- **Pulse amplitude modulation (PAM)**: In PAM, wireless data is transmitted by modulating the amplitude of light pulses. *M*-PAM (e.g., 4-PAM, 8-PAM) can improve spectral efficiency by allowing more than one bit per single symbol transmission. It is identified that PAM is more bandwidth-efficient than OOK and PPM schemes. But it is highly sensitive to receiver noise and non-linearity in the LED's response. This effect is highly dominant for higher-order signal constellations. To address this, techniques like equalization and pre-distortion are often employed. PAM is a preferred choice for certain VLC systems and applications that require a balance between throughput communication/hardware complexity.
- **Carrier-less amplitude and phase modulation (CAP)**: CAP is a bandpass modulation technique that encodes data in both the amplitude and phase of the communication signal without using an explicit carrier signal. In VLC, CAP offers better spectral efficiency than PAM. It also has a simpler implementation than OFDM, since it avoids FFT/IFFT operations in the modulation/demodulation. However, it is more vulnerable to the non-linear effect of the LED, and it requires complex equalization at the receiver to combat the channel dispersion effect. CAP's ability to support high data rates with very low computational complexity makes it a promising alternative to OFDM in medium-to-high-speed VLC applications.

12.2.5.2 Multi-carrier modulation schemes

- **DC-biased optical OFDM (DCO-OFDM)**: DCO-OFDM is one of the most common types of OFDM schemes adapted for VLC systems that use IM and DD. Due to the bipolar nature of standard OFDM signals, DCO-OFDM formation applies Hermitian symmetry to produce a real-valued signal. Then it adds a DC bias to shift all signal values into the non-negative range suitable for VLC transmission. Even though this technique achieves a high spectral efficiency and allows the full use of subcarriers for data transmission, DCO-OFDM suffers from a very high-power consumption due to the added DC bias and is sensitive to LED non-linearity and clipping distortion effects. Advanced signal bias optimization and predistortion techniques are often used to mitigate these drawbacks inherent in these systems. DCO-OFDM is appropriate for high-speed VLC applications, such as video streaming, and indoor broadcast wireless networks.
- **Asymmetrically clipped optical OFDM (ACO-OFDM)**: ACO-OFDM is also another commonly used unipolar OFDM variant specifically designed to address the above-mentioned drawbacks of DCO-OFDM. ACO-OFDM transmits only the odd part of subcarriers of the OFDM signal and clips the negative portion of the signal. This technique ensures that the clipping noise does not interfere with data-bearing subcarriers, by exploiting the orthogonality property of OFDM to ensure that the clipping noise does not interfere with data-bearing subcarriers. These designs and techniques remove the need for a DC bias for OFDM signals, making ACO-OFDM more energy-efficient and a better-suited version of OFDM for power-constrained VLC applications. However, this adaptation sacrifices spectral efficiency of the communication since it uses only a half of the subcarriers for transmission. ACO-OFDM is preferred for VLC applications where energy efficiency and linear mode of operation are more critical than high data rate, such as mobile, wearable, and IoT VLC systems.
- **Flipped OFDM (F-OFDM)**: Similar to ACO-OFDM, F-OFDM is also another unipolar OFDM variant applicable for VLC systems. It splits the real-valued bipolar OFDM signal into positive and negative parts and transmits each in separate time

slots, flipping the negative component to make it positive. The receiver reconstructs the original signal by subtracting the flipped signal from the positive one. Flip-OFDM has lower complexity than ACO-OFDM and avoids the power inefficiencies of DCO-OFDM. However, it effectively doubles the time required to transmit each symbol, reducing spectral efficiency.

- **Hadamard-coded modulation (HCM):** HCM is also a multi-carrier modulation scheme that uses orthogonal Hadamard matrices to encode data into optical light pulses. Unlike OFDM, HCM generates non-negative signals inherently from its operation, avoiding the need for DC bias or clipping as in OFDM schemes. The use of HCM is resilient to the non-linearity of LEDs and offers good energy efficiency. However, HCM's spectral efficiency is generally lower than that of OFDM, and implementation complexity can also be increased with Hadamard matrix size. It is noted that HCM is well-suited for low-to-medium data-rate VLC systems where link robustness and low power consumption are the critical factors.
- **Multi-band CAP:** Multi-band CAP is a high-speed modulation technique that extends traditional CAP by dividing the signal into multiple frequency bands, each carrying an independent CAP signal. Unlike OFDM techniques, CAP avoids the use of FFT/IFFT operations by using orthogonal pulse-shaping filters to generate subcarriers. This criterion results in lower computational complexity and hardware requirements. Multi-band CAP is highly suitable for VLC systems with practical bandwidth-limited LEDs, as it allows flexible sub-band allocation and better resistance to LED frequency response roll-off. However, such VLC systems require careful filter design and equalization to minimize inter-band interference. Studies have demonstrated multi-Gbps performance using multi-band CAP in VLC systems, showing its potential as a computationally efficient alternative to OFDM.

12.2.6 Multiple access techniques

- **Time division multiple access (TDMA):** TDMA is considered the simplest multiple access technique in which each user is assigned a specific time slot for communication. It ensures that there is no interference between users, as only one user transmits at a time. TDMA is simple to implement and is well-suited for VLC systems with a small number of users or applications with predictable traffic patterns. However, its efficiency degrades in extremely dynamic environments where user demands vary, leading to potential under-utilization of communication bandwidth during idle periods.
- **Frequency division multiple access (FDMA):** In FDMA, the available optical bandwidth is divided into several frequency bands, and each band is assigned to a different user. Hence, this technique enables simultaneous data transmission among users without time-sharing and is especially relevant in VLC systems employing multi-color LEDs or LDs, where different wavelengths (colors) can act as distinct user channels. However, FDMA's use in VLC is limited by the inherently narrow communication bandwidth of most LEDs and requires complex optical filtering to avoid inter-channel interference.
- **Code division multiple access (CDMA):** This multiple access method allows multiple users to occupy the same time and frequency resources by assigning them a unique spreading code. Such codes help the receiver to distinguish between different user signals, enabling asynchronous communication with improved security and robustness against interference. In VLC systems, implementing CDMA poses significant challenges due to the need for precise synchronization and complex decoding mechanisms. Despite this, it remains a strong candidate for secure or interference-limited indoor VLC applications.

- **Non-orthogonal multiple access (NOMA):** NOMA is a modern multiple access approach gaining popularity in VLC research community due to its ability to serve multiple users on the same frequency and time resource. It differentiates users based on their channel gains and assigns them different power levels. At the receiver, stronger users perform successive interference cancellation (SIC) to decode their own and weaker users' signals. The inherent directive nature, extremely different path losses in VLC systems make it well-suited for NOMA schemes. NOMA significantly enhances spectral efficiency and supports massive connectivity, but its performance heavily depends on accurate channel estimation, power allocation, and effective SIC implementation. Hence, the use of NOMA can be challenging in highly dynamic channel conditions.
- **Spatial division multiple access (SDMA):** SDMA exploits the spatial properties of VLC by serving multiple users located in different physical locations. Through the use of LED arrays, beam steering, or angle diversity receivers, SDMA allows VLC systems to simultaneously transmit to multiple users without overlapping beams. This method aligns naturally with VLC's directionality and is particularly beneficial in dense indoor scenarios like offices, classrooms, and also with IoT applications. However, SDMA requires accurate user positioning and advanced optics to dynamically control the spatial separation of signals, and hence, this requires advanced hardware structures.

12.3 APPLICATIONS OF VLC

Due to the above-mentioned characteristics and promising advances of VLC, it has been proposed as an attractive communication technology that can be used in a wide range of wireless communication applications emerging in different areas, including indoor, vehicular, underwater, and unmanned aerial vehicle (UAV) communication systems as shown in Figure 12.3.

12.3.1 Indoor wireless communication

VLC is a highly effective communication method in indoor environments where LED lighting is already in use for indoor illumination. By modulating the optical intensity level of the light emitted by LEDs, which is imperceptible to the human eye, data can be transmitted at high speed. This makes VLC a promising alternative or complement to traditional

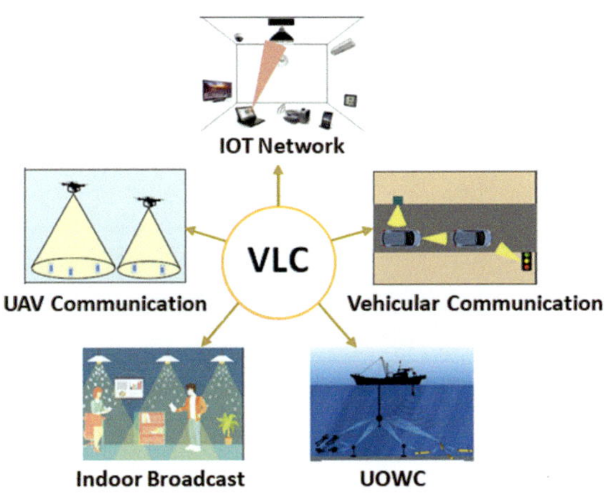

Figure 12.3 Applications of VLC systems.

Wi-Fi, particularly in environments with high RF congestion or sensitive RF environments. Moreover, VLC's LoS nature, coupled with its inability to penetrate walls, enhances communication security and minimizes the risk of unauthorized access [2]. Such attractive features of indoor VLC systems have garnered the attention of system designers to promote LiFi as a complementary technology to conventional wireless local area network (WLAN) technologies such as wireless-fidelity (Wi-Fi) networks. Modern research on VLC has shown that by using VLC in indoor communication systems, up to Gb/s data rate can be achieved [19].

12.3.2 Vehicular communication

LEDs are employed in street lights, traffic lights and signs, and vehicles for illumination and indication purposes. Hence, existing vehicular lighting infrastructures can be used for vehicular communication without additional cost. Vehicular communication is categorized into two, namely vehicle-to-infrastructure (V2I) [20] communication and vehicle-to-vehicle (V2V) communication [21]. Existing traffic light systems and spotlights on streets can be adapted to deliver information in vehicle-to-infrastructure communication systems where information collected by vehicles using sensors can be communicated to traffic infrastructures for decision-making. In V2V communication, LED headlights in vehicles can be used as VLC transmitters. This enables the communication among vehicles for traffic management as well as local communication. The use of VLC for vehicular communication applications enables high-speed data transfer, and better traffic management solutions can be implemented.

12.3.3 Underwater communication

Traditionally, RF and acoustic wave-based communication systems have been used for underwater wireless communication (UWC) applications. Recently, underwater VLC (UVLC) has been studied for maritime communication systems as a promising solution due to impairments in RF and acoustic wave-based UWC systems such as limited bandwidth and high attenuation through water [4]. In order to show the critical advantages of UVLC systems, Table 12.1 compares UVLC systems with underwater RF and acoustic communication-based systems. In general, UVLC systems are capable of achieving a very high data rate for

Table 12.1 Comparison of UWC systems

System parameters	*Optical waves*	*RF waves*	*Acoustic waves*
Max. data rate	10 Gbps	0.1 Gbps	10 Kbps
Max. commun. range	50 km	100 m	20 km
Latency	Low	Moderate	High
Max Tx. power	5 W	100 W	1 kW
Antenna size (m)	0.1	0.5	0.1
Cost	Low	High	High
Speed (m/s)	2.24×10^8	$2-3\times10^8$	1500
Attenuation	Distance	Freq. and conductivity	Conduc.
Modulation	OOK, PAM	ASK, FSK, PSK	FSK, DPSK
Reliability	Medium	Low	High
Beam	High direc.	Omni-direc.	Omni-direc.
Multipath	Low	Medium	high

short-range UWC applications. Due to extremely low water attenuation conditions, most UVLC systems use blue–green range of the light signals for communication [4]. In addition, narrow light beams emitted from lasers can be applied to achieve longer communication distances. Modern UVLC systems can be applied for ship-to-submarine, submarine-to-submarine, diver-to-diver communication, data collection from seabed sensors using autonomous underwater vehicles (AUVs), and AUV-to-AUV communication.

12.3.4 IoT applications

With the rapid expansion of the IoT applications, VLC systems offer an efficient communication medium that leverages existing lighting infrastructures to connect various IoT devices [22]. In smart home environments, VLC can enable secure and interference-free communication between devices such as sensors, thermostats, and appliances. Similarly, in industrial settings, VLC supports real-time monitoring and control of automated processes in industry lines, contributing to operational efficiency and reducing energy consumption. Its low power requirements and the reuse of existing hardware make VLC a cost-effective solution for massive IoT networks [22]. The internet of underwater things is an emerging area where underwater sensors communicate with AUVs or over-the-air APs to send their data collected underwater.

12.3.5 UAV-based communications

UAVs can be deployed for several scenarios such as smart agriculture, transportation, fault inspection, and tracking. In addition, to providing high-speed wireless communication access in outdoor environments, UAVs can be used as APs for a large number of users [23]. Especially, environments with dense users can be covered using such UAVs and can serve as dynamic APs studying the dynamic nature of users. Recently, researchers have focused on VLC-based UAV communication systems in which UAVs act as VLC APs. The directive light patterns of LEDs are suitable as transmitters on the UAVs to provide ground coverage. In the context of UAV-based VLC systems, several interesting problems have gained the attention of researchers. Some of them are UAV-user association, UAV trajectory planning, cell formation, pointing errors, and resource allocation problems [23, 24].

12.4 CHALLENGES OF VLC SYSTEMS

VLC systems can undergo several challenges and require the attention of the research community toward proving solutions to mitigate the performance losses inherent from such impairments. In the below section, we discuss some of these challenges such as blockage effect, user mobility, random receiver orientation, and non-linearity of transmitters/receivers.

12.4.1 Blockages

Blockages in VLC systems present a significant challenge, particularly in dynamic indoor environments where the LoS channel between the transmitter and the receiver can be frequently interrupted fully or partially. Such blockage conditions may result from human movement (due to mobile users or non-users), furniture inside the room, or other opaque objects temporarily or permanently obstructing the optical path (such as UAVs). Different from RF-based systems, VLC systems rely heavily on direct or reflected light paths, making VLC highly susceptible to shadowing and occlusion effects. Recent studies have shown that even brief/small blockage conditions can lead to considerable signal attenuation or complete wireless communication loss, depending on the system's design and the availability of alternative light paths (e.g., diffuse reflections) [25]. This has been identified as a major issue in VLC systems.

Modeling of blockage conditions in VLC systems helps identify the critical impact of LoS interruptions on performance, reliability, and applicability of blockage-inherent VLC systems. To this end, several blockage modeling approaches have been considered, such as geometrical models, closed-form probabilistic models, and stochastic geometry models. In geometry-based models, the blockages are models by transmitter/receiver position, blockage position, and the geometry of the blockages. This modeling is more accurate for a given instance of time. Closed-form probabilistic models estimate blockage probability based on spatial positioning, showing that minor device adjustments can reduce LoS risk [8]. This approach is more helpful for wireless systems' performance analysis, since derived distributions are critical for performance analysis. Stochastic geometry models use spatial point processes to represent random blocker distribution and movement, allowing generalized performance analysis.

Now, we present a geometry-based blockage model. In Figure 12.4, we show the geometry of a transmitter, a receiver, and a human user approximated by a cylinder which is used to model the existence of a blockage [8]. Consider i-th LED transmitter is placed at the (x_i, y_i, z_i) on the room ceiling and j-th PD receiver is placed at the (x_j, y_j, z_j) location inside the room. We calculate whether a human blockage exists for the (i, j)-th LED/PD pair due to the k-th non-user cylindrical blockage following the steps below. Assume (x_k, y_k, z_k) is the center of the base of the cylindrical blockage, h_k is the height of the cylindrical blockage, and r_k is the radius of the cylinder considered. We see this scenario from different views, i.e., the top view and side view as shown in Figure 12.4. Now, first applying simple trigonometric geometrical relations to the top view shown in Figure 12.4(a), $d_{k,i,j}$ can be expressed by

$$d_{k,i,j} = \left(\left|(y_i - y_j)x_k - (x_i - x_j)y_k + x_i y_j - x_j y_i\right|\right) / \sqrt{(y_i - y_j)^2 + (x_i - x_j)^2}. \tag{12.8}$$

Here $d_{k,i,j}$ is defined as the perpendicular distance between the trajectory line connecting the i-th transmitter and the j-th receiver and the center of the k-th cylindrical blockage seen from the top view. We see that for the case $d_{k,i,j} > r_k$, no intersection of the cylindrical blockage with the LoS channel, and hence, no blockage condition is observed. However, otherwise it is not straightforward to calculate the blockage and other views need to be

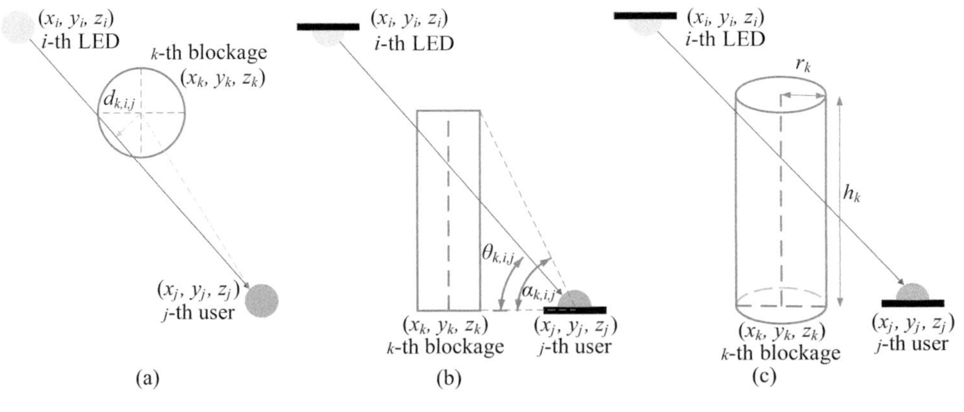

Figure 12.4 Geometric blockage modeling in VLC systems: (a) top view, (b) side view, and (c) perspective view.

considered. In the event of $d_{k,i,j} \leq r_k$, we study the side view to calculate the blockage condition. Using the geometric relations, we can see $m_k = \sqrt{r_k^2 - d_{k,i,j}^2}$, where $2m_k$ is the width of the cross-section of the intersecting plane of the cylindrical blockage on the side view. Hence, $\alpha_{k,i,j} < \theta_{k,i,j}$ corresponds to no-blockage conditions, while $\alpha_{k,i,j} \geq \theta_{b,i,k}$ implies blockage exists, where $\alpha_{k,i,j}$ and $\theta_{k,i,j}$ are the angles as shown in Figure 12.4(b) from the side view. In this case, $\alpha_{k,i,j}$ and $\theta_{k,i,j}$ angles can be mathematically expressed as

$$\alpha_{k,i,j} = tan^{-1}\left((z_i - z_j) / \sqrt{(x_i - x_j)^2 + (y_i - y_j)^2}\right), \tag{12.9}$$

and

$$-8mm\theta_{k,i,j} = tan^{-1}\left(h_j / \left(\sqrt{(x_i - x_j)^2 + (y_i - y_j)^2 - f^2} - m_k\right)\right). \tag{12.10}$$

In order to obtain a more realistic model, we create random human blockages that move around the room space to create time-varying blockage conditions. For each time instance, we calculate the blockage in the channel gain.

To mitigate the performance losses due to these blockage conditions, researchers have proposed several solutions, including guiding users, and using optical IRSs, and ML-based adaptive optimization of system parameters to improve the system performance under blockage conditions [8, 26–29]. In [29], the downlink communication performance of an indoor VLC system undergoing human blockage conditions has been analyzed. A VLC system that can improve communication robustness by guiding mobile users to avoid blockage conditions has been studied in [28]. The VLC system studied in the paper [26] uses mirror-based optical IRSs to improve the communication sum data rate of indoor VLC systems. Additionally, optical IRS has been used to improve the communication link reliability of NOMA-VLC system in [27]. An ML-based system parameter optimization framework for VLC/RF hybrid system under human blockage conditions has been studied in [8].

12.4.2 User mobility

User mobility in VLC systems presents significant challenges due to the highly directional and LoS-dependent nature of optical wireless links. As users move within indoor environments, variations in receiver orientation, transmitter-to-receiver distance, and blocking objects can lead to frequent link disruptions, rapid fluctuations in received signal strength, increased outage probability, and performance losses. To account for and address these challenges, VLC researchers have developed dynamic channel models and mobility-aware handover mechanisms in the literature. For example, the user mobility in indoor VLC systems can be modeled using random way point (RWP) model, random walk model, Manhattan grid model, Gass–Markov mobility model, obstacle aware models, pedestrian mobility models, and group mobility models. More recently, the researchers focus on closed-form mathematical models such as optimal control and clothoid function models which fit well with user mobility measurements [8].

In the recent research works, clothoid functions and optimal control-based models have been used to model the human motion in an indoor environment [30]. We first explain the clothoid function–based human mobility model. The clothoid is a trajectory which satisfies the set of differential equations expressed by

$$\dot{x}(t) = cos(\theta(t)), \tag{12.11}$$

$$\dot{y}(t) = sin(\theta(t)), \tag{12.12}$$

and

$$\dot{\theta}(t) = k_1 + k_2 t. \tag{12.13}$$

The initial conditions can be expressed as

$$x(0) = x_0, \quad y(0) = y_0, \quad \theta(0) = \theta_0, \tag{12.14}$$

where (x_0, y_0) is the coordinates of the initial human user position, and θ_0 is the initial azimuth angle of the user, respectively. k_1 denotes the initial curvature of the trajectory, k_2 the sharpness of the curved trajectory, and t the curvilinear abscissa. The parametric expressions for the clothoid function-based trajectory coordinates can be expressed as

$$x(s) = x_0 + \int_0^t \cos\left(\theta_0 + k_1 l + \frac{1}{2} k_2 l^2\right) dl, \tag{12.15}$$

and

$$y(s) = y_0 + \int_0^t \sin\left(\theta_0 + k_1 l + \frac{1}{2} k_2 l^2\right) dl. \tag{12.16}$$

where l represents the integration variable which is used to evaluate integration from the initial point to the trajectory length t. In order to account for the randomness of the user mobility patterns and obtain multiple trajectories, it is assumed that user mobility trajectories are generated with random Gaussian noise with zero mean and variance σ_p^2 added to the θ_0.

The key concept of optimal control-based mobility patterns is to find a set of random points and fill the intermediate points to complete the user mobility trajectories. These intermediate points can be accurately obtained using the human walking trajectory model presented in [30]. Consider a set of random points along the user mobility trajectory, $\mathbf{q}_i = (x_i, y_i)$, where $i = \{1, \ldots, M\}$. Here, M is defined as the number of points on the room area that can be on the human trajectory pattern. We then obtain the intermediate points of the trajectory as follows. The points between $\mathbf{q}_i$ and $\mathbf{q}_{i+1}$ can be found by solving the following optimization problem [30].

$$\min_{\mathbf{X}(.), \mathbf{U}(.), K_{i+1}} \int_{K_i}^{K_{i+1}} \Phi_k(x(t), \mathbf{u}(t)) dt. \tag{12.17}$$

There are three constraints in this problem, namely (i) dynamic constraint: $\dot{\mathbf{x}}(t) = f(t, \mathbf{x}(t), \mathbf{u}(t))$, (ii) initial condition: $\mathbf{x}(K_i) = \mathbf{x}_i$, and (iii) final condition: $\mathbf{x}(K_{i+1}) = \mathbf{x}_{i+1}$. Here, $\mathbf{x}(t) = (x(t), y(t), \beta(t), v_f, v_o, \omega)^T$ is the state variable and $\mathbf{u}(t) = (u_1(t), u_2(t), u_3(t))^T$ is the control variable. $(x(t), y(t))$ is denoted as the (x, y) coordinates of the human user at the time instance t. $\beta(t)$ is the azimuth angle of the user moving at the time instance t, v_f is the velocity of the human for the forward direction, v_o is the velocity of the human for the orthogonal direction to the movement, ω is the angular velocity of the user, K_i is the travel time until the i-th point considered, respectively. In this setup, $\Phi_k(\mathbf{x}(t), \mathbf{u}(t))$ can be expressed as [30]

$$\begin{aligned}\Phi_k(x(t), u(t)) &= b_0 + b_1 u_1^2(t) + b_2 u_2^2(t) + b_3 u_3^2(t) + b_4 \Psi\left(x(t), x(K_{i+1})\right)^2 \\ &+ b_5\left((x(K_{i+1}) - x(t))^2 + (y(K_{i+1}) - y(t))^2\right) + b_6\left(\beta(K_{i+1}) - \beta(t)\right)^2.\end{aligned} \tag{12.18}$$

Here, we considered $b_0 = 1$, $b_1 = 1.2$, $b_2 = 1.7$, $b_3 = 0.7$, $b_4 = 5.2$, $b_5 = 5$, $b_6 = 8$ in our considered setup. The dynamic constraint equation of the human motion trajectory can be expressed $\dot{\mathbf{x}}(t) = (\dot{x}(t), \dot{y}(t), \dot{\beta}(t), \dot{v}_f, \dot{v}_o, \dot{\omega})^T$. Here, $\dot{x}(t) = cos(\beta(t))v_f - sin(\beta(t))v_o$, $\dot{y}(t) = sin(\beta(t))v_f + cos(\beta(t))v_o$, $\dot{\beta}(t) = \omega$, $\dot{v}_f = u_1$, $\dot{v}_o = u_2$, and $\dot{\omega} = u_3$. Moreover, $(K_{i+1} - K_i)$ is not fixed with a predefined value. However, we obtain K_{i+1} for each $\mathbf{q}_{i+1}$ point, using an iterative search approach. This is followed due to the observation of variable time consumption of movement between two random points on the trajectory. We start this process by setting

$$K_{i+1} = \left(\sqrt{(x(K_{i+1}) - x(K_i))^2 + (y(K_{i+1}) - y(K_i))^2}\right) / v_n + K_i. \tag{12.19}$$

Here, v_n is defined as the nominal speed of the user. Next, we increase $K_{i+1} - K_i$ difference until $\Phi_k(\mathbf{x}(t), \mathbf{u}(t))$ is smaller than a pre-defined threshold value. The azimuth angle $\beta(t)$ and user coordinates $(x(t), y(t))$ for time instance t are recorded while this mobility conditions are maintained. These signals are sampled at ik_cK_s times, where K_s is the time duration between two sampling time slots to obtain the i-th human trajectory coordinates, (x^i, y^i), and azimuth angle of the user at i-th instance, β^i throughout the whole movement. An example use case of the user mobility patterns obtained using above mentioned process and predicted user position values using a neural network-based ML approach is shown in Figure 12.5.

In addition, various methods have been proposed by researchers to combat the performance losses caused by the user mobility in VLC systems. For example, in the research work [31], authors proposed a mobility-aware VLC system that predicts user trajectories to facilitate proactive handover techniques and maintain stable connectivity. On the other hand, adaptive beam steering and receiver diversity approaches have been used to mitigate the effects of receiver orientation, and user mobility-induced misalignment effects in VLC systems [32]. Furthermore, mobility modeling frameworks, such as the random waypoint and pedestrian mobility models, have been integrated into VLC simulations to better understand link stability and latency under realistic movement patterns [33]. These efforts are proposed to enhance the robustness and applicability of VLC systems in dynamic indoor environments.

12.4.3 Random receiver orientation

Random receiver orientation significantly contributes to the performance loss in VLC systems due to LEDs' highly directional optical beam patterns and VLC's dependence on LoS alignment. In indoor environments, mobile users often hold or place their devices at arbitrary angles (when placed on their palm), leading to fluctuations in the angle of light incidence on

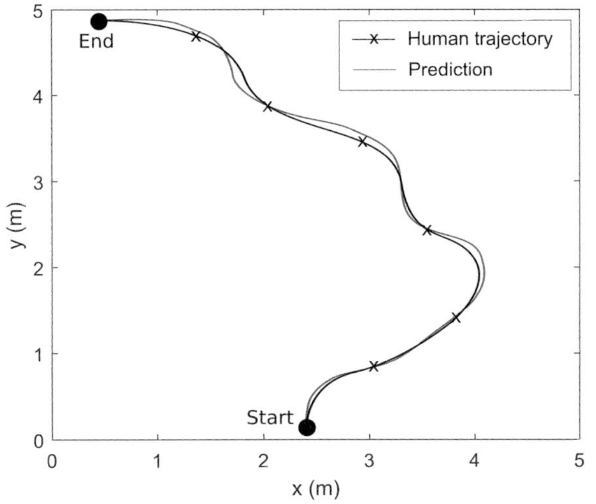

Figure 12.5 Example scenario of an indoor human mobility pattern and its path prediction using neural network-based ML approach.

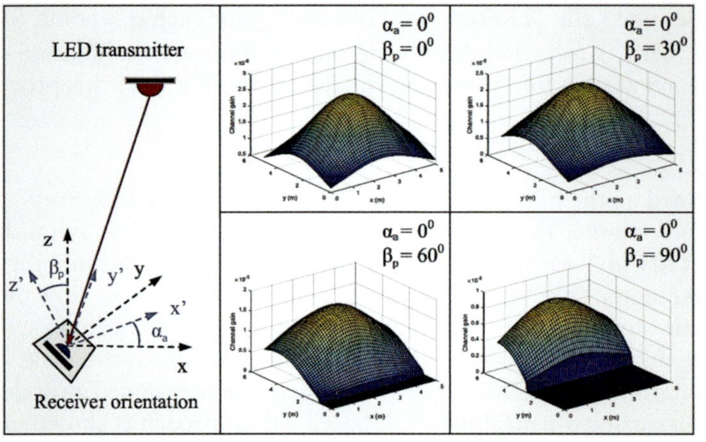

Figure 12.6 Variation of the channel gain on the 2D indoor area for different receiver orientation conditions.

PD. Also, the effective area exposed to the incoming light fluctuates with this scenario. This phenomenon can result in considerable signal attenuation, increase bit error rates, and even complete link outages in practical systems. Some studies have shown that even small deviations in device orientation, such as a smartphone tilting away from the transmitter, can reduce received signal strength by several decibels when highly directional transmitters and low FoV receivers are used. In general, the VLC receiver can be subjected to two independent angles such as azimuth angle and polar angle. The azimuth angle aligns with the user's walking direction (when the mobile device is used in portrait mode), while the polar angle is dependent on the activity (depends on sitting or walking differently). From measurements obtained in the literature, it is verified that azimuth angle is uniform distributed; however, the polar angle is approximated by the Laplacian distributed in sitting activity while approximated by the Gaussian distribution while walking. An example use case for the effect of azimuth (α_a) and polar (β_p) angles on the VLC channel gain is shown in Figure 12.6.

The random receiver orientation modeling based on realistic measurements obtained in an indoor room setup is presented in [34]. Based on the considered measurement setup, the polar angle and the azimuth angle of the mobile device are independent to each other as well as to the user position. The polar angle of the j-th receiver for the walking activity in the i-th time instance, β_j^i, assumed to follow a Gaussian distribution, and adjacent samples are time correlated due to the nature of walking activity. To account for this effect and to model this effect, a correlated Gaussian random process (RP) can be used to generate samples of the polar angle. To this end a first-order linear auto-regressive (AR) model is used to generate the i-th sample of the correlated Gaussian RP explained above, which is expressed as

$$\beta_j^i = c_1 + c_2\beta_j^{i-1} + w_n^i. \tag{12.20}$$

Here, c_1 represents the biased level of the polar angle, c_2 is defined as the constant factor of the AR model, and w_n^i is the AWGN with variance $\sigma_{w_n}^2$ [34]. We calculate the parameters of the AR-based model in Eq. (12.20) as

$$c_2 = \mathbb{R}_{\beta_j}\left(\frac{T_{c,\beta_j}}{T_{sam}}\right)^{\frac{T_{sam}}{T_{c,\beta_j}}}, \tag{12.21}$$

$$c_1 = (1 - c_2)\mathbb{E}\left[\beta_j^i\right], \tag{12.22}$$

and

$$\sigma_{w_n}^2 = (1 - c_2^2)\sigma_{\beta_j}^2. \tag{12.23}$$

Here, $\mathbb{E}(\beta_j^i)$ can be denoted as the mean of the β_j^i, T_{c,β_j} represents the coherence time of the channel, T_{sam} is the sampling time, $\mathbb{R}_\beta(\cdot)$ is defined as the auto-correlation function, and $\sigma_{\beta_j}^2$ it represents the variance of β_j^i. The azimuth orientation angle at the j-th VLC receiver for the i-th time instance, α_j^i, can be calculated using the geometric relations to obtain the angle between the user moving direction and the x-axis in the room's local coordinate frame.

To address this effect of random receiver orientation, researchers have recently proposed orientation-aware channel models and adaptive techniques such as multiple PD arrangements, hemispherical receivers, and real-time orientation tracking to maintain robust connectivity. Incorporating and investigating the challenges inherent from random orientation effects into VLC system design is thus essential for accurately predicting performance and ensuring reliable communication in real-world usage scenarios.

12.4.4 Non-linearity of the LED transmitters

The non-linearity of VLC transmitters, particularly in LEDs, presents a critical challenge to VLC system performance. VLC transmitters typically use IM and DD is used at the receiver, where the transmitted optical signal modulates the LED's optical intensity level. However, LED transmitters exhibit non-linear transfer characteristics, especially outside their optimal dynamic range, which distorts the transmitted signal. When the driving signal exceeds the LED's linear region, it leads to signal clipping or compression, resulting in harmonic distortion, inter-modulation products, and spectral broadening. These distortions degrade system performance by increasing BER, reducing achievable data rates, and complicating signal detection at the receiver.

To mitigate these non-linear effects, several solutions have been proposed in the literature. One such common approach is using pre-distortion techniques, where the signal is pre-compensated before transmission to counteract the LED's non-linearity. Digital predistortion techniques can model and inverse the non-linear characteristics of the LEDs, effectively linearizing the overall system response. Another approach involves peak-to-average power reduction approaches, especially VLC in systems that use OFDM, which is highly susceptible to LED non-linearity. Additionally, bias optimization ensures the LED operates within its linear region by carefully selecting the DC bias point. Some studies also explore modulation formats with inherently low sensitivity to non-linearity, such as ACO-OFDM or CAP. Collectively, these methods help enhance VLC performance and enable reliable high-speed communication despite the inherent limitations of LED transmitters.

12.5 EMERGING FUTURE DIRECTIONS

As a promising technology, VLC can be integrated with most of the emerging technologies to provide more superior, robust, and practical solutions to the challenges in wireless communication systems and applications. In the following section, we discuss the possible future research directions for future VLC systems. In particular, we discuss the use of artificial intelligence, RISs, semantic communication, quantum communication, and liquid lens technology in VLC systems.

12.5.1 Artificial intelligence

ML techniques are increasingly being used to tackle physical-layer challenges in VLC, such as channel estimation, non-linearity compensation, and signal detection. For example, neural-network-based channel estimation frameworks can predict VLC channel taps under varying ambient conditions and device orientation, reducing error compared to traditional linear estimators. Similarly, CNNs, deep belief networks (DBNs), and AdaBoost classifiers have been applied to real-world VLC signal datasets to perform modulation classification and demodulation, achieving improved accuracy even at longer distances or with higher-order modulation schemes. These AI techniques offer robust solutions in non-linear and time-varying environments where conventional algorithms struggle.

At the network layer, AI in particular—reinforcement learning (RL) and multi-armed bandit algorithms—is used to optimize resource allocation and handover decisions in dynamic VLC environments. A collaborative online learning-based handover scheme (COLH), for instance, clusters users sharing similar feedback and learns which access point (VLC or RF) will maximize link duration, without needing precise location information. Deep-LSTM and other recurrent models have also been deployed to predict LoS outages and user trajectories, allowing the system to proactively adjust handovers and resource allocation in hybrid RF/VLC networks. Such AI-driven mobility prediction enhances reliability and reduces unnecessary service interruptions in smart indoor communication systems.

12.5.2 Reconfigurable intelligent surfaces

RISs are gaining significant attention in the realm of wireless communications and are now being explored as a transformative technology for VLC systems. RIS consists of programmable surfaces composed of passive or semi-passive elements that can dynamically manipulate the direction, phase, and intensity of incident light to control the optical propagation environment. In VLC, RIS can be strategically deployed on ceilings, walls, or other indoor structures to overcome LoS limitations, to overcome signal blockages and to extend communication coverage in complex environments. Recent research has shown that liquid crystal (LC)-based RIS and micro-electromechanical systems (MEMS)-based RIS can provide beam steering and beamforming capabilities without requiring active power-hungry components. By adapting to user mobility and environmental dynamics, RIS enables flexible and energy-efficient channel reconfiguration, significantly enhancing system reliability and spectral efficiency. Furthermore, RIS can be integrated with multiple-input multiple-output (MIMO)-based wireless systems and NLOS VLC systems to support high user density and dynamic traffic conditions. Despite its promise, RIS for VLC remains in early development stages, and key challenges such as precise optical modeling, real-time control, and scalability must be addressed to fully exploit its potential in next-generation VLC networks.

12.5.3 Semantic communication

Semantic communication, which focuses on communicating the meaning and relevance of information rather than exact data reproduction, is emerging as a promising approach in wireless communication systems, as it can drastically reduce the resource requirement, especially the bandwidth requirement. Recently, there is a growing interest toward using semantic communication for enhancing VLC efficiency. By leveraging semantic-aware encoding, ML, and context-driven inference, VLC systems can prioritize the delivery of critical information, reduce bandwidth consumption, and improve robustness under challenging channel conditions such as mobility, blockage, or misalignment. This paradigm shift is particularly relevant for IoT and edge intelligence scenarios, where devices may have limited computational and communication resources. Integrating semantic communication into VLC opens up new avenues for low-latency and energy-efficient transmission, especially in human-centric and mission-critical applications. In particular, semantic communication can be promising for resource-limited UVLC systems in which limited transmit energy is available in underwater sensor nodes. Moreover, OCC as a special sector of VLC can be combined with semantic communication and beneficial for underwater systems, especially for highly turbid, and misalignment-expected underwater environments as OCC provides robust communication while semantic autoencoders provide a way to extract useful information for underwater communication. Such systems are critical for underwater exploration, and seabed mapping activities.

Despite these advances, realizing semantic VLC systems requires interdisciplinary research that bridges information theory, natural language processing, and optical communication, along with the development of new performance metrics that go beyond traditional bit error rate and throughput. Moreover, the optimized design of semantic autoencoders that suits the VLC applications is another research direction. In particular, researchers are focusing on generative adversarial networks, and advanced deep learning solutions. In this regard, architectures should be carefully selected depending on the selected application, the type of semantic information to be transmitted, and the inputs/outputs.

12.5.4 Quantum communication

Quantum communication in VLC systems is an emerging interdisciplinary research area that explores integrating quantum principles—such as quantum key distribution (QKD)—into optical wireless environments using visible spectrum light sources like LEDs and lasers. Unlike traditional quantum communication that typically operates in the near-infrared, VLC-based quantum systems seek to leverage the ubiquity and low-cost infrastructure of LEDs for secure data transmission. However, implementing quantum protocols over visible light poses unique challenges, including high background noise from ambient lighting, limited photon-counting sensitivity in commercial receivers, and strict requirements for LoS alignment. Despite these obstacles, preliminary studies have demonstrated the feasibility of using polarization or IM of single photons for QKD in controlled indoor VLC settings. With advancements in single-photon detectors, modulation techniques, and noise-resilient protocols, quantum-enhanced VLC holds promise for delivering highly secure wireless communication in future smart environments.

12.5.5 Liquid lens technology

Recent advancements in optical lens technology, particularly those findings that use adaptive optics, have led to the emergence of reconfigurable and tunable liquid lens architectures. These innovations offer significant potential to enhance the performance of VLC systems by enabling dynamic optimization of lens parameters for improved signal transmission and reception efficiency [17, 35–39]. Among these approaches, LC-based lens structures have garnered attention for their ability to modulate the refractive index and manipulate light propagation in VLC applications. In [35], LC-based IRSs were implemented at the transmitter side to improve data rate uniformity while maintaining appropriate illumination levels. Conversely, in [37], LC-based IRSs positioned at the receiver can dynamically steer incoming light beams toward the PD, exactly to the effective area of the PD, enhancing overall system performance. Additionally, mechanical structure-based liquid lenses have been studied for potential uses in dynamic and robust beam steering capacity. For example, in the research work of [17], an adaptive/tunable liquid lens with three degrees of freedom, including focal length, azimuth angle, and polar angle was proposed for suppressing the inter-channel interference in MIMO VLC systems, thus improving communication reliability. While mechanical liquid lenses have shown promise, non-mechanical tunable liquid lenses (TLLs) have emerged as a compelling alternative due to their ability to alter the shape and orientation of the liquid interface without mechanical components. These lenses offer smooth and precise control over light direction, crucial for high-performance VLC systems. In [36], a 3D-capable beam steering method using an electrowetting surface-based TLL structure combined with a liquid prism was introduced, offering fine-grained control of beam orientation. Similarly, [39] demonstrated effective one/two-dimensional beam steering using various multiple liquid lenses, showing the enhanced beam control capabilities of this technology. More recently, the applicability of TLLs in VLC systems has been explored in [38], where an electrowetting-based cuboid-shaped TLL was integrated into a receiver to counteract the performance losses inherent in dynamic user conditions such as user mobility and random receiver orientation. These developments mark a promising step toward practical and adaptive VLC system implementations.

Despite these initial efforts, the integration of liquid lens technology into VLC systems remains in its nascent stage. Significant research gaps persist, particularly regarding the deployment of TLLs to counter challenges in VLC systems such as random user mobility, random receiver orientation, transmitter-receiver misalignment, and LoS blockage effects. Furthermore, TLLs hold the potential to work in synergy with wall-mounted IRSs, improving system resilience under dynamic and obstructive environments. In MIMO and multiuser VLC scenarios, TLLs can be strategically employed at the transmitter and/or receiver ends to reduce interference even before digital signal processing, thus elevating system throughput and reliability. However, to realize these benefits, deeper investigations into the physical characteristics and control mechanisms of TLLs are essential. Research is needed to determine optimal lens shapes, suitable liquid materials, and effective mechanical versus non-mechanical actuation techniques. Moreover, to support analytical and simulation-based studies, simplified yet realistic channel models must be developed. These models should account for the physical and operational limitations of TLLs, enabling the theoretical evaluation of TLL-assisted VLC systems under various practical constraints.

12.5.6 Optical ISAC

Integrated Sensing and Communication (ISAC) in the context of VLC is an emerging research direction that seeks to combine data transmission and environmental sensing into a unified optical system. While ISAC has been widely studied in the RF domain, its extension to the optical domain referred to as Optical ISAC (O-ISAC) leverages the unique properties of light, such as high directionality, spatial resolution, and immunity to RF interference, to simultaneously support high-speed data transmission and precise indoor sensing or localization.

In VLC-based O-ISAC, the light emitted by LEDs or LDs is used not only for transmitting data but also to sense parameters like user position, gesture, mobility patterns, and even surface reflectivity. For example, by analyzing the reflected light intensity and time-of-flight (ToF) or phase shift, the system can estimate the location or movement of users or objects, while still maintaining reliable communication links. This enables applications like human activity recognition, occupancy detection, or gesture-based control in smart homes and offices. Recent works propose using modulated light patterns and PD arrays or image sensors to extract both communication data and spatial information in real time, with minimal hardware overhead [40]. Moreover, pinhole camera-based O-ISAC systems have been proposed as more robust solutions to improve the sensing accuracy. Further, O-ISAC can be a promising solution for underwater communication systems where RF-based sensing is extremely challenging.

However, integrating ISAC into VLC comes with challenges such as trade-offs between data rate and sensing resolution, managing ambient light interference, and ensuring synchronization between sensing and communication tasks. Research is ongoing to develop joint waveform designs, sensing-aware modulation schemes, and deep learning-based signal processing to balance these objectives. As VLC infrastructure expands in indoor environments, Optical ISAC is poised to become a key enabler of context-aware, intelligent wireless systems that go beyond communication to interact with and interpret the environment.

12.6 CONCLUSION

The recent interest in high data rate, robustness, low latency, and high security in wireless communication systems has led to VLC systems. Although the applicability of VLC systems is limited in certain applications, VLC can enable a wide range of novel wireless applications, opening the way to next-generation networks such as 6G. In this book chapter, the fundamentals of VLC systems were discussed, focusing on transceiver architecture, channel models, modulation techniques, and multiple access. We also categorized the applications of VLC systems into indoor, underwater, vehicular, and aerial scenarios. In addition, the practical challenges and limitations of VLC and possible solution approaches were also highlighted. As an emerging technology, possible research directions and integrations with modern technologies and technologies that are expected to emerge were also studied. Studies reveal that VLC can be a promising solution for an extremely wide range of emerging wireless communication applications for future wireless systems, particularly for short-range high data rate applications.

REFERENCES

1. Hany Elgala, Raed Mesleh, and Harald Haas. Indoor optical wireless communication: potential and state-of-the-art. *IEEE Commun. Mag.*, 49:56–62, Sept. 2011.
2. Parth H. Pathak, Xiaotao Feng, Pengfei Hu, and Prasant Mohapatra. Visible light communication, networking, and sensing: a survey, potential and challenges. *IEEE Commun. Surveys Tutor.*, 17:2047–2077, Sept. 2015.
3. Toshihiko Komine and Masao Nakagawa. Fundamental analysis for visible-light communication system using LED lights. *IEEE Trans. Consum. Electron.*, 50:100–107, Feb. 2004.
4. Hemani Kaushal and Georges Kaddoum. Underwater optical wireless communication. *IEEE Access*, 4:1518–1547, Apr. 2016.
5. Harald Haas, Liang Yin, Yunlu Wang, and Cheng Chen. What is LiFi? *J. Lightwave Technol.*, 34:1533–1544, Mar. 2016.
6. Nan Chi, Yingjun Zhou, Yiran Wei, and Fangchen Hu. Visible light communication in 6G: Advances, challenges, and prospects. *IEEE Veh. Technol. Mag.*, 15:93–102, Dec. 2020.
7. Yusuf Said Eroğlu, Yavuz Yapıcı, and İsmail Güvenç. Impact of random receiver orientation on visible light communications channel. *IEEE Trans. Commun.*, 67:1313–1325, Feb. 2019.
8. Kapila W. S. Palitharathna, Himal A. Suraweera, Roshan I. Godaliyadda, Vijitha R. Herath, and Zhiguo Ding. Neural-network-based blockage prediction and optimization in lightwave power-transfer-enabled hybrid VLC/RF systems. *IEEE Internet Things J.*, 11:5237–5248, Feb. 2024.
9. Anas Chaaban, Zouheir Rezki, and Mohamed-Slim Alouini. On the capacity of intensity-modulation direct-detection gaussian optical wireless communication channels: a tutorial. *IEEE Commun. Surv. Tutor.*, 24:455–491, 1st Quat. 2022.
10. Panagiotis D. Diamantoulakis, George K. Karagiannidis, and Zhiguo Ding. Simultaneous lightwave information and power transfer (SLIPT). *IEEE Trans. Green Commun. Netw.*, 2:764–773, Sep. 2018.

11. Vasilis K. Papanikolaou et al. Simultaneous lightwave information and power transfer in 6G networks. *IEEE Commun. Mag.*, 62:16–22, Mar. 2024.
12. Sander Bastiaens, Morteza Alijani, Wout Joseph, and David Plets. Visible light positioning as a next-generation indoor positioning technology: a tutorial. *IEEE Commun. Surv. Tutor.*, 26:2867–2913, 4th Quat. 2024.
13. Lucas Teixeira et al. A review of visible light communication LED drivers. *IEEE J. Emerg. Sel. Top. Power Electron.*, 10:919–933, June 2022.
14. Tai-Cheng Yu et al. Visible light communication system technology review: devices, architectures, and applications. *Crystals*, 11, Sep. 2021.
15. Mehmet A. Kucuk, Cenk Albayrak, Kadir Turk, and Halim Yanikomeroglu. Optical filter design for multi-color VLC in reflective environments with multiple lighting sources. *IEEE Access*, 12:169344–169355, Nov. 2024.
16. Cuiwei He and Chen Chen. A review of advanced transceiver technologies in visible light communications. *Photon.*, 10, June 2023.
17. Kapila W. S. Palitharathna, Christodoulos Skouroumounis, and Ioannis Krikidis. Liquid lens-based imaging receiver for MIMO VLC systems," *IEEE Trans. Commun.*, 73:11663–11678, 2025.
18. Kapila W. S. Palitharathna, Himal A. Suraweera, Roshan I. Godaliyadda, Vijitha R. Herath, and John S. Thompson. Average rate analysis of cooperative NOMA aided underwater optical wireless systems. *IEEE Open J. Commun. Soc.*, 2:2292–2310, Sept. 2021.
19. Changmin Lee et al. 26 Gbit/s LiFi system with laser-based white light transmitter. *J. Lightwave Technol.*, 40:1432–1439, Mar. 2022.
20. Wen-Hsuan Shen and Hsin-Mu Tsai. Testing vehicle-to-vehicle visible light communications in real-world driving scenarios. In *Proc. IEEE Veh. Netw. Conf. (VNC 2017)*, pages 187–194, Feb. 2017.
21. Masayuki Kinoshita et al. Motion modeling of mobile transmitter for image sensor based I2V-VLC, V2I-VLC, and V2V-VLC. In *Proc. IEEE Globecom Works. (GC Wkshps 2014)*, pages 450–455, Mar. 2014.
22. Giorgio Corbellini, Kaan Aksit, Stefan Schmid, Stefan Mangold, and Thomas R. Gross. Connecting networks of toys and smartphones with visible light communication. *IEEE Commun. Mag.*, 52:72–78, July 2014.
23. Yining Wang, Mingzhe Chen, Zhaohui Yang, Tao Luo, and Walid Saad. Deep learning for optimal deployment of UAVs with visible light communications. *IEEE Trans. Wirel. Commun.*, 19:7049–7063, Nov. 2020.
24. Mohammad Reza Maleki, Mohammad Robat Mili, Mohammad Reza Javan, Nader Mokari, and Eduard A. Jorswieck. Multi-agent reinforcement learning trajectory design and two-stage resource management in CoMP UAV VLC networks. *IEEE Trans. Commun.*, 70:7464–7476, Nov. 2022.
25. Xiping Wu. Performance analysis of visible light communications with channel blockage caused by human bodies. In *Proc. IEEE Int. Conf. Commun. (ICC 2023)*, pages 6541–6545, Rome, Italy, May 2023.
26. Sylvester Aboagye, Telex M. N. Ngatched, Octavia A. Dobre, and Alain R. Ndjiongue. Intelligent reflecting surface-aided indoor visible light communication systems. *IEEE Commun. Lett.*, 25:3913–3917, Dec. 2021.
27. Hanaa Abumarshoud, Bassant Selim, Mallik Tatipamula, and Harald Haas. Intelligent reflecting surfaces for enhanced NOMA-based visible light communications. In *Proc. IEEE ICC 2022*, pages 571–576, Seoul, South Korea, 2022.
28. Jona Beysens, Qing Wang, and Sofie Pollin. Exploiting blockage in VLC networks through user rotations. *IEEE Open J. Commun. Soc.*, 1:1084–1099, July 2020.
29. Anand Singh, Gourab Ghatak, Anand Srivastava, Vivek Ashok Bohara, and Anand Kumar Jagadeesan. Performance analysis of indoor communication system using off-the-shelf LEDs with human blockages. *IEEE Open J. Commun. Soc.*, 2:187–198, Jan. 2021.
30. Isabelle Maroger, Olivier Stasse, and Bruno Watier. Walking human trajectory models and their application to humanoid robot locomotion. In *IEEE/RSJ Int. Conf. Intell. Robots Syst. (IROS 2020)*, pages 3465–3472, Las Vegas, USA, Feb. 2020.
31. Mohammad Amir Dastgheib, Hamzeh Beyranvand, Jawad A. Salehi, and Martin Maier. Mobility-aware resource allocation in VLC networks using t-step look-ahead policy. *J. Lightwave Technol.*, 36:5358–5370, Dec. 2018.

32. Ahmed Taha Hussein, Mohammed T. Alresheedi, and Jaafar M. H. Elmirghani. Fast and efficient adaptation techniques for visible light communication systems. *J. Optical Commun. Netw.*, 8:382–397, June 2016.
33. Rong Zhang, Mauro Biagi, Lutz Lampe, Thomas D. C. Little, Stefan Mangold, and Zhengyuan Xu. Guest editorial localisation, communication and networking with VLC. *IEEE J. Sel. Areas Commun.*, 36:1–7, Jan. 2018.
34. Mohammad Dehghani Soltani, Ardimas Andi Purwita, Zhihong Zeng, Harald Haas, and Majid Safari. Modeling the random orientation of mobile devices: measurement, analysis and LiFi use case. *IEEE Trans. Commun.*, 67:2157–2172, Mar. 2019.
35. Sylvester Aboagye, Telex M. N. Ngatched, Alain R. Ndjiongue, Octavia A. Dobre, and Hyundong Shin. Liquid crystal-based RIS for VLC transmitters: performance analysis, challenges, and opportunities. *IEEE Wirel. Commun.*, 31:98–105, Aug. 2024.
36. Junsik Lee, Jooho Lee, and Yong Hyub Won. Nonmechanical three-dimensional beam steering using electrowetting-based liquid lens and liquid prism. *Opt. Express*, 27:36757–36766, Dec. 2019.
37. Alain R. Ndjiongue, Telex M. N. Ngatched, Octavia A. Dobre, and Harald Haas. Re-configurable intelligent surface-based VLC receivers using tunable liquid-crystals: the concept. *J. Lightwave Technol.*, 39:3193–3200, May 2021.
38. Kapila W. S. Palitharathna, Constantinos Psomas, and Ioannis Krikidis. Outage probability analysis of tunable liquid lens-assisted VLC systems. *arXiv preprint arXiv:2505.14236*, 2025.
39. Mo Zohrabi, Robert H. Cormack, and Juliet T. Gopinath. Wide-angle nonmechanical beam steering using liquid lenses. *Opt. Express*, 24:23798–23809, Oct. 2016.
40. Yunfeng Wen, Fang Yang, Jian Song, and Zhu Han. Optical integrated sensing and communication: architectures, potentials and challenges. *IEEE Internet Things Mag.*, 7:68–74, July 2024.

Chapter 13

Green wireless communication

Baljeet Ram, Jaya Yadav, Vipin Kumar, Amit Kumar Singh, and Nikhil Kumar Rajput

13.1 INTRODUCTION

Green wireless communication focuses on establishing wireless networks that are as efficient as well as eco-friendly. The core idea of this concept is to reduce the carbon footprint in communication systems while guaranteeing better performance and reliability [1]. It adopts diverse technologies such as 5G and B5G communication networks, Wi-Fi systems, network systems, sensor networks, and satellite communications. This initiative covers improvements in hardware, software, network design, and operational strategies to minimize energy consumption throughout the entire system, from user equipment (UE) to core network infrastructure [2].

The massive growth of mobile devices, smartphones, and internet of things (IoT) devices, and emerging services like autonomous vehicles, Augmented Reality/ Virtual Reality (AR/ VR) are ever-increasing. Some of these services require high data rates (e.g., HD Video streaming), some require low latency (e.g., medical operations), while some require both high data rates and low latency (e.g., AR/VR). Meeting these diverse requirements with traditional wireless systems is challenging. Furthermore, these systems consume more energy and hence are not suitable for a green environment. The existing algorithms and approaches are unable to fulfill users' demands while consuming less energy [3]. To overcome the limitations of the existing approach, green wireless communication mainly focuses on achieving high data rates, low latency, and simultaneously addressing the issue of making the environment green using energy efficiently.

The objective of green wireless communication is to create a wireless system that emits less CO_2 and consumes less energy. In addition, it focuses to on facilitating QoS and QoE (Quality of Experience) services to the intended users. The emerging technologies, such as 5G and 6G, are designed mainly to be environmentally friendly. For green wireless communication, researchers are looking at numerous approaches to provide optimal solutions to reduce the power consumption. Such approaches are optimizing RAN technologies, using energy-efficient hardware, and introducing smart sleep modes for base stations [4]. However, the impact of making wireless systems green extends beyond environmental benefits. It cuts down operational expenses (OPEX) for service providers and encourages vendors to provide optimum and sustainable solutions. It becomes possible because of multi-vendor acceptance of solutions.

Open-Radio Access Networks (O-RAN) is based on the fundamentals of openness, disaggregation, intelligence, and virtualization. These fundamentals provide flexibility and scalability in designing and developing functions involved in O-RAN. The openness encourages the research community to participate and innovate for next-generation wireless systems. Using the existing methods in O-RAN may cause more energy consumption because of disaggregation. Without synchronization between components, they consume

DOI: 10.1201/9781003637721-13

more energy [5, 6]. Even if synchronization is achieved, these methods are not energy-saver and energy-efficient. Therefore, new energy-efficient methods and techniques are required to attain the objective of green communication. Furthermore, the 5G and B5G technologies also focus on high data rate, low latency, and fulfilling diverse users' demands. The energy-efficient methods should also satisfy users' requirements. Dynamic resource allocation, smart power scaling, virtualization, and artificial intelligence (AI)-driven traffic prediction are some of the techniques that can make network components work and use energy efficiently based on current demands [7]. The energy can also be saved by allowing components to switch to low-power or sleep modes when they are not in use. Such techniques should minimize Capital Expenditure (CAPEX), OPEX, and achieve global sustainability goals [8].

In this chapter, we address the importance of green communication, rising energy costs, fundamental architecture of O-RAN, different methods for energy efficiency, and serving users' requirements. The aim of green wireless communication is to lower energy use, cut carbon emissions, improve energy efficiency at all network levels, and hence create environmentally friendly and cost-effective wireless networks. The requirements for fast, uninterrupted connectivity increase energy consumption in communication systems. The architecture of O-RAN exhibits different components and efforts to make green and sustainable communication. The foremost efforts include optimizing network design, using renewable energy sources, and implementing smart power management strategies [9]. The AI/machine learning (ML) algorithms on RAN Intelligence Control (RIC) are exploited to make policies that help to make decisions for allocating resources to users, handover, network congestion, and switching between ON and OFF. These algorithms are gaining momentum because of their suitability for efficiently providing solutions to challenges prevailing in wireless networks. However, they require more research for reducing energy consumption and the complexities of real network environments. These AI/ML models may work well under ideal conditions, but overlook issues like user mobility and changing traffic patterns. Moreover, they are designed for a RAN component or specific network layer, but when they are integrated with the system, their performance decreases [10]. The usage of renewable energy resources is drawing attention for the green and sustainable development of wireless networks.

13.2 GREEN 5G AND BEYOND

The exponential growth of mobile devices and emerging services like AR/VR, tactile Internet has created the need to evolve wireless communication and think differently to fulfill users' demands. While 5G and the emerging 6G systems promise high speed, low latency, and ultra-reliability, there is growing concern about their energy consumption. The green architectural network is becoming more pivotal in 5G and 6G due to global concern about sustainable wireless communication, energy saving, and protecting the environment. In 5G communications, efforts to improve energy efficiency mainly focus on making the network denser. This is done by deploying small cells to reduce transmission distances and power consumption. The cells can be placed strategically to increase the data rate while using less energy than traditional macro cell base stations [11]. However, the large-scale MIMO has a large number of antennas, which have the potential to control precisely the direction of radio signals using beamforming. It provides better spectral and energy efficiency to reduce power consumption. Another significant aspect of 5G is network slicing, which allows the creation of multiple virtual networks over a single physical infrastructure. In this, each slice can be customized for different services, such as enhanced mobile broadband (eMBB), ultra-reliable low latency communications (uRLLC), and massive machine-type communications (mMTC). Meanwhile, software-defined network (SDN) and network function virtualization (NFV) further contribute to reducing unnecessary energy consumption by controlling idle resources centrally during off-peak times [12].

The 6G technology focuses even more on eMBB, uRLLC, and mMTC with sustainable energy efficiency and cost-effectiveness. This technology is expected to provide data speeds of up to 1 terabit per second, ultra-fast communication, and a wide range of connected devices. With the increasing attention toward green communications, cell-free massive MIMO has been identified as one of the most promising solutions. Cell-free networks differ from traditional cellular systems, which are divided into cells, causing interference, signal overlap, and more energy consumption. Intelligent reflecting surfaces (IRS) are another development for 6G architecture. These surfaces represent a powerful paradigm shift as they enable the radio environment to be altered by reconfiguring its electromagnetic properties, allowing signals to be reflected throughout dense urban spaces and guide them in the desired direction so that users can receive all intended information without consuming power to boost the signal. AI or ML applications are well-suited for 5G and beyond energy-efficient designs [13]. Of these, AI can be used for predictive and adaptive energy management – for example, turning base stations on or off according to traffic patterns, and optimizing user scheduling to balance load while minimizing energy consumption. Furthermore, edge computing is key to energy consumption reduction by bringing data processing closer to the end user. These decrease latency, and as the traffic of data in the core is offloaded, it reduces energy for long-distance data moving. Figure 13.1 shows the typical diagram of Green 5G and B5G technology.

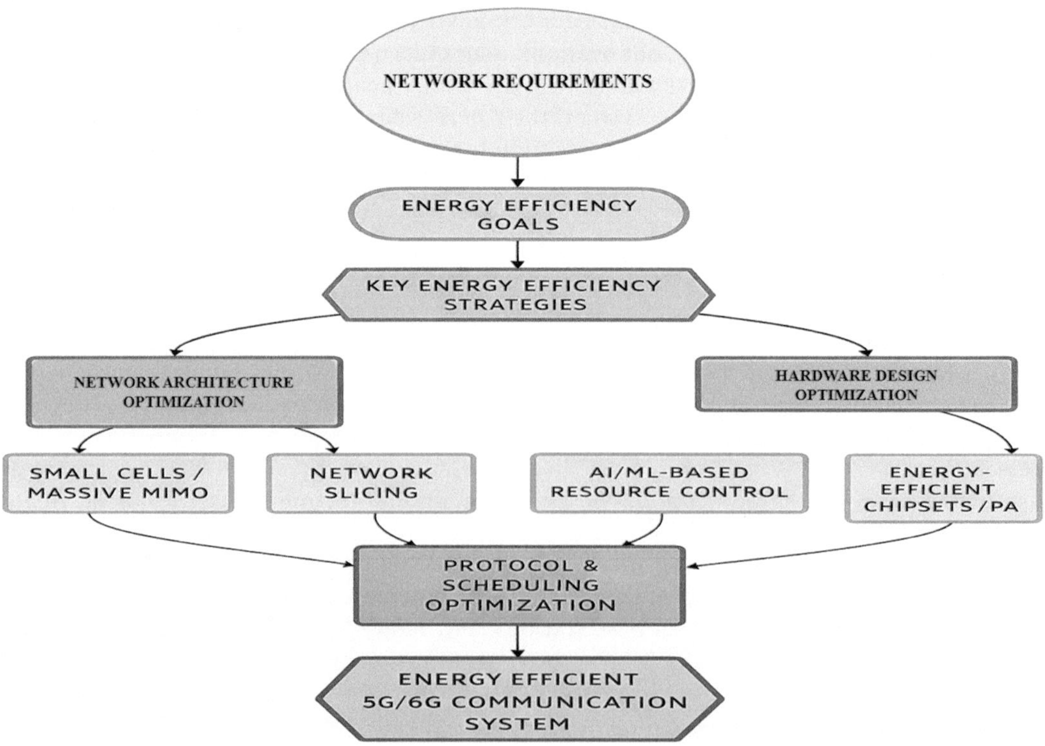

Figure 13.1 Typical diagram of green 5G and its beyond.

13.3 O-RAN IN NEXT-GENERATION WIRELESS NETWORKS

The O-RAN Alliance is a consortium focused on transforming traditional RAN into an open, intelligent, and virtualized architecture aimed to deliver diverse user demands in a cost-effective and energy-efficient manner for future mobile networks. It can achieve dynamic and optimum allocation of resources and supports real-time energy management through open, interoperable, and virtualized functions. Using AI/ML algorithms, O-RAN can manage and predict traffic patterns, which can lead to switching off inactive network elements. Furthermore, its disaggregated components of multi-vendors, including open energy-efficient hardware and software solutions promise to support scalable and cost-effective network deployments while lowering energy consumption. This multi-vendor and open solution helps to make environmentally friendly wireless communications, which makes O-RAN a key framework for sustainable and innovative 5G and 6G networks [14].

13.3.1 Fundamental architecture of O-RAN

The O-RAN is built on the principles of disaggregation, openness, and virtualization. It is crucial enabler of implementing the potential of 5G and B5G technologies by introducing intelligence and flexibility. It disaggregates traditional RAN (i.e., radio head and band-based unit) into three components connected by standardized open interfaces: open radio unit (O-RU), open distributed unit (O-DU), and open central unit (O-CU), as shown in Figure 13.2. O-RU handles lower physical layer and RF processing, whereas O-DU, connected by an open front haul interface with O-RU, manages real-time protocols, which are higher physical layer (High-PHY), medium access control (MAC), and radio link control (RLC) protocols. O-CU connected with O-DU through an open mid-haul interface

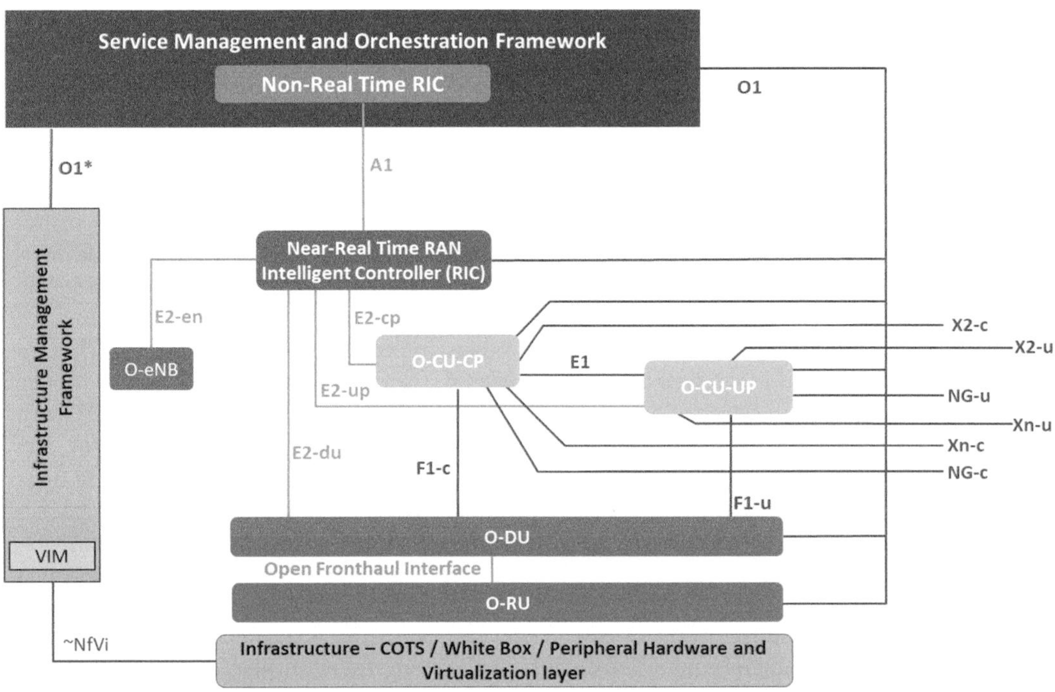

Figure 13.2 Logical architecture of O-RAN [31].

(F1 interface), which handles higher layer protocols, namely radio resource control (RRC), packet data convergence protocol (PDCP), and service data adaptation protocol (SDAP). O-CU is further divided into O-CU-CP (control plane) and O-CU-UP (user plane) [31].

The service management and orchestration (SMO) framework is an intelligent components of O-RAN. It is an intelligent cloud native framework for automating, managing, and orchestrating the entire O-RAN network. Its primary functions include orchestration and lifecycle management, non-RT RIC, operations, administration and maintenance (OAM), and data collection and management. Its objective is improving network performance, enhancing user experience, reducing operational cost, and increasing energy efficiency. The flexibility, openness, and sustainability objectives envisioned for 5G and 6G networks are built into SMO. AI/ML-based algorithms running on non-RT RIC forecast network traffic and generate policies [15]. SMO communicates with the non-RT RIC to run AI/ML applications through rApps. These rApps include strategies for saving energy, forecasting traffic, and controlling the network based on policies that support eco-friendly wireless communication. This enrichment information and real-time traces help SMO to make strategic decisions like shutting down underused network components, adjusting transmission power levels, and reallocating resources to meet demands while focusing on energy efficiency [16].

SMO also opens the door for multiple vendors to integrate network components. Academicians small-scale industries can design and develop different network functions and models. They can provide different solutions to problems like traffic prediction, energy optimization, and controlling network components by developing rApps. The operator can choose a solution that fits their requirement and user satisfaction. In the future, it can choose another solution without affecting the remaining system. These solutions are not only QoS effective but they are also energy-efficient and hence promote the development of power-optimized hardware and software solutions. SMO offers detailed control over power consumption in the RAN and allows precise control of energy-aware scheduling. It controls power usage according to the load on the network. Integration with edge computing reduces latency and energy use by analyzing data closer to the user. This also enables smart, scalable, and eco-friendly operations for the O-RAN architecture [17]. Thus, SMO plays an important role in reducing operational costs and carbon footprints. Its central management, AI-driven optimization, and vendor-neutral orchestration provide technical efficiency.

Another intelligent component of O-RAN is near-RT RIC. It aims to improve the intelligence, flexibility, and efficiency, operating within a time range of 10 milliseconds to 1 second, the near-RT RIC manages and improves radio network functions almost instantly using customizable, component-based apps called xApps. The near-RT RIC interacts with the underlying RAN elements, including the O-DU and O-CU, through standardized E2 interfaces. This feature enables policy-based control, dynamic resource allocation, and efficient QoS control across the network. One of its benefits is its use of AI/ML algorithms to evaluate a network's condition quickly, which helps decision-makers improve performance, mobility, and energy efficiency [18]. In green wireless communications, the most important energy-efficient functions are load balancing, on/off BS sleep mode, power control, and interference control. For instance, xApps supervising a user and control plane traffic flows can command a RAN element, like an RU, to reduce transmit power or shut off a part of the RAN that is not used, resulting in energy savings without a deterioration in quality of service. Near-RT RIC allows operators to connect different RAN components and applications easily [19].

Thus, the O-RAN architecture is designed and developed for enhancing QoS, improving network performance, optimizing energy, and encouraging small industries through vendor neutrality. This enables the goal of building sustainable and eco-friendly communication infrastructure in 5G and B5G.

13.3.2 O-RAN for network sustainability

Wireless communication networks are now recognized to have significant economic and ecological effects in the forthcoming years [20] due to 5G and B5G emergence, which facilitates heterogeneous services and provides on-demand services. This vast connectivity is acknowledged as a concern for both the environment and financial resources. In light of this, a novel approach to "green wireless networks", which focuses on the environmental effects on cellular networks, has inspired many researchers. The O-RAN is a promising technology that provides a new way to tackle such issues, introducing various concepts such as open, disaggregated, virtualized, and intelligent network structure, as existing RAN architecture is typically built with proprietary-based solutions from particular vendors. This leads to vendor lock-in, which restricts interoperability, reduces flexibility, and leads to higher deployment and operating costs [21]. The O-RAN architecture enables flexible resource allocation and network utilization via open and standardized interfaces that allow vendor neutrality. Due to this, small operators can participate to provide high-quality solutions tailored to the network's needs. They can propose more energy-efficient solutions and boost competition. Now, the network provider can integrate hardware and software components from various providers to achieve two main objectives. It comes under the growing interest in developing sustainable networks; the first one is reducing the operational expenditure for operators, and the other one is protecting the environment by minimizing CO_2 emissions. It is easy for operators to independently replace or update single components with new energy-efficient technologies and hardware without affecting the overall network infrastructure.

Simultaneously, RAN virtualization reduces the need for dedicated physical infrastructure, resulting in higher resource utilization without any hardware deployment. In addition, this reduces energy consumption and maintenance costs [22]. Even more, the cloud-based approach makes it easy to adjust the network functions based on the traffic pattern, which works toward higher energy efficiency and a minimal utilization of idle resources. At the same time, unlocking intelligent components inside the SMO framework automates tasks such as configuration, monitoring, and fault detection, which cuts down manual work and reduces labor costs [23]. Through SMO, traffic prediction, smart scheduling, and energy consumption management can drastically reduce the energy usage. The scalability and flexibility attributes of O-RAN design allow for the automated scaling of respective network elements by using accurate coverage and capacity requirements. For example, if traffic goes up in a particular region, only those E2 nodes need to be updated or modified. This also saves unnecessary expenditure on the overall system [24]. O-RAN design employs AI and ML algorithms to improve network efficiency and decrease energy consumption. Near-RT RIC and non-RT RIC are part of the architecture. These controllers provide configurable programs known as xApps and rApps. They cater to various responsibilities related to energy saving, which is explained below [25].

Predictive Load Management: AI plays a vital role in upgrading network management performance and efficiency using its predictive and adaptive capabilities. AI algorithms learn the traffic patterns and user mobility that enable them to predict real-time demand. This allows the network to switch cell activation/deactivation and power levels based on current demand and predicted demand. By analyzing operational data, AI systems learn to facilitate better decision-making and improve the management of complex network configurations. This further results in improved stability and capacity of the network through adaptation to changing conditions, early detection of issues, thereby allowing corrective measures before deterioration in service quality occurs. This makes networks more resilient and increases their capacity to handle traffic surges [26].

Intelligent Resource Scheduling: To utilize the network resources optimally AI/ML algorithms monitor the traffic pattern and real-time demands. Resources are then allocated in a way to maximize performance with the least energy consumption. This is an intelligent allocation for the efficient and effective usage of resources by constant monitoring and evaluation of demands for spectrum. It avoids unnecessary allocations, hence saving cost and the environment. Such a proactive approach ensures that resources will be utilized effectively and efficiently by dynamically adjusting usage according to demand in real time. Long-term, it improves energy efficiency and enhances performance [27].

Interference Mitigation: To reduce any interference, AI contributes in controlling communication between cells in proximity. Thus, the signal transmission process becomes smoother and more effective. It reduces power utilization and enhances total performance. AI/ML applications can use several energy-saving techniques and satisfy QoS requirements to promote an eco-friendly environment and improve performance. This promotes the larger objective of sustainable wireless communication and raises overall energy efficiency.

13.4 ENERGY-EFFICIENT TECHNOLOGIES IN O-RAN

Understanding the power consumption implications of O-RAN architectures is important to realize their potential for green communication techniques. O-RAN network slicing allows for assigning resources in the optimal way to fulfill user demands. It never turned on the services for low priority to cut down on wasted energy. AI can handle this more effectively by continuously analyzing traffic patterns and optimizing power consumption by virtualized network components. It basically performs smart sleep mode in which RUs and DUs can activate only when they are needed to save energy. These advancements in O-RAN's architecture provide more scalable and sustainable network operations by adding openness and intelligence. This makes energy-efficient and eco-friendly network usage in 5G and B5G wireless networks [27].

13.4.1 Network slicing

One of the prominent technologies that O-RAN enables is network slicing. It is a key evolution introduced in 5G and B5G networks. Network slicing allows us to share a single physical network into multiple virtualized and isolated logical networks within O-RAN. All of these networks can be tailor-made to the needs of various services like eMBB, URLLC, and mMTC to optimize resource usage in multiple services.

Network slicing allows us to dynamically allocate resources based on the real-time requirements of every slice. As different services have different needs to be considered in terms of application-specific latency, throughput, and reliability. For instance, eMBB services may have high data rates, but mMTC may need to be energy-efficient and require wide coverage. Operators can use network slicing to modify the size of specific slices in response to changes in traffic and service importance. This guides the better utilization of computing resources, bandwidth, and power consumption. In the context of energy efficiency, network slicing helps to cut down the unnecessary energy use by avoiding the usual "*one-size-fits-all*" resource allocation method [28]. As the existing technique is based on static setups which offer more resources than required. This leads to the waste of resources and energy due to idle devices get resources that consume energy. Network slicing enables operators to disable or reduce the priority of lower-priority slices when utilization is low or in regions with minimal traffic.

In addition, network slicing capabilities were further extended by the RIC component due to its role as a centralized intelligence controller. Multiple RAN components- Centralized Units (CUs) Distributed Units (DUs), and Radio Units (RUs)-work together for the enhancement of every network slice, thus enabling real-time optimization. This ensures that each UE gets a proper number of resources and energy in each slice. Besides this, AI-driven analytics further enhance network slicing through the ability to anticipate various service requests. This allows the networks to make changes in advance and choose energy-efficient options. It further helps to achieve green communication goals by minimizing the environmental impact of mobile networks in large cities during off-peak hours.

13.4.2 AI-Optimized dynamic resource allocation

At present, conventional wireless system replaces through smart, innovative, and energy-efficient resource management. By using adaptive algorithms that operate in real time, energy consumption can be reduced without slowing down the network performance. To make the NextGen of wireless networks more energy-efficient, AI-driven resource allocation is necessary-it ensures that resources are allocated with maximum efficiency. AI-based solutions constantly monitor network conditions and adjust resource usage in real time by reducing them when traffic is low and increasing them when it is high. This flexibility increases spectrum efficiency and significantly reduces energy consumption [29].

These AI models use past data to learn traffic patterns and predict peak consumption of UEs demand and other network factors. Specifically, in energy efficiency mode, when base stations predict low traffic, they may enter sleep or low power modes without impacting the quality of the service. This type of energy management saves a great quantity of energy and is quite useful in highly populated cities or when energy consumption is low. Real-time network settings adjustments inspired by an AI model to maximize power by monitoring key performance indicators such as QoS, user requests, the strength of the signals, and interference levels. In order to comply with UE demands, policies, and AI/ML models assist in determining the appropriate beamforming directions, scheduling techniques, and transmission power levels. By determining the optimal antenna configuration and shutting off unused devices, AI can improve energy efficiency. AI-driven technologies offer a unified approach to improving the environmental sustainability of communication systems. These intelligent networks help to switch from conventional, reactive approaches to proactive, context-aware, and energy-efficient ones [30].

13.4.3 Sleep modes for power management

Sleep modes are associated with a gradual shutdown of the base stations' components. In these different levels, components are considered to activate and deactivate for reducing energy consumption while maintaining service quality. The traffic demand determines which sleep mode the RAN elements RU, DU, and CU can enter. More components are deactivated as you move through the levels, which lowers the BS's power consumption. However, this results in a longer delay as it takes a longer time to reactivate the BS. This subsection discusses different types of sleep modes for reducing energy consumption. But this also introduced challenges to coordinate sleep transitions across a multi-vendor environment [32].

13.4.3.1 Types of sleep modes

Sleep modes help save energy when network components are not being used. They are particularly effective during low-traffic periods or when traffic demands are intermittent. In wireless networks, base stations, UE, and other network components can fully operate, go idle, or enter sleep modes. The primary goal of these modes is to eliminate power consumption when networks are idle or experiencing low demands, while simultaneously ensuring the networks are responsive to traffic variations [33, 34]:

Base Station Sleep Modes: High amounts of energy are used by base stations, which are particularly used in advanced networks like 5G and beyond and play a critical role. For instance, during low utilization times or off-peak times, the operations of the base stations can be optimized by putting them in sleep or low power modes. This can be done through disabling some components of the base station, such as the radio frequency chains, and leaving only the essential functions to operate and ensure that they are active so that they can handle incoming traffic. This may include shutting down excess equipment or utilizing an extensive power idle manner.

Cell Sleep Modes: In cellular networks, sleep mode can also be implemented at the cellular level. For example, in a small cell network or in a heterogeneous network, inactive cells may be put to sleep mode, and only a few active cells would be available to cater to user needs. This would ensure that resources are utilized efficiently and only active cells could be brought up or dynamically woken up to serve users' needs in real-time. See Figure 13.3.

UE Sleep Modes: In addition to the base stations and cells, the UE, such as a smartphone, can utilize sleep modes to save energy. For instance, smartphones can turn off or decline to a lower-power state if not used for a moment. But if UE needed to access any service, it wakes up automatically.

13.4.3.2 Implementation of sleep modes in low-traffic periods

In a wireless cellular network, implementing sleep modes is a complex dynamic scheduling technique. The sleep mode system should intelligently identify times of low traffic by assessing current traffic conditions and user demand. It consists of several steps outlined below [35]:

Traffic Monitoring: To enable sleep mode, it is critical to determine when traffic demand is low. This is done through the constant monitoring of user behavior, data transfer rates, and the number of users within a location. The monitoring helps in learning when the demand for traffic is low or high to optimize network resources and ensure QoS.

State Transition Logic: In this, with the detection of low traffic by the system, it needs to determine which components among base stations, cells, or network interfaces

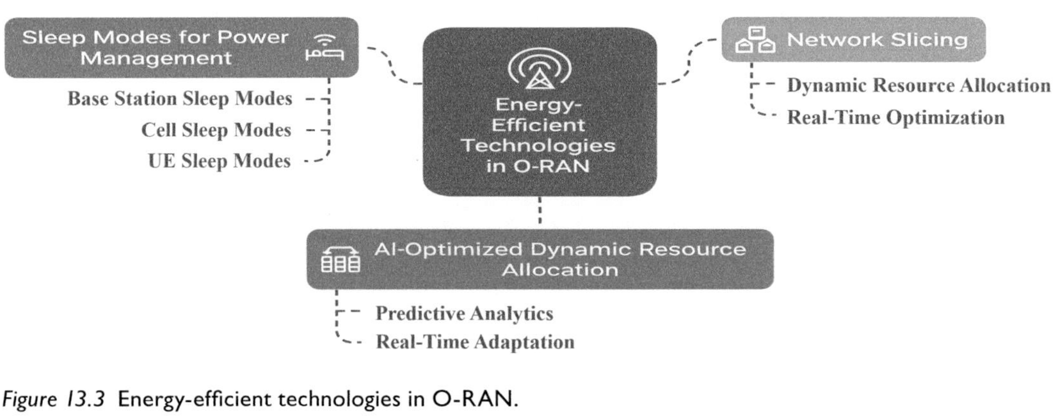

Figure 13.3 Energy-efficient technologies in O-RAN.

can go to sleep. This needs complex algorithms that must make a trade-off between energy saving and service maintenance. For instance, in partial sleep mode, one is able to conserve energy while maintaining essential coverage to prevent service interruptions.

Energy-Efficient Scheduling: Dynamic scheduling approaches, such as demand prediction models and traffic forecasting, are necessary for finding out when network components should sleep or wake up. This process maintains the system's efficiency and seamless operation, hence reducing the number of service interruptions. These technologies help determine when network components should wake up to prevent delays in offering services through the prediction of demand and traffic patterns.

Intelligent Wake-Up Mechanisms: A reliable wake-up mechanism is, thus, a critical requirement for such approaches in wireless communication systems. In this case, a more advanced wake-up mechanism would integrate more complicated and high-level algorithms that offer less wake-up time while ensuring more energy efficiency. As such, they maintain a delicate balance between minimizing the energy wasted during idle time and ensuring that the overall framework performs the intended functions without interruptions. Several approaches could be used to enable intelligent wake-up mechanisms in wireless communication systems.

Traffic Prediction Algorithms: Traffic prediction can help to optimize performance and energy consumption [36, 37]. Traffic modeling is used to predict traffic patterns and compute various performance measures. Intelligent algorithms use historical data to predict traffic patterns for future traffic demands. These algorithms enable them to determine in advance when a sudden increase in traffic can be anticipated and encourage them to take pre-emptive actions and activate the necessary network components to meet the anticipated demand.

Event-Driven Wake-Up: Activating the wake-up mechanisms relies on event-driven triggers. These triggers respond to various network events, such as incoming calls, data requests, or specific user activities. If an event happens or is seen, the network components are placed in a normal mode from their dormant position, and the service continues without any impediments.

Low-Latency Wake-Up: For 5G and B5G networks, low latency is paramount for the QoS user experience. Enhanced and optimized wake-up protocols need to be established to seamlessly transition from a sleep state into full operation with as little delay as possible. This involves minor updates to the wake-up process and ensuring that necessary resources are available to handle network traffic efficiently.

Collaborative Wake-Up Mechanisms: The purpose of collaborative wake-up solutions is to make the wake-up process consistent across many parts of a heterogeneous network, such as macro cells, small cells, and edge devices. This cooperation aims at better utilization of network resources, less energy consumption, and faster user response.

13.5 CHALLENGES IN BALANCING PERFORMANCE AND SUSTAINABILITY

The openness and disaggregation principles of O-RAN architecture bring many advantages, like improved efficiency, flexibility, reduced COPEX/OPEX, and programmability. Despite these advantages, it also brings sustainability issues, especially in terms of energy consumption. These present challenges for researchers and industry to achieve optimal performance and well allocation of resources. eMBB and uRLLC are two types of heterogeneous services with distinct requirements: high data throughput and low latency with a reliable connection.

Reduction of energy consumption and carbon emission in heterogeneous applications is one of the most prominent challenges. As discussed earlier, finding a balance between sustainability and performance is more difficult because of the multiple service types and traffic. This can be handled very well by integrating intelligence through AI and ML models. Another challenge is that the rise of renewable energy sources and novel energy-saving techniques may raise financial and technological concerns. Overcoming these challenges, wireless communication systems need accurate resource management, network management, and power-efficient equipment to be operable and friendly to the environment [38].

13.5.1 Overcoming the trade-off between energy-efficiency and performance

Reduction of energy consumption combined with enhancing QoS and QoE is quite challenging. Therefore, striking an ideal balance between energy efficiency and performance necessitates coming up with new approaches to increase both while retaining QoE. Among the priorities is the utilization of intelligent context-aware network management approaches that guarantee top network performance and reduced energy consumption. These methods rely on current data on user needs, mobility patterns, and traffic distribution, and frequently readjust resource allocation in real-time to adapt to the areas of highest demand on this basis. The use of ML methods is especially important, as it allows using predictive models to predict the network usage and change power use modes on that prediction basis. Sleep mode for some base stations would involve having them enter an idling state, engaging only those nodes that guarantee continuous coverage due to predictive models that ensure quick reactivation as required at all times, which means they should not be put to sleep in the first place. Moreover, the possibility of network densification, supported by small cells, has been proven. Deployment of a great number of low-power small cells in addition to a smaller number of high-power macro base stations significantly reduces the load for each individual station and, therefore, the need to work at full power around the clock. When combined with power harvesting, such as solar or wind energy used by base stations, this principle could theoretically allow for a significantly greater than now reduction in energy consumption without affecting the required QoS level [39].

However, it is essential not to allow service disruption due to aggressive power-saving methods. For instance, one of the aggressive power-saving methods is sudden shutdowns of network elements, which can lead to considerable service disruption. For applications where low latency is essential, such as video conferences and self-driving cars, this can be unacceptable. The resolution we suggest here is based on the concept of graceful degradation. Instead of working abruptly in an on-off mode, network elements could function in an energy-efficient mode, which still provides essential service. In such a model, even though the device is saving energy, it can still maintain vital links. Another possible idea is to conduct optimization in replicated network layers. Instead of concentrating solely on the physical layer, i.e., the transmission power, cross-layer optimization considers how the changes in the physical layer affect higher layers, including routing, congestion control, and application performance. For instance, Adaptive Modulation and Coding (AMC) can adjust its transmission settings based on channel conditions, saving energy while keeping the error rate low [40]. Moreover, the development of the user-centric network design continues. Instead of seeking to provide all users with the same optimal high-performance coverage at all times, networks can be designed to serve client needs. Critical services such as emergency communication and financial transactions are given maximal resource access, while non-critical services are directed to energy-efficient paths that may sacrifice a tiny measure of performance gain for a considerable amount of energy savings.

13.5.2 Implementation challenges in disaggregated architecture

Wireless networks are evolving to become more sustainable and efficient, and a disaggregated network architecture is becoming increasingly viable. The approach of decoupling hardware and software and achieving integration with multiple vendors to cut operating expenditures, boost flexibility, and implement eco-friendly network deployments is another high-ranking approach. Despite this, it would be inaccurate to expect that the proposed strategy is without obstacles. Among the principal difficulties are successful implementation achieved through real-time data interchange and interoperability, which is one of the principal conditions for achieving energy efficiency and operating energy efficiency [41].

Managing the interoperability of equipment from different vendors is one of the critical areas of networking. Traditional networks are built using equipment from a single vendor, are very well integrated, but are tied to supplier standards, limiting the ability to make changes. Disaggregated networks provide the opportunity to go beyond the framework of a single technology, and individual components, such as RU, DU, and CU, "run away" from the standards of vendor "locks". This high level of openness stimulates the appearance of additional opportunities and savings, but at the same time, it brings complexity in the verification of work questions of interoperability. Although there are interfaces or protocols prescribed in standards or LSA, for example, O-RAN, Third Generation Partnership Project (3GPP), equipment from different providers can implement them differently in their equipment. Moreover, even small differences in the configuration of the software itself, the timing of interaction, or the hierarchy of abstraction of hardware will immediately cause performance bottlenecks or the impossibility of connecting. In the case of green communication through wireless communication, one must also remember that energy consumption is a question of coordinating resources. Errors occur not only in the form of retransmissions and signaling overhead but also in the form of redundant and meaningless energy consumption. To ensure such interaction, thorough testing is needed, certification systems of certification, and above all, a reliable configuration tool that is capable of managing configuration among heterogeneous vendors. Primarily, the main organizations that are busy with this are the O-RAN Alliance, Telecom Infra Project (TIP), and a number of open-source initiatives. They strive to create reference designs, open interfaces, and organize events to test the work of multiple vendors, but the problem is that not everyone adheres to their initiated policies, and tools for profiling and debugging are not enough.

In disaggregated and eco-friendly networks, it is a considerable challenge to maintain seamless real-time data exchange without having control overhead rise to high levels. First, the disaggregation implies an increase in the number of control points and interfaces between network units. As a result, disaggregation causes delays and high signaling overhead, especially for time-critical operations such as beamforming, resource allocation, and handover processing: the lower the control overhead is, the less rescheduling will happen for the computation resources. Second, in efficient communication systems, where energy-saving features such as sleep modes or adaptive transmission strategies are critical, quick exchange is a requirement. In the event of the late or high-overhead coordinator of network elements or a lack of one, the energy-saving alternative will be offered too late to be utilized or not be proposed at all, which makes green optimization senseless. Edge computing and the optimal function of the control plane are the most important measures in this regard. If computation intelligence is located closer to the network edge, localized decisions can be made faster, avoiding the necessity to send all data through a central controller. It can use lightweight orchestration frameworks powered by AI to position important control messages first and eliminate unwanted overloading, enabling real-time responses without losing energy. Moreover, SDN and NFV can help by abstracting the control overflows, efficiency, and physicality in flow control. However, each of these instruments needs to be managed resourcefully to not add more overhead [42].

13.6 RENEWABLE ENERGY INTEGRATION IN O-RAN NETWORKS

The motivation for integrating renewable energy with O-RAN networks aims to enhance eco-friendly wireless communication, specifically powered using solar, wind, and hybrid power to operate the disaggregated network elements. Renewable-powered O-RAN operates on open and virtualized systems and uses sophisticated energy management modules, such as AI-based RICs, which use processing to consume power corresponding to the power availability, which reduces the dependency on pollutant fossil fuel power sources, promoting sustainable decentralized deployments. Despite the remaining challenges of energy fluctuations and instant coordination factors, the advancing energy storage modules and corresponding coordination make renewable-powered O-RAN a feasible platform for constructing energy-efficient, new-generation wireless networks [43].

13.6.1 Solar and wind energy for green wireless infrastructure

The demands from the wireless communication sector are increasing as the mobile data usage is rapidly growing worldwide. This trend raises the pressure on the whole sector to minimize the ecological footprint. Since most of the network energy consumption occurs in the base stations, reducing the carbon footprint is a priority that these stations can help to attain. One such strategy is feeding O-RAN infrastructure utilizing solar and wind power to boost the sustainability and energy efficiency of the wireless systems. Using sustainable sources to power O-RAN base stations will allow network operators to install RUs, DUs, and CUs at various locations. For example, they can set up RUs, DUs, and CUs in remote and rural areas where it might be expensive or inaccessible to access the traditional power grid. Consequently, solar panels and wind turbines can be employed to provide sustainable, on-site power generation for these nodes. Solar energy is the most suitable energy source for powering O-RAN base stations in locations with high solar radiation. Photovoltaic installations may generate constant energy during the day. When combined with energy storage solutions such as lithium-ion batteries, the installation can store some of the energy to power the base station at night or during a power outage. In contrast, wind may be the energy source for powering O-RAN base stations along the coast, in mountainous regions, and wherever else wind intensity is sufficient. By using local renewable sources, the network operators can be less reliant on the centralized grid, slashing modern power costs and reducing carbon dioxide emissions [42].

Moreover, O-RAN, designed on cloud technology, allows network intelligence to supervise this consumption dynamically, addressing even the RIC. For instance, AI algorithms can predict how much energy will be produced and adjust network operations, including traffic management or sleep modes, based on how much renewable power is available. As a result, green energy is utilized most effectively while maintaining the quality of service. This necessitates hybrid energy systems that combine grid power with renewable sources. While the sustainability advantages of renewable energy sources are enormous, such energy sources are not available at all times. For example, the power from a solar system is heavily reliant on the weather and day-night cycle, while a wind system's output may not be consistent throughout the day and in different areas of the world throughout the year. Such a lack of consistency can undermine the power grid's consistency, which is vital in supporting the communication infrastructure. Therefore, more hybrid energy systems are finding application in this area. Hybrid energy systems combine renewable resources with the conventional grid power and storage to ensure constant and reliable power flow. The system utilizes solar or wind power in an event where an excess amount of power is generated by renewables. When the amount of power from the renewables is low, the system effortlessly switches to grid power or stored battery power. Such systems also enable smart energy use, in which some of the less critical network operations can be shut off during the times when renewable energy is scarce, while the more critical or essential services are always on. As time goes on, ML algorithms develop, and they improve the balancing efficiency in grid and renewables consumption, which will eventually drive the consumption of renewable energy. Research is also ongoing with the support of microgrid technologies and advanced energy management platforms intended to enable renewable energy sharing among the many O-RAN nodes from a specific location to enhance the sustainability of wireless infrastructure.

13.6.2 Battery storage and smart grids for network operations

The advances in the wireless communication sector toward sustainable and energy-efficient options, battery storage, and smart grid technology are becoming equally important (see Figure 13.4). In energy storage, distribution, and usage, battery systems and smart grids significantly contribute to the overall objective of eco-friendly wireless communication. Energy storage solutions are critical in network infrastructure powered by renewable sources, such as solar and wind power. These sources experience inconsistencies, mainly because they may work best during non-peak hours for the users. This calls for the need to utilize excess power when generation is optimal, as is seen in smart grids. Battery systems, including traditional lithium-ion to controlled lithium iron phosphate and emerging solid-state batteries, offer efficiencies such as the optimal amount of power, longevity, and speed when charging. Integrating such into the energy management layer of the network allows the operator to make decisions such as storing power from batteries to utilize during hours of peak tariff,

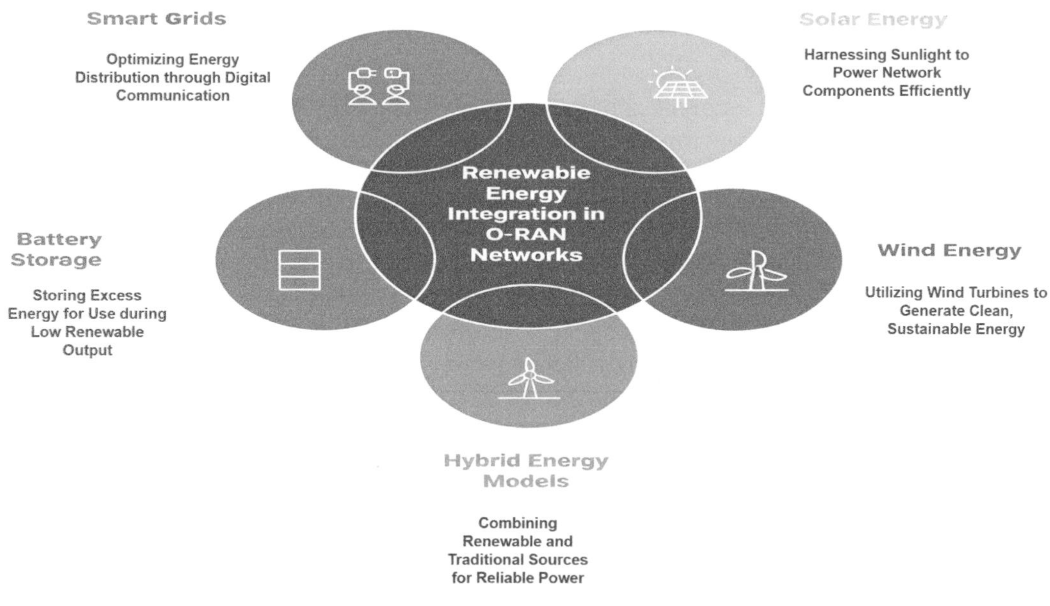

Figure 13.4 Sustainable wireless networks through renewable energy integration.

and allows devices to run directly from the grid at other times. Additionally, integrating AI-powered analytics and energy forecasting technology in the battery may result in sustainable and affordable usage. The network, through the system, can observe the flow of power within the cells to minimize wastage through AI-engineered energy production forecasting based on AI historical charging patterns [40].

Smart grids are advanced power systems that employ digital communication and automation to improve electricity generation, distribution, and consumption efficiency. One of the grid's functions during wireless network operations is real-time energy balancing by monitoring energy demand in the bulk power system and modulating the energy supply. Smart grids enable energy balancing in real time by monitoring energy demand at the base station and modulating the energy supply. Real-time two-way communication between the grid and network infrastructure allows the wireless operator to respond to changes in power availability or network load nearly instantly. For instance, if the mobile traffic increases unexpectedly, the smart grid may reassign energy resources to meet this new demand. On the other hand, if mobile traffic decreases, the grid may reduce the power allocation by activating energy-saving mode or diverting surplus renewable power into energy storage. The intelligent grid enables integrating several energy sources like solar, wind, and grid electricity within one system. Multi-source integration allows the grid to pick the most cost-effective and environmentally friendly energy source available at any moment. It also allows the intelligent grid to promote energy sharing in a city setting where the energy infrastructure is densely distributed. For instance, if an O-RAN base station generates more solar power than it requires, it may redistribute the electricity to neighboring base stations that require more energy or have less solar energy access. The intelligent grid supports an integration approach. Smart grids enable demand response programs, where the operator receives monetary compensation for adjusting energy consumption to reduce peak power demand. Telecommunication operators lower the demand, hence lowering the cost for end consumers, using energy storage and load management technology to contribute to the traditional grid's reliability and reduce their operation cost [43].

13.7 CONCLUSION

This chapter delves into the potential of green wireless communication in reducing energy consumption to overcome climate change while simultaneously increasing QoS. To achieve this, multiple techniques were discussed to improve energy efficiency and reduce energy consumption in ecological systems by upgrading hardware, software, network design, and operations. It highlights the approaches to increase power efficiency of next-generation wireless networks O-RAN architecture autonomously achieves this vision by enabling disaggregation, openness, incorporating intelligence, and making it easier to customize applications based on demand. SMO automates, orchestrates, and manages the entire O-RAN through intelligence. Non-RT RIC provides long-term policies and ML models, and near-RT RIC applies them for real-time optimization of RAN functions. Some of the intelligent, automated energy-efficient strategies are network slicing and AI-based resource allocation, which customize resource usage. This chapter discusses several methods to improve resource usage, including turning off the base station during off-peak hours, using sleep mode during low-traffic hours, and switching to active mode during peak hours, among others. In conclusion, O-RAN significantly contributes to the development of environmentally sustainable wireless networks. O-RAN must utilize renewable energy sources such as solar and wind power to diminish reliance on fossil fuels.

REFERENCES

1. Wang, L.-C. and Rangapillai, S. (2012) 'A survey on green 5G cellular networks', *Proceedings of the 2012 International Conference on Signal Processing and Communications (SPCOM)*, IEEE, pp. 1–5.
2. Li, G.Y., Xu, Z., Xiong, C., Yang, C., Zhang, S., Chen, Y. and Xu, S. (2011) 'Energy-efficient wireless communications: Tutorial, survey, and open issues', *IEEE Wireless Communications*, 18(6), pp. 28–35.
3. Wu, J., Rangan, S. and Zhang, H. (2012) *Green Communications: Theoretical Fundamentals, Algorithms and Applications*. Boca Raton, FL: CRC Press.
4. Mabrouk, A. and Zayani, R. (2023) 'Toward energy-efficient 6G networks: Uplink cell-free massive MIMO with NLD cancellation technique of hardware impairments', *IEEE Access*, 11, pp. 105314–105329.
5. Bhargava, V. (2017) 'Energy efficiency in massive MIMO-based 5G networks: Opportunities and challenges', *IEEE Wireless Communications*, 24(3), pp. 86–94.
6. Han, C., Harrold, T., Armour, S., Krikidis, I., Videv, S., Grant, P.M., Haas, H. *et al.* (2011) 'Green radio: Radio techniques to enable energy-efficient wireless networks', *IEEE Communications Magazine*, 49(6), pp. 46–54.
7. Yang, Y., You, X., Juntti, M., Wang, C.-X., Leib, H. and Ding, Z. (2013) 'Guest editorial—Spectrum and energy efficient design of wireless communication networks: Part I', *IEEE Journal on Selected Areas in Communications*, 31(5), pp. 825–828.
8. Auer, M., Giannini, V., Gódor, I., Imran, M.A., Malmodin, J. and Queseth, O. (2011) 'How much energy is needed to run a wireless network?', *IEEE Wireless Communications*, 18(5), pp. 40–49.
9. Malmodin, J. and Lundén, D. (2018) 'The energy and carbon footprint of the global ICT and E&M sectors 2010–2015', *Journal of Industrial Ecology*, 22(4), pp. 728–740.
10. Lakshmi, G.M., Mahmood, M.S., Kusuma, V. and Basha, D.K. (2022) 'Analysis of TG-based content addressable memory circuits', *Proceedings of the IEEE International Conference on Applied Artificial Intelligence and Computing (ICAAIC)*, pp. 1772–1777.
11. Björnson, E., Sanguinetti, L. and Kountouris, M. (2016) 'Deploying dense networks for maximal energy efficiency: Small cells meet massive MIMO', *IEEE Journal on Selected Areas in Communications*, 34(4), pp. 832–847.
12. Messaoudi, F., Farhani, F. and Zaafouri, A. (2022) 'Control strategy of a single-phase two-stage transformerless grid-connected PV system', *Proceedings of the IEEE International Conference on Electrical Sciences and Technologies in Maghreb (CISTEM)*, pp. 1–6.
13. Dahlman, E., Parkvall, S. and Sköld, J. (2017) 'NR: The new 5G radio access technology', *IEEE Communications Standards Magazine*, 1(4), pp. 24–30.
14. Chen, Y., Li, T., Cui, W., Tuo, G., Liu, X. and Wang, Y. (2022) 'A deep reinforcement learning-based efficient optimisation solution for inverse kinematics of hyper-redundant robots', *Proceedings of the IEEE International Conference on Robotics and Biomimetics (ROBIO)*, pp. 501–506.
15. Polese, M., Mezzavilla, M., Dutta, S., Rangan, S. and Zorzi, M. (2022) 'Understanding O-RAN: Architecture, interfaces, algorithms, security, and research challenges', *IEEE Communications Standards Magazine*, 6(4), pp. 36–43.
16. Khosravi, S., Riggio, M. and Mahmoodi, T. (2022) 'AI-driven RAN management in O-RAN: Opportunities and challenges', *IEEE Access*, 10, pp. 123456–123470.
17. Chen, M., Challita, U., Saad, W., Yin, C. and Debbah, M. (2019) 'Artificial neural networks-based machine learning for wireless networks: A tutorial', *IEEE Communications Surveys & Tutorials*, 21(4), pp. 3039–3071.
18. Bianzino, A.P., Chaudet, C., Rossi, D. and Rougier, J. (2012) 'A survey of green networking research', *IEEE Communications Surveys & Tutorials*, 14(1), pp. 3–20.
19. Riggio, M., Khosravi, S. and Mahmoodi, T. (2023) 'Real-time RAN control with xApps in O-RAN: Design and evaluation', *IEEE Transactions on Network and Service Management*, 19(4), pp. 4567–4580.
20. Hasan, Z., Boostanimehr, H. and Bhargava, V.K. (2011) 'Green cellular networks: A survey, some research issues and challenges', *IEEE Communications Surveys & Tutorials*, 13(4), pp. 524–540.
21. Mao, Q., Hu, F. and Hao, Q. (2018) 'Deep learning for intelligent wireless networks: A comprehensive survey', *IEEE Communications Surveys & Tutorials*, 20(4), pp. 2595–2621.
22. Salem, F.E., Saker, L., Woungang, I., Cheriet, M. and Cherkaoui, S. (2019) 'Optimal policies of advanced sleep modes for energy-efficient 5G networks', *Proceedings of the IEEE International Symposium on Network Computing and Applications (NCA)*.

23. Salem, F.E., Saker, L., Woungang, I., Cheriet, M. and Cherkaoui, S. (2019) 'Traffic-aware advanced sleep modes management in 5G networks', *Proceedings of the IEEE Wireless Communications and Networking Conference (WCNC)*.
24. Mahapatra, R., Chandra, A., Gupta, L. and Misra, S. (2016) 'Energy efficiency trade-off mechanism towards wireless green communication: A survey', *IEEE Communications Surveys & Tutorials*, 18(1), pp. 686–705.
25. Alsharif, M.H., Kelechi, A.H., Kim, J. and Kim, J.H. (2019) 'Energy efficiency and coverage trade-off in 5G for eco-friendly and sustainable cellular networks', *Symmetry*, 11(3), p. 408.
26. Ali, M.Y., Hossain, T.H. and Mowla, M.M. (2019) 'A trade-off between energy and spectral efficiency in massive MIMO 5G systems', *Proceedings of the International Conference on Electrical, Computer and Telecommunication Engineering (ICECTE)*.
27. Abou-Of, M.A., Taha, A.H. and Sedky, A.A. (2016) 'Trade-off between low power and energy efficiency in benchmarking', *Proceedings of the International Conference on Information and Communication Systems (ICICS)*, pp. 322–326.
28. Friedman, R., Kogan, A. and Krivolapov, Y. (2013) 'On power and throughput tradeoffs of Wi-Fi and Bluetooth in smartphones', *IEEE Transactions on Mobile Computing*, 12(7), pp. 1363–1376.
29. Li, Y., Liu, X., Zhang, H. and Zhu, P. (2018) 'Power optimization for energy-efficient wireless communications using hybrid-ARQ', *Proceedings of MILCOM*, pp. 1062–1067.
30. Mao, Y., You, C., Zhang, J., Huang, K. and Letaief, K.B. (2017) 'A survey on mobile edge computing: The communication perspective', *IEEE Communications Surveys & Tutorials*, 19(4), pp. 2322–2358.
31. O-RAN Alliance. (2022) *O-RAN Near-Real-Time RAN Intelligent Controller: E2 Application Protocol (E2AP) Specification*, Version 02.00.
32. Ashfaq, K. and Safdar, G.A. (2024) 'Energy efficiency and interoperability through O-RAN rapid transition protocol (ORTP)', *Computer Networks*, 252, p. 110658.
33. Zhang, H., Chu, X., Guo, W. and Wang, S. (2015) 'Coexistence of Wi-Fi and heterogeneous small cell networks sharing unlicensed spectrum', *IEEE Communications Magazine*, 53(3), pp. 158–164.
34. Han, S., Yang, C., Wang, G. and Lei, M. (2012) 'On the energy efficiency of base station sleeping with multicell cooperation', *IEEE Transactions on Vehicular Technology*, 61(2), pp. 1013–1024.
35. Kord, A., Coder, J.B. and Le, V. (2024) 'Evolving open RAN interoperability: A large-scale definition', *Proceedings of the IEEE International Conference on Communications Workshops (ICC Workshops)*.
36. Bebortta, S., Pati, B., Panigrahi, C.R. *et al.* (2024) 'Dynamic performance modeling framework for QoS-aware 5G-based IoT-edge systems', *International Journal of Information Technology*, 16, pp. 2149–2160.
37. Bebortta, S., Singh, A.K. and Senapati, D. (2022) 'Performance analysis of multi-access edge computing networks for heterogeneous IoT systems', *Materials Today: Proceedings*, 58, pp. 267–272.
38. Bennis, M., Debbah, M. and Poor, H.V. (2018) 'Ultra-reliable and low-latency wireless communication: Tail, risk, and scale', *Proceedings of the IEEE*, 106(10), pp. 1834–1853.
39. Sarigiannidis, P., Karapistoli, E. and Economides, A.A. (2017) 'Modeling the Internet of Things under attack: A G-network approach', *IEEE Internet of Things Journal*, 4(6), pp. 1964–1977.
40. Liu, X., Sun, S.X. and Huang, G. (2020) 'Decentralized services computing paradigm for blockchain-based data governance', *IEEE Transactions on Services Computing*, 13(2), pp. 343–355.
41. Alaba, F.A., Al-Amrani, M.A. and Hameed, I.A. (2024) 'AIoT-enabled smart grids', in *Artificial Intelligence of Things for Sustainable Development Goals*. Cham: Springer Nature, pp. 59–79.
42. Aktar, M.R., Hossain, M.S., Abedin, S.F. and Hasan, M.K. (2023) 'Energy-efficient hybrid powered cloud radio access network (C-RAN) for 5G', *IEEE Access*, 11, pp. 3208–3220.
43. Wang, Y., Wang, J., Li, J., Zhu, P., Wang, D. and You, X. (2022) 'MADDPG-based RAN network slicing for smart grid with NOMA', *Proceedings of the IEEE International Conference on Communication Technology (ICCT)*, pp. 1224–1229.

Chapter 14

Future trends in wireless communication technologies

Neha Singh, Sasmita Dash, and Supriya Aggarwal

14.1 INTRODUCTION

There is a paradigm shift in the wireless communication world. Although the fifth-generation networks have already started changing the way we interrelate devices and use information, the telecommunications industry is already beyond these feats. The 5G–6G and 6G-and-beyond journey is not only a speedy and latency improvement, though it is certainly a paradigm shift in how wireless systems are designed, operated, and exploited throughout the society. The fifth-generation wireless networks have radically changed the connection capabilities, and there are applications that could only be discussed in theory before the era of wireless networks of the fifth generation. Such networks offer fast connectivity with low latency and can accommodate the types of communication of megamachines over the Internet of Things, smart cities, virtual and augmented reality, and Industry 4.0. However, as various organizations around the world are rolling out 5G technology, the inherent flaws of this disruptive technology are now more evident. The wireless evolution is taking a definite trend. Every generation has been coming around every decade with disruptive ability, and at the same time, the technical and business environment requires the next evolution. The movers behind wireless technologies have evolved, although they are still pegged on the desire of a man to have a faster means of communication, capacity, and more integrated wireless systems within his day-to-day life. Figure 14.1 shows the focus areas of future wireless technologies.

14.1.1 5G: Current impact and limitations

The 5G networks have brought about concrete benefits that can already be realized in contemporary telecommunications. The technology has been able to offer much higher data rates than the past generations, with high-band 5G offering rates of over 1 Gb/s, and mid-band 5G delivering rates of between 100 and 900 Mb/s. With this dramatic scale of capacity, new uses have been unlocked in remote surgery, autonomous vehicle testing, and immersive entertainment experiences.

Nevertheless, the implementation of 5G has demonstrated severe challenges that limit its additional development [1]. The cost of infrastructure 5G networks is high and has restricted usage to the economically disadvantaged areas, which in effect further enhances the digital divide between developed and developing countries. Security vulnerabilities have arisen as operators have increased the network surface to support additional connected devices and edge computing functions.

Networks of 5G consume high levels of energy since the infrastructure needs a lot of power to run, and this raises environmental concerns as well as a continuous cost of operation to the service providers. Also, the scarcity of high-frequency spectrum has compelled operators to take tough decisions concerning the trade-off between coverage and capacity. The interoperability

DOI: 10.1201/9781003637721-14

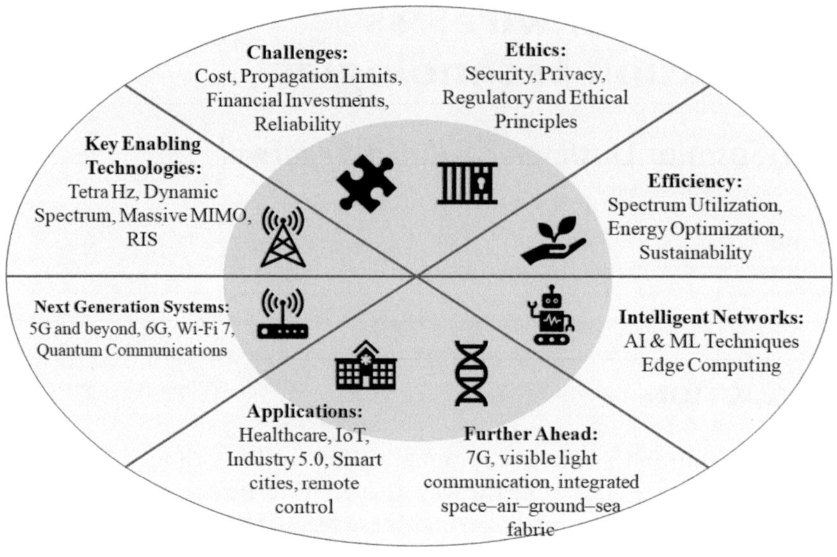

Figure 14.1 Focus areas in future wireless technologies.

issues of the 5G systems have turned out to be more problematic than they were expected. Regional differences and vendor-specific implementation have resulted in areas of incompatibility that restrain the efficiency of the network. Moreover, the general design of 5G, even though groundbreaking when compared to 4G, is based on the cellular principles of design that are currently seen as a possible limitation of next-generation services. Quite on the contrary, these limitations have not slowed down the use of 5G. Instead, they have indicated what the next-generation systems need to achieve. The collective understanding by network operators, equipment vendors, and research laboratories is that the way to go is not only to make a series of marginal improvements but to conceptualize the way wireless systems function differently.

14.1.2 Drivers for future wireless technologies

There are a number of formidable drivers of research and development of wireless systems beyond 5G. The former is uncomplicated: the number of connected devices and consumed data is growing explosively. Internet of Things devices are ever-increasing, and it is estimated that the number of connected devices across the globe will reach more than 15 billion by the end of the 2020s. Such devices produce enormous amounts of data that should be transferred, stored, and processed. Even 5G networks are strained by this load, especially when it comes to cases of extremely large data collection using distributed sensors. The second force is the appearance of radically novel use cases that stretch the capabilities of the existing networks to the limit. Not only is a high bandwidth required, but also an extremely low latency and a reliable connection are necessary for the extended reality applications that span a virtual reality, augmented reality, and mixed reality experience. The transportation and manufacturing autonomous systems require real-time communications and a latency of milliseconds in the single digits. Holographic communication and digital twin technologies can transform the way humans engage with information, but they need the nature of connectivity, which is not achievable at present. Sustainability of the environment has also become a rather significant motivator. Direct effects of power consumption of wireless networks are influences on carbon emissions and the cost of operation. The networks in the future should be more efficient yet provide better performance, which is a difficult task that has made the use of new methods of transmitting a signal and network structure a subject of research.

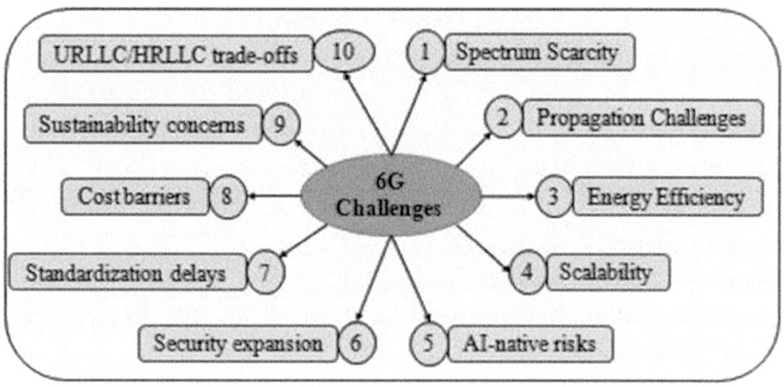

Figure 14.2 Challenges in 6G.

14.1.3 Vision of 6G and beyond

The sixth generation of wireless networks is projected to be commercially applicable at the beginning of the 2030s [2]. Although the 5G standardization was aimed at improving mobile broadband, ultra-reliable communications, and massive connectivity, the 6G is being projected at a broader range. The vision does not only cover better communication abilities but also integration of sensing, computing, and intelligence, directly to the network itself. The 6G networks are projected to achieve much higher data rates than 5G, and some studies indicate that the transmission rates may be 1000 times higher. What is more important is that the architecture will support the ability to support completely new types of services, such as synchronized distributed massive MIMO and extreme MIMO configurations, non-terrestrial networks with global coverage, and integrated communication combined with advanced sensing functions. The geographical scope of 6G services will be unmatched, and area coverage will be by ground infrastructure, low and medium Earth orbit satellites, and high-altitude platform systems. The key challenging areas that the development of 6G will encounter are shown in Figure 14.2.

The vision is further to the services that are, at the moment, in science fiction. Expansive mixed reality experiences would enable users spread around continents to interact in shared online spaces with low latency. Self-directed mobility systems would make use of ubiquitous connections through wireless links to organize vehicle movements. The massive digital twinning would generate synchronized digital versions of physical infrastructure where predictive maintenance and optimization of them could be done on scales never known before. Even after 6G, research groups are already brainstorming on 7G and beyond. These talks center around some basic abilities, such as built-in sensing and communication, in which the wireless signals are used both to communicate as well as to do environmental sensing, and the transparent integration of the non-terrestrial and terrestrial network space into a truely global one [3].

14.2 NEXT-GENERATION WIRELESS NETWORKS

14.2.1 6G networks

The 6G research community is developing technically radically new foundations in contrast to the way other generations of wires were envisioned. The development of 6G is not happening in a vacuum, and instead, it is the result of a level of learning that has been amassed throughout the development and deployment of 5G. This wireless evolution process has been useful because the field experience of implementation has helped to determine the theoretical concepts that can be applied to practical benefits and those that need critical modification. The architecture of 6G network is being imagined as a multi-dimensional network which can simultaneously work in a wide variety of spectrum bands. Although 5G focused on frequencies as high as about 100 GHz, 6G studies also dwell on lower centimetric frequencies (with bands of 7–15 GHz) as well as much higher sub-terahertz (THz) frequencies (92–300 GHz). This multi-band design gives the opportunity to optimize network deployment to the desired services, with lower frequencies being used in the wide area due to the wide area and higher frequencies being used in the localized and high-capacity connections. The intelligence architecture of 6G is very different than the intelligence architecture of 5G. As 5G started to consider machine learning (ML) as a network optimization tool, 6G will introduce the concept of intelligence at the network design level. The network will have self-healing features, self-organizing resources, and predictive algorithms that keep changing in accordance with the changing times. This is in contrast to the networks that react to problems once they arise, to networks that predict and avert problems before they affect the quality of services. Another distinguishing feature of 6G vision is the global coverage. In contrast to 5G, which is practically based on a terrestrial infrastructure, 6G networks are strictly being developed in such a way that they incorporate satellite systems, high-altitude platform systems, and unmanned aerial vehicles into one comprehensive whole. It should not be seen as the process of imposing non-terrestrial components onto terrestrial networks, but instead forms a single architecture in which these various network domains can be used as complementary parts of a single system. The satellite constellations in low-earth orbit will offer backhaul and access connectivity to end users in remote areas, and ground systems will serve to provide dense connectivity in urban areas. The first standardization studies on 6G started in 2025, and now the official standardisation is in process and commercial applications are expected to be to be available in the early 2030s. This timeline gives a duration of ten years to develop technologies out of research ideas to workable commercial systems, which is approximately parallel to the wireless generation cycles.

14.2.2 Wi-Fi 7

Although 6G is the long-term development of cellular technology, the wireless ecosystem has complementary technologies, which are also evolving simultaneously. The latest development of unlicensed spectrum communications is Wi-Fi 7, also known as 802.11be, which is showing great advancements [4]. Wi-Fi 7 is a gigantic increase in capacity over its predecessors, and it has the highest data rates of about 46.1 Gb/s, which is about four times higher than Wi-Fi 6, which could achieve about 9.6 Gb/s. This is a dramatic gain in the performance of Wi-Fi due to a number of complementary technologies acting together. Multi-link operation also enables devices to use several frequency bands at the same time, such as the older 2.4 GHz band, 5 GHz band, and the newer 6 GHz band. Higher-order modulation schemes, namely 4096-QAM modulation, allow each data symbol to feature 12 bits of information instead of 10 bits afforded by the 1024-QAM scheme used by Wi-Fi 6, which in effect doubles the theoretical transmission rates by a quarter. Wi-Fi 7 can fit a lot more data in the spectrum than its predecessors because of the adoption of ultra-wide bandwidth with channel widths of up to 320 MHz. The consequences of Wi-Fi 7 go beyond the simple speed boosters. With lower latency and less overhead, it is reliably able to support a variety of new applications such as 8K video streaming, real-time multiplayer gaming, and immersive augmented reality experiences. The updated protocols also hold the promise of better power efficiency in the form of better resource allocation and optimized transmission approaches, establishing Wi-Fi 7 as the option to support battery-powered gadgets that were once forced to use cellular connections. Wi-Fi 7 plays a very important role within the larger ecosystem of wireless technologies. Wi-Fi is used in high-capacity, localized areas where 5G and future 6G are concentrated on large areas. The overlapping of these technologies enables networks to maximise the deployment of the various standards, whereby Wi-Fi can serve extensive high-density demand areas such as offices and homes, whilst cellular networks can serve mobility and wide coverage. It is predicted that this complementary relationship will continue due to the development of wireless technologies, and each standard will be optimized to suit its specific purpose.

14.2.3 Quantum communication

One of the most radically different paths in the research of wireless technology has been the exploration of quantum mechanics as a means of communication. Instead of applying the electromagnetic waves to transmit information in traditional forms, quantum communication takes advantage of the strange behavior of quantum mechanics by developing a channel that would be theoretically impossible to breach and inherently secure. The best quantum communication technology that is the most developed is quantum key distribution, which is approaching the feasibility of being real. The Bennett Brassard 1984 protocol, usually referred to as BB84, defines a means for two parties to share encryption keys where complete certainty is ensured by the laws of physics. In an effort of a third party to intercept or measure quantum states during key exchange, measuring the quantum state itself disrupts the quantum state in such a way that it is instantly observable to legitimate parties. This produces a communication medium wherein eavesdropping is not only computationally hard, but it is also physically impossible to carry it out without being detected. Existing quantum key distribution (QKD) systems [5–7] have distances of up to about 120 km based on optical fiber, which is enough to connect major nodes of infrastructure in metropolitan regions. Recent breakthroughs have led to QKD distances of more than 380 km in the laboratory and toward continental-scale distribution networks. These systems are compatible with existing network infrastructure by the use of quantum-enabled encryptors, which receive keys provided on QKD appliances and use them to encrypt and decrypt Ethernet and fiber channel traffic up to 100 Gb/s. The major catalyst behind the deployment of quantum communication is securing the future against the prospective quantum computers. Encryption algorithms that secure sensitive information nowadays, such as the popular RSA protocol, are based on the hardness of factoring large numbers. A quantum computer powerful enough would be able to factorize in a reasonable time, which would make all data encrypted by the current methods decryptable. Companies that have long-term data security needs are starting to adopt quantum-safe encryption schemes, as they have realized that the data encrypted today may be intercepted and decrypted in the future by an opponent with a quantum computer. Implementation of quantum communication in 6G networks is still mostly in the research phase, although there is an initial effort toward how quantum key distribution may be implemented into core network capabilities. This is due to the fact that the current QKD implementations are limited, and it appears that quantum communication will be used by a limited set of high-security applications, instead of offering widespread and ubiquitous quantum-secured connectivity to entire networks.

14.3 KEY ENABLING TECHNOLOGIES

14.3.1 Terahertz (THz) and sub-THz communication

The gap in frequencies between infrared and microwave radiation, i.e., between about 0.1 and 10 THz, is an unexploited wireless communication frontier. THz communication has become a key technology to realize the very high data rates required by 6G use cases [8].

Compared to lower frequencies, which have already been heavily allocated and utilized, THz bands are quite lightly occupied, providing vast potential bandwidth in future communications systems. THz communication has a big potential in sheer numbers. Theoretical projections have suggested that the transmission rates of 100 Gb/s or more have been achieved over distances of a few meters, with experimental demonstrations making higher transmission rates, reaching toward terabits per second. This phenomenal rate can be achieved simply because of the huge bandwidth at THz frequencies; a single THz band can support tens of gigahertz of useful bandwidth compared to megahertz ranges in conventional cellular bands. Shorter wavelengths at THz frequencies allow smaller antennas to be used, and more directional transmission can be made, which opens up the potential of dense spatial multiplexing, whereby multiple data streams can be transmitted simultaneously in close proximity without causing much interference.

Nonetheless, THz communication is associated with considerable practical difficulties that are to be overcome before the mass implementation is possible. The propagation loss is very high at THz frequencies, so signals thin out quickly with range, so communications can only be done over a fairly short distance, perhaps meters, not kilometers. This is a natural restriction, and thus THz communication is only applicable in a few special cases, such as the ultra-fast campus network, wireless backhaul inside the data centers, and limited-range device-to-device communication. The breakthroughs in various fields at the same time are needed to build practical THz systems. Sources and teraflop lasers need to be efficient at ambient temperature, and sufficiently coherent to be able to modulate data, which remained only recently within reach of success. Enhanced sensitivity will be required to receive weak THz signals, and this has not yet been achieved, even with the advances in photonic integrated circuits. Large-scale designs and fabrication of THz antennas and waveguides is a complicated process, although scientists are working on remedies through silicon photonics technology. The introduction of THz communication in any 6G network will be widespread around 2030; however, this is subject to successful completion of engineering issues and standardizations. The first deployment will probably be targeted at niche applications where the range constraints are tolerable in favor of a very high bandwidth, and then expanded to more general applications as technology becomes more mature.

14.3.2 Cognitive and dynamic spectrum access

Spectrum scarcity is the basic limitation to wireless capacity, which is the lack of frequency bands that can be used to communicate. The conventional spectrum allocation allocates bands to certain services by licensing by the government, creating an inefficient scenario where some frequency is not fully utilized, and some frequency is overloaded. The idea of cognitive radio and dynamic spectrum access technologies focuses on solving this issue so that the wireless systems can intelligently detect the environment and opportunistically utilize the spectrum that is temporarily free. Dynamic spectrum access works by coordinating three functions: spectrum awareness, spectrum access, and cognitive processing. Spectrum awareness entails constantly monitoring the radio frequency environment to determine which frequency bands are occupied and which ones are free [9].

This information is processed in a manner that gives smart decisions regarding how the spectrum can be allocated to be used, and what opportunities are available to secondary access. Spectrum access establishes measures to use the existing spectrum opportunities without disrupting the primary users with a license on the bands. This practical application of the cognitive radio systems needs a few technical abilities. Spectrum sensing should well identify the existence of licensed primary users without false arms that are very high; an error in believing that a band is free can lead to severe interference to licensed services. This detection has to be fast enough to counter the varying spectrum availability before there is deterioration in the quality of service. The decision-making layer of the cognitive processing should consider several aspects, such as whether and how to utilize the available spectrum, given not only what the available frequencies are, but also propagation characteristics, traffic requirements, and the possibilities of interference.

ML algorithms have also become especially promising in the machine optimization of cognitive radio. Instead of applying predetermined sensing algorithms and spectrum allocation algorithms, ML models may learn by observing the past trends in using the spectrum and adjust their own behavior. Such techniques are able to forecast spectrum availability, find the optimal frequencies to use given the current traffic needs, and dynamically adapt transmission parameters to changing conditions. ML-based cognitive systems are more resilient than unchanging approaches due to the capacity to continually learn and thus improve [10]. Cognitive radio integration into 5G networks has already started in advanced deployments, and full standardization of the technology is anticipated once 6G specifications are finalised. The digital spectrum access with sub-THz bands above can be combined with cognitive spectrum access, in which case the cognitive systems can efficiently use these resources without wasting them, even if they are miles distant.

14.3.3 Cell-free massive MIMO and ultra-dense connectivity

The standard cellular network topology divides the coverage into cells that are covered by base stations, forming boundary effects, where the quality of service drops at the edges of the cells, and possible interference between neighboring cells. Cell-free massive MIMO architectures suggest a much more radical solution: instead of having many antennas in one place, distribute the antenna elements over a large geographic extent, and have all the distributed elements serving all the users cooperatively by centralized processing. Ultra-dense cell-free massive MIMO is an advancement of this idea, scaled to 6G, in which the density of connections is more than 100 million devices per square kilometer, 100 times higher than a 5G target. The densely populated environment is inherently supported by this architecture with ubiquitous connectivity, whereas the distributed antenna deployment is superior to the traditional cellular systems in sparse regions. The positive propagation properties of massive MIMO, where access point channel vectors between different users are roughly orthogonal, permit the use of simple processing algorithms to attain excellent performance with high spectral and energy efficiency [11].

There are significant technical benefits of cell-free massive MIMO. The array gains and multiplexing gains increase with the number of antennas within the system, and hence, there is a potential of having a very high spectral efficiency with proper signal processing. Channel hardening, in which the squared norm of channel vectors approaches determinism, implies that no longer does performance depend on instantaneous fading, but on average channel statistics, and as such, other receiver designs are simpler than those of a system lacking the property of hardening. Distributed architectures eliminate the cell boundary effect that plagues traditional cellular networks by eliminating the cell boundaries.

However, it is very difficult to make an ultra-dense cell-free structure work. The large populations of distributed antenna elements and users need to be coordinated by a fronthaul network between access points and centralized processing, and this type of backhaul link has significant bandwidth demands. The computational requirements to centralize on all the user signals increase with the number of users. The stricter the requirements on synchronization, the greater the number of distributed elements. The earliest practical implementations of cell-free architectures have been in research testbeds and small commercial pilots, and operators have seen the prospect of much greater capacity and coverage. It is believed that the technology will be at the center of 6G networks, especially in the dense urban areas that require high connection rates.

14.3.4 Reconfigurable intelligent surfaces

Reconfigurable smart surfaces are a new means of managing the wireless propagation environment. Instead of accepting the propagation channel being a given constraint and developing communication systems to operate despite this constraint, reconfigurable intelligent surfaces (RIS) technology provides active control of the way in which electromagnetic waves propagate in the environment. RIS is a collection of numerous small passive reflecting elements, the characteristics of which can be dynamically adjusted to channel the waves that come into it to their desired destination.

The basic idea behind RIS is simply beautiful, and it has a potent implication. Individual phases can be set on each element of the surface to add a certain phase shift to received signals. With these phase shifts organized on a large scale over a large number of elements, the surface is able to essentially act as a directional mirror to wireless signals, directing energy emitted by a transmitter to a specific receiver without shadowing of obstacles and multiple paths fading that otherwise would impair performance [12].

The advantages of RIS implementation are varied. RIS can also enhance coverage into deep shadow regions as well as propagate signals by enhancing signal propagation paths. This additional coverage is possible without the energy usage that active amplification would entail, as RIS elements work in the passive mode, reflecting the existing signals with a modified phase. The enhanced propagation also enhances the signal quality in most cases, and this also decreases the transmit power needed to sustain the quality of the service. It is this coverage extension and reduced power that can be drawn to indicate that RIS can meet the issue of capacity, coverage, and energy efficiency in wireless networks concurrently.

Of interest are RIS security applications. RIS could eliminate eavesdropping of wireless communications by dynamically changing signal paths to guarantee that signals reach unwanted receivers with destructive interference and those that are intended with constructive interference. This establishes some type of physical-layer security that is hard to crack by the adversaries. RIS has a number of challenges in its practical implementation. The individual surface control demands either a wired connection of the control signals or wirelessly controlled control channels, which complicates the design of RIS systems. The optimization of the surface configuration demands an understanding of the channel conditions, which have to be estimated or measured. This overhead may be enormous in high-change environments. It is not yet easy to produce RIS on a large scale, because it would take innovations in fabrication technology to produce large surfaces with elements that can be controlled individually.

The existing studies in the field of RIS are addressing the theoretical boundaries as well as the practical implementation strategies. RIS has already been integrated with 5G networks in advanced deployments, and standardization dialogues in 6G applications have already started. It is projected that the technology will gain more significance as the density of the spectrum grows, and high-frequency propagation gets more difficult.

14.3.5 Convergence of terrestrial and non-terrestrial networks

The amalgamation of terrestrial wireless infrastructure with non-terrestrial networks such as satellite communications, high-altitude platform systems, and aerial networks, with the help of unmanned aircraft, is one of the most unique features of the 6G vision [3]. Instead of considering these domains as discrete ones, 6G architecture sees them as a part of a single network that can easily combine the resources of various domains to fulfil users.

Terrestrial systems alone cannot offer the capabilities that non-terrestrial networks can offer. Low-earth-orbit satellite constellations are used to cover geographic regions in which installing ground-based infrastructure would be economically impractical or technically impractical. The orbital paths of the satellite systems are predictable and produce a regular coverage pattern that is useful in planning the network. There are several satellite planes in varying altitudes, which offer architectural flexibility, enabling networks to decide on which orbital layer is most suited to specific services.

There are various challenges that need solutions to achieve the technical integration of the terrestrial and non-terrestrial networks. Propagation characteristics are vastly different between terrestrial and satellite connections, and it can be seen that waveforms and coding schemes should be adjusted to ensure compatibility. Topology management in the case of the movement of satellites in orbit necessitates dynamic topology, where network connections keep changing as satellites relocate with respect to the ground stations and users. Satellite networks combined with the current ground infrastructure will require prudent spectrum coexistence planning in order to prevent interference.

Carrier aggregation is one potential architecture that could be applied to non-terrestrial space, whereby users can simultaneously use both terrestrial base stations and satellite systems and superimpose the bandwidth of multiple connections. This involves strict power control and coordination, yet allows optimum use of the available resources. The long-term vision goes further than mere integration to what has been called unification, whereby terrestrial and non-terrestrial systems are brought to work as one smoothly integrated network in which resource allocation, handover, and service provision are coordinated across domain boundaries as easily as inter-base station handovers in a terrestrial network. This vision cannot be reached only through the development of technologies but also through international collaboration in spectrum allocation and regulation systems.

14.4 INTELLIGENT NETWORKS

14.4.1 AI and ML integration

The introduction of artificial intelligence (AI) and ML in wireless networks is a fundamental change in the design, deployment, and operation of networks. In contrast to the earlier versions of networks that added learning and optimization, 6G networks are being designed with intelligence as a functional element and not a feature. This integration is through virtually all the network functions, such as radio access optimization and security management.

ML is useful at capturing patterns of a complex set of data and predicting based on the patterns. This is feasible in wireless networks, which allow optimization of the network that could not be achieved by manually configuring the network. Through past traffic trends, ML systems can anticipate future traffic requirements with reasonable precision, which enables the network to pre-empt future resources well before a traffic jam sets in, instead of responding to quality-of-life situations [13]. Likewise, a model based on ML trained on propagation data is capable of estimating channel conditions sufficiently to help make beamforming and power control decisions. ML algorithms classification and their possible use cases in wireless communication are depicted in Figure 14.3.

This application of the deep neural network to the allocation of resources to wireless networks is among the most promising fields of ML studies. The traditional resource allocation algorithms are based on a set of fixed rules that optimize under average conditions. Deep learning methods can alternatively be trained on adaptable allocation policies that are sensitive to certain network states, user patterns, and service demands. These trained policies can

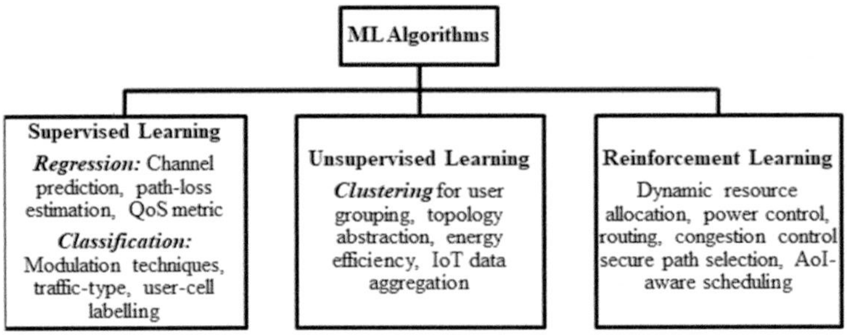

Figure 14.3 Application of machine learning algorithms in wireless communication.

be more effective than hand-written algorithms as they can use small patterns in network data that a human would be unable to detect.

ML optimization makes such virtual partitioning of one physical network into many logical networks (called network slicing) more feasible, having different properties. The algorithms Opera ML can adapt the service allocations of the slice resources in response to service demands to maximize the network utilization overall when different services share the same underlying radio and computing resources, necessitating competition between network slices. This is necessary to support the various services expected in the 6G networks. Currently, in wireless communication systems, AI and ML are used for task-oriented activities to enhance network efficiency, like resource allocation based on channel observations, proactive resource management to prevent congestion, and predictive analytics to forecast future requirements. AI and ML techniques [14] are trained from operational data collected from the networks and then deployed for prediction, control, and optimization across different layers in the communication system. Some use cases for the physical layer are tabulated in Table 14.1, while for security features, use cases are shown in Figure 14.4.

Federated learning (also known as distributed ML) is a method through which models can be trained in a way that data privacy is maintained by the collaborating network elements. Instead of collecting all network data in a single location and analysing it, federated learning

Table 14.1 Physical layer use cases for AI and ML techniques

Physical layer	*Use cases*
Channel characterization and modeling	1. Path loss, delay spreads, and angular speeds from Tx/Rx coordinates, distance, and frequency 2. Realistic design of geometry-based stochastic channel models using clustering techniques
Channel estimation and prediction	1. SVR in channel estimation in time-varying multipath and FBMCOQAM systems. Improved BER by better modeling nonlinearities and intrinsic interference. 2. DNNs in beamspace mmWave massive MIMO 3. Gaussian process regression for Age-of-Information (AoI)-aware scheduling and power control
Signal detection and modulation recognition	1. Code-aided modulation classification in multiuser SC-FDMA systems 2. Continuous phase modulation classification via likelihood-based methods 3. Autoencoder–extreme learning machine (AE-ELM) architectures 4. Denoising diffusion probabilistic models (DDPM) for offline interference signal classification

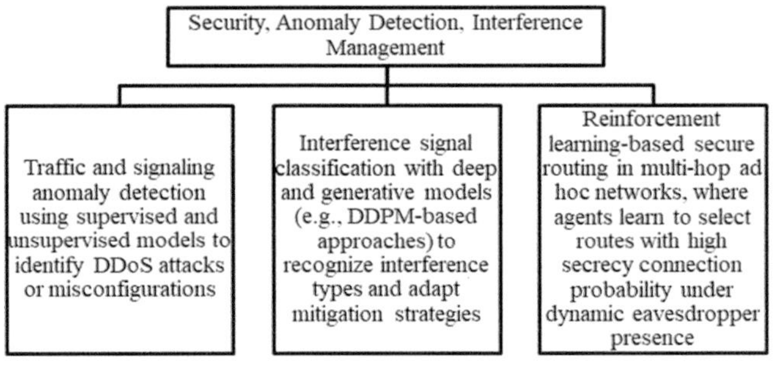

Figure 14.4 Possible use cases of AI and ML algorithms in security, anomaly detection, and interference management.

allows each node to collect models based on its data and send the trained models to be combined into a model with all the nodes. This methodology preserves the privacy of data, and yet the advantages of learning from different data sources. There are still significant implementation issues of ML in wireless networks. Neural networks take enormous amounts of data to train, posing privacy concerns in situations where networks hold sensitive user data. The calculation cost of executing ML models in real time should be kept to a minimum so as not to cause latency or unreasonable power usage. ML models are highly resistant to adversarial attacks, in which attackers purposefully formulate inputs specifically to deceive learned models, demanding defensive mechanisms that remain subjects of ongoing research.

14.4.2 Edge computing

The conventional structure of wireless networks puts the computation and storage in the central network, and end devices act as the sole communication devices. The architecture was more problematic as the applications became more computationally intensive and latency-sensitive. Edge computing [15, 16] works in reverse and brings the computation to the point of data origin, to the network edge near end devices and base stations.

Edge computing has some great advantages for wireless networks. The presence of edge computing eliminates latency by processing data on the edge rather than transmitting it on the network to remote cloud resources. Applications such as autonomous vehicles and industrial control systems are absolute when it comes to low latency that must be within single-digit milliseconds, and therefore, edge computing is required and not optional. The reduced network traffic due to the local processing relieves congestion in the backhaul, allowing other services to be carried.

Edge computing makes it possible to deploy ML in a wireless network in a practical way as well. Whereas the computational resources needed to train an extremely large model can be large, inference with already-trained models can be done on edge devices with limited resources. This enables optimization choices of ML to be taken in low latency at the network edge and consequently transmitted to network components to be implemented.

Especially the combination of ML and edge computing opens up especially strong opportunities. Edge devices have the ability to make intelligent decisions by doing local ML inference without having to wait to be responded to by remote cloud systems. As an illustration, network status information can be processed by an edge server at a cellular base station, ML inference can be performed on that server to predict optimal resource allocations, and those allocations may be implemented in the minimum possible time. There are many edge servers that may jointly train distributed models without jeopardizing data privacy by using federated learning.

The issues surrounding the deployment of edge computing are the limited resources of edge hardware, which in many cases need to be reliable, require low maintenance, and work in harsh outdoor conditions. The workload is distributed among heterogeneous edge resources, necessitating complex orchestration in order to sustain the quality of service. All edge computing requires a well-thought-over architecture to achieve smooth functionality between these different environments and cloud-based resources, as well as core network resources.

14.5 CHALLENGES, OPPORTUNITIES, AND RESEARCH DIRECTIONS

14.5.1 Spectrum scarcity and efficient utilization

Although spectrum research and more advanced wireless technologies have been achieved over the past decades, spectrum scarcity is still a major constraint that impacts the capacity of a wireless network. The spectrum of frequencies available to wireless communication that are useful, that is, frequencies with suitable propagation properties and a regulatory license permitting practical implementation, is limited and saturated with already existing services. Wireless capacity remains in high demand, and supply remains low, making the price of spectrum licenses in most countries an all-time high [17].

The classical spectrum allocation method is the spectrum allocation, where a frequency band is allocated to a particular service to offer long-term licenses, and therefore creating an ownership structure that offers certainty to investors but removes flexibility to repurpose the spectrum as services and technologies change. With this system, licensed spectrum may end up idle whilst adjacent frequencies may be overloaded. It has been found that there are high levels of inefficiencies in the use of the spectrum, with a large number of licensed bands going unused during the majority of the time, when other primary services could use extra spectrum.

Spectrum allocation economics is paradoxical. Physically speaking, according to Cooper's Law, wireless capacity has been doubled every 2.5 years in a mix of frequency reuse improvement, spectrum efficiency improvement, and density increase in deployment. However, with this human effort of improving the spectrum, the feeling of scarcity of the spectrum exists, where the demand has always kept increasing more than the supply.

There are several strategies that are being followed to enhance spectrum efficiency. Such technologies as cognitive radio and dynamic spectrum access introduced above are one of the directions in which secondary users can opportunistically access the spectrum during the periods when the primary users are not using it. New modulation schemes and coding methods are still achieving increases in information capacity delivered in each unit spectrum, although the improvements are becoming more and more incremental. A move to higher frequencies, such as millimeter-wave and THz bands opens, lightly used spectrum, which is counterbalanced by propagation issues in most applications.

Another area of improvement is the spectrum sharing policies. Allocations are no longer done exclusively for some frequency bands, but the frequency bands have become shared spectrum where many users or services are allowed to operate simultaneously under spectrum sharing regulations. This method involves making sure it is well designed to avoid undesirable interference effects, yet it is much more efficiently utilized than exclusive allocation when done correctly.

Spectrum efficiency spectrum efficiency in 6G has spectrum efficiency refinement of all these methods in addition to new avenues such as joint communication and sensing, where spectrum is used in both modes and may potentially increase spectrum utilization by a factor of two. Formulation of global harmonized spectrum policies is still necessary, but not an easy task due to the national sovereignty implications of spectrum policy.

14.5.2 Energy-efficient and environmentally sustainable designs

Energy consumption of wireless networks has increased with the scale and the enhancement of capability at the same rate. The current cellular networks are high-power-consuming and have led to the cost of operation to the network operators, as well as the environmental effects in terms of carbon emissions. With the growth in the networks to offer more coverage and greater capacity, the management of energy consumption becomes more crucial [18].

There are several complementary methods of making use of energy efficiency. On the hardware level, more efficient power amplifiers and radio elements are able to decrease energy wastage in signal transmission. The advances in antenna technology, such as the massive MIMO designs, allow more of the radio energy to be directed into the intended receivers instead of radiating the power in every direction. The several antennas used at the transmitter and receiver enable less power to be used during transmission to achieve the same coverage by combining the signals coherently.

On the network structure level, it can be optimized by achieving energy efficiency by means of smarter resource allocation. Conventional methods are a continuous transmission at constant power levels. Adaptive power control adapts transmit power according to the channel conditions, such that less energy is used in situations where propagation is good. Network slicing and edge computing would enable a decrease in the use of energy as data stored in the area can be processed there and not sent all over. Another such energy leverage is sleep mode optimization. Base stations, when not in traffic are able to go into low-power sleep states, which use very little energy and lower the processing load on the station when capacity is not required. This is especially useful in cellular networks where traffic is highly fluctuating based on time of the day, peak traffic during business hours, and very low during the nights.

The solution to the carbon emission problem is through the integration of renewable energy sources into the wireless network infrastructure. Base stations can be powered by solar panels in a large number of geographical locations, especially areas with good access to the sun. Wind turbines are applicable to remote base stations that cannot be connected to the grid. Hybrid systems that incorporate the use of several sources of renewable energy with battery backup offer a more reliable power source as compared to a single renewable source.

The environmental impact of wireless networks is not limited to their use of energy. The material used in the equipment of the network, production processes, and the overall disposal of equipment are all environmental factors. The need to design a sustainable network should consider these entire life-cycle effects and not only the operational energy usage.

One 6G network opportunity is the possibility of greatly lowering energy usage than in 5G due to more efficient technologies and architectures. The cell-free massive MIMO systems mentioned above have the potential to be more energy efficient than the traditional cellular systems. The transition to higher frequencies with more directive antennas has a smaller amount of wasted radio. Intelligence integration in networks eliminates redundant processing and allows the allotment of resources to be more accurate.

14.5.3 Security, privacy, and trust in AI-driven networks

The growing awareness of AI in wireless networks brings both security and privacy issues, as well as advantages that AI brings. In contrast to the existing models of network security, where the attack pattern is reasonably well studied based on many years of experience, AI-based security offers new vulnerabilities that are currently being explored and defined [19].

Adversarial ML is one of them, whereby attackers develop inputs with a specific aim of making the ML models make wrong judgments. One of the prototypical examples is to create radio signals that can deceive signal classifiers based on deep learning to incorrectly label channel conditions. These attacks would lead to poor resource allocation decisions taken by networks, poor quality of services, or security vulnerabilities.

Another challenge of significant magnitude is data privacy. Good performance with ML models requires large amounts of training data, which is commonly sensitive to user activities, locations, and even communications patterns, which are available in network data. By having a centralized repository of this information to be used in ML training, there is a privacy risk in case this data is breached. Although federated learning and differential privacy methods have potential solutions, they have model accuracy and computational overhead costs.

The reliability of AI-based informed decisions is not limited to the scope of preventing attacks and maintaining privacy, but also to the ability to make sure that the choices that the ML systems make can be explained and are compliant with human values. An excellent prediction system can be made by an ML system about how best to allocate resources, but only in cases where the operators are able to interpret why the prediction was made and that it is likely to be used as intended. This trust is hard to build because the internal logic of decisions of some neural networks is opaque, i.e., a black box.

The sophistication of AI-based network structures poses security management issues. As the intelligence is spread among many edge nodes and cloud systems with many network elements, it becomes harder to maintain the consistency of security policies and identify intrusions. When combining various AI systems, the interaction between them has to be thought over to avoid the appearance of unintended behavior that might result from the combination of the systems.

The academic community is working on such challenges in various ways. Strong ML algorithms are the ones that aim to be resistant to adversarial attacks by training models that explicitly take into consideration possible attacks. ML methods that preserve privacy such as federated learning minimize the necessity to centralize sensitive data. The formal verification methods seek to offer some mathematical assurances regarding the behavior of AI systems in the most vital applications.

14.5.4 Ethical and regulatory considerations for intelligent wireless systems

Introduction of smart wireless systems poses key ethical and regulatory challenges that are not limited to technical factors. The potential possibilities offered by the 6G networks' features such as ubiquitous connectivity and highly developed sensing and intelligence opens both a huge potential benefit and a veritable threat to society when used without proper protection. Governance issues of who governs vital infrastructure and the ways of exercising control over the infrastructure also exist because of the concentration of smart decision-making power in wireless networks that are run by either a private corporation or a government organization. The major issue in the past wireless generations was to provide universal access to service and avoid monopolistic tendencies of the incumbent operators. The concerns now extend to the use of data, the authority to make decisions automatically, and the visibility of algorithmic systems because of the intelligence built into networks.

AI-based systems are a grave source of algorithmic bias. When trained on prejudiced data or applying methods that enable the preservation of existing disparities, ML models will amplify these disparities when trained at a network scale. As an example, a spectrum allocation algorithm implemented through ML with mostly urban network data could fail in the low-density areas, or the algorithm can intentionally assign poorer service quality to one group of users. The policies that regulate wireless networks are changing with the AI issues. The AI Act of the European Union creates risk-based regulatory conditions for AI systems depending on the amount of harmful consequences they might bring. There are other jurisdictions where similar regulatory initiatives are being developed. Such frameworks tend to insinuate the relevance of transparency, accountability, and human supervision of autonomous systems. The more the methods of integration of wireless networks, the more international coordination will be essential. International harmonization of spectrum regulations, security standards, privacy requirements, and safety standards in the electromagnetic field is necessary to facilitate smooth connectivity throughout the globe. Different stakeholders, such as government regulators, manufacturers, and service providers, as well as civil society organizations, are needed to influence the development of policies that would balance the interests of people on one hand, with innovation on the other hand.

14.5.5 Future research roadmap toward 7G and beyond

The direction in the development of 7G and the further generations of wireless networks does not rely entirely on technological achievement but also involves more liberal concerns related to network sustainability, equity, and integration into society. The industry is just starting to appreciate the fact that wireless generations will one day reach their physical limits to the point where the next significant enhancement of speed using conventional methods becomes impractical. The basic research questions that will guide 7G are how to successfully integrate maritime, aerospace, and terrestrial communication fields into one coherent network that caters to users in all environments. Another frontier is the exploration of new spectrum bands, which can go as far as visible light communication and beyond. The increased integration of sensing, positioning, and communication functions into networks may provide synergies that allow functions that cannot be achieved when these functions are used individually. Another thing that the roadmap highlights is the need for networks to be more responsive in terms of their operating environment. Networks ought to be configured to reconfigure dynamically to particular scenarios and uses, instead of being configured to a set of typical conditions. This demands not only technological developments but also developments in our theoretical comprehension of how networks ought to be essentially designed. The preservation of the environment and conservation of energy are likely to continue being the key issues when wireless networks continue their proliferation. The networks that can be sustainable over decades and serve growing numbers of users with more and more data demand constant innovation in the field of power efficiency, the integration of renewable energy sources, and the concepts of a circular economy to manufacture and dispose of equipment. Another long-term research direction is the combination of wireless networks with quantum computing and quantum communication. As much as quantum computing is potentially a threat to existing encryption, it also presents the potential of radically new applications to optimization, simulation, and communication that could open up new possibilities to network capabilities.

14.6 CONCLUSION

The transformation of the existing 5G networks to 6G and beyond is one of the boldest technological projects in the history of telecommunication. The technical issues are great, the time scales are steep, and the financial investments needed are colossal. But the vision that has developed through the research community illustrates that there are strong possibilities of networks that offer beyond previous connectivity, intelligence, and capability, and can work sustainably in environmental and economic limits.

The successor of the wireless network will not merely offer higher data rates on the current 5G infrastructure. Instead, they will radically transform the way networks are designed, operated, and exploited in society. Non-terrestrial infrastructure integration increases network coverage in areas that have not been connected before. The intelligence incorporated in networks allows autonomous optimization and adaptation that has never existed before. Integrating communication with sensing allows networks to transmit and at the same time sense the environment.

The enabling technologies mentioned in this chapter, such as the terahertz communication, RIS, cell-free massive MIMO, quantum communication, and others, are various facets of a consistent vision of wireless networks that is both faster, more efficient, more secure, and more accessible over the globe in comparison to the past. These technologies are becoming well-proven in terms of feasibility through research demonstrations and prototype deployments.

Nevertheless, technology itself will not show whether 6G and beyond will reach their potential. For implementing the next-generation networks, it is necessary to go beyond technical innovation, but also there is a need for regulatory change, international cooperation, equal access policy, and strong security and privacy management. This is a responsibility of the research community to ensure that wireless technologies are used to the benefit of the wider society as opposed to accumulating power in the hands of the rich and influential bodies.

It has a history of approximately a ten-year span between the development of research ideas and the commercial implementation of a given wireless generation. In the case of 6G, this window is open, and standardization studies started in 2025, and commercial applications will be expected in the early 2030s. It is imperative that the research fraternity, industry partners, and policy makers take stern action within this era so that next-generation wireless technologies can be responsibly created and deployed fairly.

The world is already working on research laboratories on the next generation of 7G and beyond. The visionary ideas that are merely spectators of today, such as holographic communication across continents, autonomous systems that are coordinated by wireless networks and have microsecond latency, and the introduction of digital twins of entire cities that are synchronized in real time, could be deployed on networks in the 2030s and even later. The future technical and operational issues are daunting, though the experience of the past wireless generations shows that the telecommunications industry can address them all.

REFERENCES

1. Tezergil, B., & Onur, E. (2022). Wireless backhaul in 5G and beyond: Issues, challenges and opportunities. IEEE Communications Surveys & Tutorials, 24(4), 2579–2632.
2. Semaan, E., Tejedor, E., Kumar-Kochhar, R., Magnusson, S., & Parkvall, S. (2023). 6G spectrum-enabling the future mobile life beyond 2030. White paper GFTL-23, 243.
3. Giordani, M., & Zorzi, M. (2020). Non-terrestrial networks in the 6G era: Challenges and opportunities. IEEE Network, 35(2), 244–251.
4. Liu, X., Dong, Y., Li, Y., Lin, Y., Yang, X., & Gan, M. (2023). IEEE 802.11be Wi-Fi 7: Feature summary and performance evaluation. ArXiv abs/2309.15951.
5. Tsai, C.-W., Yang, C.-W., Lin, J., Chang, Y.-C., & Chang, R.-S. (2021). Quantum key distribution networks: Challenges and future research issues in security. Applied Sciences, 11, 3767. doi: 10.3390/app11093767
6. Scarani, V., & Kurtsiefer, C. (2014). The black paper of quantum cryptography: real implementation problems. Theoretical Computer Science, 560, 27–32.

7. Ferenczi, A., Grangier, P., & Grosshans, F. (2007, June). Calibration attack and defense in continuous variable quantum key distribution. In *International Quantum Electronics Conference* (p. IC_13). Optica Publishing Group.
8. Li, W., Yu, J., Zhu, B., Wang, F., Ding, J., Zhang, J., ..., & Yu, J. (2023). Photonic terahertz wireless communication: Towards the goal of high-speed kilometer-level transmission. Journal of Lightwave Technology, 42(3), 1159–1172.
9. Tan, X., Zhou, L., Wang, H., Sun, Y., Zhao, H., Seet, B. C., ..., & Leung, V. C. (2022). Cooperative multi-agent reinforcement-learning-based distributed dynamic spectrum access in cognitive radio networks. IEEE Internet of Things Journal, 9(19), 19477–19488.
10. Muzaffar, M. U., & Sharqi, R. (2024). A review of spectrum sensing in modern cognitive radio networks. Telecommunication Systems, 85(2), 347–363.
11. Ngo, H. Q., Interdonato, G., Larsson, E. G., Caire, G., & Andrews, J. G. (2024). Ultradense cell-free massive MIMO for 6G: Technical overview and open questions. Proceedings of the IEEE, 112(7), 805–831.
12. Aboagye, S., Ndjiongue, A. R., Ngatched, T. M., Dobre, O. A., & Poor, H. V. (2022). RIS-assisted visible light communication systems: A tutorial. IEEE Communications Surveys & Tutorials, 25(1), 251–288.
13. Pivoto, D. G. S., de Figueiredo, F. A. P., Cavdar, C., Tejerina, G. R. d. L., & Mendes, L. L. (2025). A comprehensive survey of machine learning applied to resource allocation in wireless communications. IEEE Communications Surveys & Tutorials. doi: 10.1109/COMST.2025.3552370
14. Wang, C. -X., Di Renzo, M., Stanczak, S., Wang, S., & Larsson, E. G. (February 2020). Artificial intelligence enabled wireless networking for 5G and beyond: Recent advances and future challenges. IEEE Wireless Communications, 27(1), 16–23. doi: 10.1109/MWC.001.1900292
15. Merenda, M., Porcaro, C., & Iero, D. (2020). Edge machine learning for AI-enabled IoT devices: A review. Sensors., 20(9), 2533. doi: 10.3390/s20092533
16. Ergen, M., Saoud, B., Shayea, I., El-Saleh, A. A., Ergen, O., Inan, F., & Tuysuz, M. F. (2024). Edge computing in future wireless networks: A comprehensive evaluation and vision for 6G and beyond. ICT Express, 10(5), 1151–1173.
17. Aboagye, S., Saeidi, M. A., Tabassum, H., Tayyar, Y., Hossain, E., Yang, H. C., & Alouini, M. S. (2024). Multi-band wireless communication networks: Fundamentals, challenges, and resource allocation. IEEE Transactions on Communications, 72(7), 4333–4383.
18. Anandaraj, M., Ganeshkumar, P., Naganandhini, S., & Selvaraj, K. (2024). Greening the wireless world: Energy-efficient and sustainable communication. In Intelligent Networks (pp. 216–234). CRC Press.
19. Rifa-Pous, H., Garcia-Font, V., Nunez-Gomez, C., & Salas, J. (2024). Security, trust and privacy challenges in AI-driven 6G networks. arXiv preprint arXiv:2409.10337.

Index

L

M

N

O

P

Q

R

S

T

U

V

W

Z